Lecture Notes of the Institute for Computer Sciences, Social Informatics and Telecommunications Engineering 622

Editorial Board Members

Ozgur Akan, *Middle East Technical University, Ankara, Türkiye*
Paolo Bellavista, *University of Bologna, Bologna, Italy*
Jiannong Cao, *Hong Kong Polytechnic University, Hong Kong, China*
Geoffrey Coulson, *Lancaster University, Lancaster, UK*
Falko Dressler, *University of Erlangen, Erlangen, Germany*
Domenico Ferrari, *Università Cattolica Piacenza, Piacenza, Italy*
Mario Gerla, *UCLA, Los Angeles, USA*
Hisashi Kobayashi, *Princeton University, Princeton, USA*
Sergio Palazzo, *University of Catania, Catania, Italy*
Sartaj Sahni, *University of Florida, Gainesville, USA*
Xuemin Shen, *University of Waterloo, Waterloo, Canada*
Mircea Stan, *University of Virginia, Charlottesville, USA*
Xiaohua Jia, *City University of Hong Kong, Kowloon, Hong Kong*
Albert Y. Zomaya, *University of Sydney, Sydney, Australia*

The LNICST series publishes ICST's conferences, symposia and workshops.
LNICST reports state-of-the-art results in areas related to the scope of the Institute.
The type of material published includes

- Proceedings (published in time for the respective event)
- Other edited monographs (such as project reports or invited volumes)

LNICST topics span the following areas:

- General Computer Science
- E-Economy
- E-Medicine
- Knowledge Management
- Multimedia
- Operations, Management and Policy
- Social Informatics
- Systems

Xiali Hei · Luis Garcia · Taegyu Kim ·
Kyungtae Kim
Editors

Security and Privacy in Cyber-Physical Systems and Smart Vehicles

Second EAI International Conference, SmartSP 2024
New Orleans, LA, USA, November 7–8, 2024
Proceedings

Editors
Xiali Hei
University of Louisana at Lafayette
Lafayette, LA, USA

Luis Garcia
University of Utah
Salt Lake City, UT, USA

Taegyu Kim
Pennsylvania State University
University Park, PA, USA

Kyungtae Kim
Dartmouth College
Hanover, NH, USA

ISSN 1867-8211　　　　　　ISSN 1867-822X (electronic)
Lecture Notes of the Institute for Computer Sciences, Social Informatics
and Telecommunications Engineering
ISBN 978-3-031-93353-0　　　ISBN 978-3-031-93354-7 (eBook)
https://doi.org/10.1007/978-3-031-93354-7

© ICST Institute for Computer Sciences, Social Informatics and Telecommunications Engineering 2025

This work is subject to copyright. All rights are solely and exclusively licensed by the Publisher, whether the whole or part of the material is concerned, specifically the rights of translation, reprinting, reuse of illustrations, recitation, broadcasting, reproduction on microfilms or in any other physical way, and transmission or information storage and retrieval, electronic adaptation, computer software, or by similar or dissimilar methodology now known or hereafter developed.
The use of general descriptive names, registered names, trademarks, service marks, etc. in this publication does not imply, even in the absence of a specific statement, that such names are exempt from the relevant protective laws and regulations and therefore free for general use.
The publisher, the authors and the editors are safe to assume that the advice and information in this book are believed to be true and accurate at the date of publication. Neither the publisher nor the authors or the editors give a warranty, expressed or implied, with respect to the material contained herein or for any errors or omissions that may have been made. The publisher remains neutral with regard to jurisdictional claims in published maps and institutional affiliations.

This Springer imprint is published by the registered company Springer Nature Switzerland AG
The registered company address is: Gewerbestrasse 11, 6330 Cham, Switzerland

If disposing of this product, please recycle the paper.

Preface

We are delighted to introduce the proceedings of the second edition of the European Alliance for Innovation (EAI) International Conference on Security and Privacy in Cyber-Physical Systems and Smart Vehicles (SmartSP 2024). This conference brought together researchers, developers, and practitioners around the world to discuss emerging ideas and trends in security and privacy of smart cyber-physical systems (CPS). The theme of SmartSP 2024 was "emerging applications, experimental studies, and social impacts of CPS and Smart Vehicles".

The technical program of SmartSP 2024 consisted of 16 full papers and 2 demo papers. Aside from the high-quality technical paper presentations, the technical program also featured three keynote speeches, one invited talk and one demo session. The three keynote speakers were Kevin Butler from University of Florida, USA, Kenneth Rohde from Idaho National Laboratory, USA, and Danfeng Yao from Virginia Tech, USA. The invited talk was presented by Yazhou Tu from Auburn University in USA.

Coordination with the steering chair, Bo Chen, was essential for the success of the conference. We sincerely appreciate his constant support and guidance. It was also a great pleasure to work with such an excellent organizing committee team for their hard work in organizing and supporting the conference. In particular, the Technical Program Committee, led by our TPC Co-Chairs, Taegyu Kim and Kyungtae Kim, completed the peer-review process of technical papers and made a high-quality technical program. We are also grateful to Conference Manager, Natasha Onofrei for her support and to all the authors who submitted their papers to the SmartSP 2024 conference.

We strongly believe that SmartSP provides a good forum for all researchers, developers, and practitioners to discuss all security and privacy aspects that are relevant to smart CPS. We also expect that future SmartSP conferences will be as successful and stimulating, as indicated by the contributions presented in this volume.

Taegyu Kim
Kyungtae Kim
Luis Garcia
Xiali Hei

Organization

Steering Committee

Alvaro Cardenas	University of California, Santa Cruz, USA
Bo Chen (Chair)	Michigan Technological University, USA
Mohamed Amine Ferrag	Technology Innovation Institute, UAE
Xiali Hei	University of Louisiana at Lafayette, USA
Hongxin Hu	University at Buffalo, SUNY, USA
Peng Liu	Pennsylvania State University, USA
Xiapu Luo	Hong Kong Polytechnic University, China
Weizhi Meng	Technical University of Denmark, Denmark
Indrajit Ray	Colorado State University, USA
Yuqing Zhang	University of Chinese Academy of Sciences, China

Organizing Committee

General Chair

Xiali Hei — University of Louisiana at Lafayette, USA

General Co-chair

Luis Garcia — University of Utah, USA

TPC Chair and Co-chair

Taegyu Kim	Pennsylvania State University, USA
Kyungtae Kim	Dartmouth College, USA

Sponsorship and Exhibit Chair

Luis Garcia — University of Utah, USA

Local Chairs

Abdullah Yasin Nur   University of New Orleans, USA
Raju Gottumukkala   University of Louisiana at Lafayette, USA

Publicity and Social Media Chairs

Yazhou Tu   Auburn University, USA
Zihao Zhan   University of Florida, USA
Soumyajit Dey   Indian Institute of Technology Kharagpur, India

Publications Chairs

Qiben Yan   Michigan State University, USA
Lan Zhang   Clemson University, USA

Web Chair

Md Imran Hossen   University of Louisiana at Lafayette, USA

Poster/Demo Chair

Lin Zhang   University of Pennsylvania, USA

Technical Program Committee

Yazhou Tu   Auburn University, USA
Mulong Luo   University of Texas at Austin, USA
Muslum Ozgur Ozmen   Purdue University, USA
Andrew Clark   Washington University in St. Louis, USA
Sriharsha Etigowni   National Renewable Energy Laboratory, USA
Ki-woong Park   Sejong University, South Korea
Fatima Anwar   UMass Amherst, USA
Aiping Xiong   Pennsylvania State University, USA
Mu Zhang   University of Utah, USA
Ruimin Sun   Florida International University, USA
Saman Zonouz   Georgia Tech, USA
Mert Pese   Clemson University, USA
Luis Garcia   University of Utah, USA
Aolin Ding   Accenture Labs, USA

Peng Liu	Pennsylvania State University, USA
Ruoyu Wu	Purdue University, USA
Jianliang Wu	Simon Fraser University, Canada
Habiba Farrukh	University of California, Irvine, USA
Sanchuan Chen	Auburn University, USA
Junghwan Junghwan Rhee	University of Central Oklahoma, USA
Agbotiname Imoize	University of Lagos, Nigeria
Xueping Susan Liang	Florida International University, USA
Ronghua Xu	Michigan Technological University, USA
Chao Wang	National Taiwan Normal University, Taiwan
Hokeun Kim	Arizona State University, USA
Baekgyu Kim	DGIST, South Korea
Deepak K. Tosh	University of Texas at El Paso, USA
Lan Zhang	Northern Arizona University, USA

Contents

Emerging Applications

Leaking Through the Physics: Covert Cyber-Physical Data Exfiltration
Through Unobserved Physics 3
   Matthew Chan, Luis Garcia, Nathaniel Snyder, Marcus Lucas,
   Aolin Ding, Amin Hass, Oleg Sokolsky, James Weimer, Paulo Tabuada,
   Saman Zonouz, and Mani Srivastava

AcousticScope: Understanding Biases in Voice Interaction via Automated
Acoustic Testing 26
   Shuhao Zhang, Mohammed Aldeen, Song Liao, Jeffery Young,
   and Long Cheng

Practitioner Paper: Decoding Intellectual Property: Acoustic and Magnetic
Side-Channel Attack on a 3D Printer 54
   Amirhossein Jamarani, Yazhou Tu, and Xiali Hei

Security Techniques for Cyber-Physical Systems

An Efficient and Applicable Physical Fingerprinting Framework
for the Controller Area Network Utilizing Deep Learning Algorithm
Trained on Recurrence Plots 77
   Rafi Ud Daula Refat, Alireza Mohammadi, and Hafiz Malik

Short Paper: Software Bill of Materials Management for Embedded
Vehicle Systems 100
   Teddy Nyambe, Rik Chatterjee, and Jeremy Daily

ProvPredictor: Utilizing Provenance Information for Real-Time IoT
Policy Enforcement 110
   Michael Norris, Patrick McDaniel, Syed Rafiul Hussain, and Gang Tan

A Case Study of API Design for Interoperability and Security
of the Internet of Things 135
   Dongha Kim, Chanhee Lee, and Hokeun Kım

ShadowConn: Breaking the Entanglement of Cross-Platform IoT
Delegation in Multi-user Environments 158
   Huan Bui and Chenglong Fu

Hardware and Firmware Security

Hardware-Assisted Runtime In-vehicle ECU Firmware Self-attestation
and Self-repair 187
    Josh Dafoe, Job Siy, Niusen Chen, and Bo Chen

Unveiling the Operation and Configuration of a Real-World Bulk
Substation Network 211
    Keerthi Koneru, Juan Lozano, John Castellanos, Emmanuele Zambon,
    and Alvaro Cardenas

RustBound: Function Boundary Detection over Rust Stripped Binaries 237
    Ryan Evans, William Hawkins, and Boyang Wang

Adversarial Attacks in Autonomous Systems

Transient Adversarial 3D Projection Attacks on Object Detection
in Autonomous Driving 259
    Ce Zhou, Qiben Yan, and Sijia Liu

Assessing Deep Learning Model Accuracy in Varied Surface Conditions
for CPS: A Comparative Study 279
    Cade Jacobson, Mathew Clutter, and Francis Akowuah

Practitioner Paper: A Real-Time Defense Against Object Vanishing
Adversarial Patch Attacks for Object Detection in Autonomous Vehicles 296
    Jaden Mu

Ethics, Privacy, and Human-Centric Considerations

Ethical Considerations and Policy Implications for Large Language
Models: Guiding Responsible Development and Deployment 311
    Ziyin Zhou, Xu Ji, Jianyi Zhang, Zhangchi Zhao, Xiali Hei,
    and Kim-Kwang Raymond Choo

Integrating Human Preferences for Moral Decision Making in Autonomous
Vehicles 330
    Bishal Thapa, Henry Griffith, and Heena Rathore

Privacy-Enrooted Car Systems (PECS): Preliminary Design 344
    Giampaolo Bella, Gianpietro Castiglione, Sergio Esposito,
    Mirko Giuseppe Mangano, Mirco Marchetti, Marcello Maugeri,
    Mario Raciti, Salvatore Riccobene, and Daniele Francesco Santamaria

Demos

Demo: All-in-One Solution for Online Abuse Research 361
   Mohammed Aldeen, Pranav Pradosh Silimkhan, Ethan Anderson,
   Taran Kavuru, Tsu-Yao Chang, Jin Ma, Feng Luo, Hongxin Hu,
   and Long Cheng

Author Index 367

Emerging Applications

Leaking Through the Physics: Covert Cyber-Physical Data Exfiltration Through Unobserved Physics

Matthew Chan[1], Luis Garcia[2(✉)], Nathaniel Snyder[3], Marcus Lucas[3], Aolin Ding[4], Amin Hass[4], Oleg Sokolsky[5], James Weimer[6], Paulo Tabuada[3], Saman Zonouz[7], and Mani Srivastava[3,8]

[1] Rutgers University, New Brunswick, NJ, USA
[2] University of Utah, Salt Lake City, UT, USA
la.garcia@utah.edu
[3] University of California, Los Angeles, CA, USA
[4] Security R&D, Accenture Labs, Accenture, Washington DC, USA
[5] University of Pennsylvania, Philadelphia, PA, USA
[6] Vanderbilt University, Nashville, TN, USA
[7] Georgia Institute of Technology, Atlanta, GA, USA
[8] Amazon, Seattle, USA

Abstract. From magnetic fields to differential power analysis, side-channel attacks have emerged as a significant threat to cyber-physical systems (CPS). Attackers often exploit out-of-band channels through signal-injection attacks to perform data exfiltration. These attacks primarily target unmonitored channels and open-loop cyber-physical systems. However, attacks through physical channels are relatively unexplored because state estimation methods are conventionally believed to be sufficient for detecting malicious physical actuation. In this paper, we propose a novel method for out-of-band data exfiltration from a cyber-physical system based on unobserved physics. We present that, despite the presence of state estimation-based intrusion detection techniques, our data exfiltration method can exploit the limitations of physical state observability to circumvent these protections. It allows for the stealthy exfiltration of sensitive data from the network using existing cyber-physical models and the infrastructure of individual devices. We evaluate the efficacy of our data exfiltration technique in the context of two real-world testbed scenarios: an industrial robotic arm controller and an autonomous surveillance drone. We discuss potential defenses against this type of attack and their limitations.

Keywords: Cyber-physical System Security · Side Channels · Edge Computing · Data Exfiltration

1 Introduction

The explosion of edge computing has called for an increased emphasis on the collateral attributes of edge networks, e.g., mobility, wide-spread geographical location, low latency, and heterogeneity [8]. As an enabling technology, edge computing drives the transition of the cyber-physical systems (CPS), from a centralized computation paradigm to distributed computing on the *edge*. An edge-enabled CPS performs data processing on devices that reside a single "hop" away from sensors and actuators, i.e., directly interfacing with sensors and actuators. We have observed that many works focus on addressing the security challenges of sensor components within the CPS, such as sensor spoofing attacks [4,5,30]. Yet, insufficient attention has been given to the actuator components and the assurances of *physical* actuation against information leakage.

Traditionally, the practice of air-gapping critical systems –i.e., physically isolating a computer from an unsecured network –is believed to mitigate network vulnerabilities and reduce the associated attack surface. However, recent attacks have overcome air-gap defenses to exfiltrate data via covert channels that exploit device peripherals, including electromagnetic emanations [10,24,28,39], magnetic fields [35], power consumption [20,26], acoustic noise [6,23], observable characteristics [36], and thermal emissions [25]. While typical countermeasures, such as the physical concealment of side channels, are feasible in static scenarios like data centers, they hardly work in the context of autonomous CPS architectures. Specifically, edge computation architectures such as autonomous CPS have evolved to perform computation on much more dynamic and adaptive edge network devices. As opposed to static data centers, these low-level devices are physically exposed as they sense and actuate in the physical environment, exhibiting more complexities than the aforementioned channels. Meanwhile, conventional countermeasures may not be able to deploy on these devices to provide cyber-physical runtime guarantees due to their resource constraints of memory, power consumption, and computational capacity. Therefore, traditional air-gapping techniques or physical concealment of side channels are not sufficient to secure information leakage for autonomous CPS on the edge.

Despite the potential vulnerabilities, exploiting physical covert channels to exfiltrate sensitive information is challenging in practice. The main reason is that these cyber-physical systems are typically monitored via state estimation techniques to ensure the system is behaving correctly [2,14,38]. By extracting and modeling the safety-critical CPS state updates, such as sensor measurements and physical dynamics, state estimation monitors can identify malicious behaviors caused by data exfiltration. Therefore, an attacker's encoding mechanism for data exfiltration would need to be designed so as not to have the state estimator raise any flags. Because distributed CPS architectures are running inferencing closer to or on the edge devices, an attacker may have access to higher-level information inferred from the data and, as such, has to encode fewer bits into an attack since more information can be encoded into each bit. For instance, a drone may be performing object detection in a privacy-sensitive area. An attacker would only have to encode low-level label information into the data exfiltration as opposed to sending the raw data.

In this paper, we show how a CPS's physical actuation can be used as a covert channel to exfiltrate sensitive data. In particular, we introduce a cyber-physical encoding technique that maintains *stealthiness* against an entity that is monitoring the cyber-physical system via state estimation techniques. We empirically demonstrate how an attacker would maximize the rate of transmission while maintaining stealthiness with respect to the physical covert channel. This also implies that our approach maintains the utility of the cyber-physical application. For instance, to encode data into the actuation of a drone, our approach would encode data into the movement of the drone while ensuring that the drone completes its waypoint navigation correctly. This approach is analogous to prior attacks that exploit the semantic models of autonomous systems for stealthiness, e.g., cyber-physical attacks that target state-estimation techniques [13, 18].

We evaluate our attack on two exemplary cyber-physical systems: a robotic arm in the context of an industrial control system as well as a drone surveilling an area of interest. For each system, we encode the data across a variety of applications and evaluate the efficacy of each attack. We use computer vision techniques to observe the physical actuation and decode the encoded bits. We also evaluate each attack against monitors with varying levels of probabilistic certainty about the estimated system states. We optimize our attacks against state-of-the-art state estimation techniques and show how we would maximize the transmission rate for each case with respect to the state estimation noise.

Our contributions are summarized as follows:

- We present a new covert data exfiltration through unobserved physics on cyber-physical edge devices.
- We propose a cyber-physical attack model that optimizes data exfiltration of covert channels against state estimation techniques to maximize the transmission rate while maintaining stealthiness.
- We evaluate the efficacy of our approach across two exemplary autonomous, safety-critical cyber-physical system applications: industrial IoT robotics and autonomous drone surveillance.

2 Background

In this section, we provide a background on "out-of-band" covert data exfiltration, to better understand the system model and its associated threat model. Currently, covert data exfiltration works have shown how physical side channels exist across many normally unmonitored modalities, usually in the context of air-gapped systems. These include electromagnetic radiation [24], magnetic fields [35], power consumption [26], acoustic channel [6, 23], optical field [36], as well as thermal emissions [25]. In all cases, these systems typically propose a cyber-physical air-gapped covert channel followed by an associated countermeasure to prevent such channels from being exploited. Subsequent works will then continue this attacker-monitor game where a new covert channel is proposed to attack the hardened system. For instance, in defense against the aforementioned attacks where data was exfiltrated via electromagnetic radiation [24, 28, 39], technical countermeasures are proposed such as physical insulation and software-based reductions of information-bearing emissions.

The procedural and technical countermeasures presented for the aforementioned attacks generally propose insulation of the physical channels in which data can be exfiltrated that are subsequently exploited. For both attacks and defenses, these approaches fail to encapsulate the physical model of these channels that stem from the memory-mapped inputs and outputs of the system [33]. Such physical models can be used to perform cyber-physical state estimation to understand what will be the physical impact of a particular action in the cyber space [14, 45]. Further, state estimation allows for providing an understanding of the mutual dependency between physical channels, e.g., the correlation between a 3D printer's object printing operations and its acoustic channel [6]. From a monitor's perspective, state estimation not only enables the cyber-physical noise models that may need to be insulated, but also can perform intrusion detection if an attacker is explicitly encoding data into a particular channel that deviates from the estimated state of the channel. From an attacker's perspective, state estimation techniques can be used to craft complex cyber-physical attacks on neglected physical channels [13]. In both cases, the respective problems are exacerbated when moving from the static, immobile systems considered in these works (e.g., data center computers) to mobile and autonomous edge devices that are difficult to physically insulate and expose even more cyber-physical channels. In this paper, we aim to study this notion in greater depth, particularly in the context of autonomous CPS on the edge.

3 Problem Formulation

In this section, we provide a system model for physical covert channels and share our insights on the limitations of prior monitors. We then define the threat model along with our adversarial assumptions and challenges in this context.

3.1 System Model

The system model we consider in this paper is depicted in Fig. 1, where the visible IoT/CPS Edge device is monitoring a sensitive application in the context of an edge computing-enabled CPS model. A supervisory controller sends high-level control commands accordingly, e.g., an air traffic controller sends a coordinate setpoint for a drone. The supervisory controller is also running a state estimation monitor [2, 14, 38] to check for anomalies (e.g., caused by cyber-physical attacks or component failures) in the system state of the CPS on the edge, excluding merely random perturbations. This monitor ensures the safe operation of the CPS by verifying that the system state is consistent with previously sent control commands, e.g., a drone's previous state has been updated according to the physical dynamics. We assume that the CPS has local control loops that convert the high-level commands from the supervisory controller to local actuation with respect to its internally maintained state estimation, e.g., a drone's stability and waypoint navigation control loops. Finally, we assume that there may be one or more humans in the same vicinity that can observe the physical characteristics

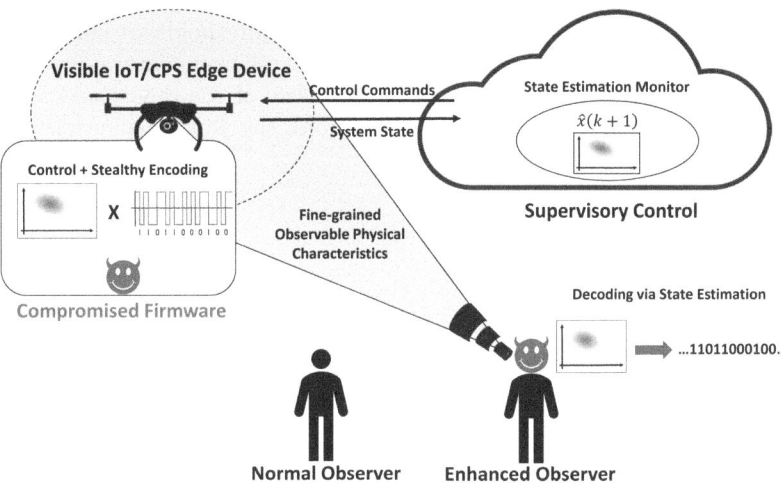

Fig. 1. Data exfiltration attack overview.

of the device from a distance. The notion of human perception is an analog to perceptibility with respect to distortion models in the context of adversarial machine learning [27,37]. We now discuss the limitations of such state estimation monitors and the threat model of how an attacker can achieve data exfiltration with respect to this system model.

3.2 Limitations of State Estimation Monitors

The typical goal of a monitor is to leverage an appropriate state estimator to detect any anomalies. Theoretically, a perfectly tuned state estimator model for all memory-mapped physical I/O and its associated physical covert channels of a CPS would detect any cyber-physical data exfiltration attack [7]. However, in practice, it is difficult to develop a perfectly robust state estimation model for real-world applications due to imprecision caused by sensor noise and environmental factors [12]. Furthermore, a state estimator can sense the observable or partially observable set of physical variables, which means that such a monitor heavily relies on the availability and quality of sensor instrumentation, as well as the level of process noise associated with the observable physical variables and channels. For instance, if a robotic arm infers its own pose and reports only the XY-coordinates of the robot's end effector before and after a movement command, then a state estimator is only able to report the posterior $\hat{x}_t$ then prior $\bar{x}_{t+1}$ state estimates and associated covariances of the end effector. An anomaly could be detected using a distance metric such as a Euclidean distance [14] to see if the current XY-coordinates, $\{(x_1, y_1), (x_2, y_2)\}$ are close enough to the $\hat{x}$ and $\hat{y}$ estimates within a certain error ϵ:

$$\sqrt{(\hat{y_1} - y_1)^2 + (\hat{x_1} - x_1)^2 + (\hat{y_2} - y_2)^2 + (\hat{x_2} - x_2)^2} < \epsilon. \tag{1}$$

However, in practical deployment, the monitor's thresholds are conservatively calculated to avoid excessive false positives due to environmental noise and system overshooting [13]. That means for an attacker, if the deviation caused by data exfiltration is within the accepted threshold, the state-estimation monitor will not be able to detect it.

3.3 Threat Model

The threat model has two components: the compromised edge-enabled CPS that is encoding sensitive information into physical actuation, and an adversarial observer that is decoding the encoded actuation. These components collude in order to perform covert data exfiltration. The goal of the attacker is to encode data into a physical covert channel while maintaining *stealthiness*. To perform stealthy data exfiltration, the attacker will observe the system's estimator output and the controller's input and use their actuation capability to influence the process noise and sensor measurement noise. The choice and encoding scheme of the attacker will be domain-specific and described in Sect. 4. We first describe how the CPS software can be compromised, then discuss the characteristics of the adversarial observer.

Compromised CPS Software. We assume that an attacker has compromised the CPS software in such a way that provides the attacker full control of one or more physical actuators of the edge CPS and, thus, can encode the sensitive information into the target covert channels. The most straightforward way of a software compromise is through a direct malicious software update, performed either remotely or locally, as shown in [12,18]. Alternatively, a compromise could be propagated using runtime attacks conducted through compromised local network devices, as in the case of Stuxnet [16].

Derived State Estimator. We also assume that the attacker has access to the physical dynamics of the edge CPS such that an attacker can derive the state estimator along with an estimated noise model for the device. This can be achieved through the use of reverse-engineering and system identification techniques presented in prior works [12,31,40]. We do not assume that the attacker can compromise the state estimation monitor itself, as the monitor is deployed on the supervisory controller and may be located on a different chip than the exploited software module [29].

Enhanced Adversarial Observer Model. For our enhanced adversarial observer, we assume that an adversary has sufficient sensing capability to adequately sense the covert physical channel on which the data exfiltration is occurring. This may require enhanced resolution ("zooming-in," in the optical case) compared to a normal observer. With the assumed knowledge of the physical dynamics, the adversary can develop an appropriate encoding scheme to meet these sensing requirements.

3.4 Challenges and Definitions

Stealthiness. We define stealthiness as the attacker's ability to deviate the CPS's state such that any threshold levels of the monitor are not crossed as a result of the attack, the attacked state(s) do not strongly correlate with non-attacked state variables, and the attack is conducted on a state which does not utilize a *colored-noise* state estimator, i.e. if the measurement noise is correlated, then computing an ensemble average for the auto-correlation of the measurement noise (empirically) would differ from the known correlation signal.

Imperceptibility. We define imperceptibility with respect to a "human observer" – or an observer that may be monitoring the CPS through a particular modality or set of modalities from sensors equivalent to a human's "sensors". This problem is analogous to the problem of adversarial machine learning where an attacker is introducing perturbations to a model's input data while minimizing some loss function such that the system will misclassify the data sample while maintaining imperceptibility of the perturbations [34, 37]. In this context, the perfect model of perceptibility is the associated decoding mechanism itself. In addition to maintaining stealthiness with respect to the state estimation model, an attacker will also minimize the encoded movements such that the decoding function will only work with a decoder that has sufficient sensing granularity, e.g., a camera equipped with an appropriate focal length to capture the encoded movements of the physical channel.

4 Cyber-Physical Data Exfiltration

In this section, we present our approach for performing covert CPS data exfiltration. We begin with a discussion of signal temporal logic and its concept of robustness. We apply it to a motivating example of a robotic arm to show the process of physical side-channel data encoding, even in the presence of a state estimation monitor. We then discuss how this encoded information is exfiltrated through a decoding process.

4.1 A Signal Temporal Logic Formulation

In order to develop a framework for performing covert data exfiltration over physical channels, we draw upon a previous signal temporal logic (STL) [11] formulation for stealthy CPS attacks. STL allows us to reason over sequences of states and values. We can model physical system constraints by combining STL predicates like ϕ and ψ:

$$\phi = y < c \mid \neg \phi \mid \phi \vee \psi \mid \phi \wedge \psi \mid \Box_I \phi \mid \Diamond_I \phi \mid \phi \, \mathrm{U}_I \, \psi$$

In addition to the normal logical operators, STL includes $\Box_I$ for always, $\Diamond_I$ for eventually, and U_I for until, for which the predicates are evaluated over an interval I. As an example, if we want to formulate in STL that a system eventually ends up at a position (x_1, y_1) by time t and its y-value never exceeds c, we can define the STL specification $\theta = \Diamond_{[0,t]}(x = x_1 \land y = y_1) \land \square_{[0,n]}(y < c)$. Then, we can check whether θ is satisfied by some signal trace $S = \{x_{1..n}, y_{1..n}\}$.

Another useful property of STL is the built-in ability to quantify the robustness $\rho(\phi, s, t)$ of signals with respect to STL specifications [15]. Notably, if ρ is positive, the signal satisfies the STL specification, and if ρ is negative, the signal does not satisfy the specification. Robustness is important in falsification and search-based test generation scenarios for finding safety property violations [11,42]. However, unlike those works, we use robustness as a metric to help investigate the stealthiness and imperceptibility of physical-channel data exfiltration. We can model both our CPS system and anomaly detector constraints using STL specifications. This can guide the development of encoding schemes, and allows us to evaluate the effectiveness of covert data exfiltration with respect to system and monitor constraints.

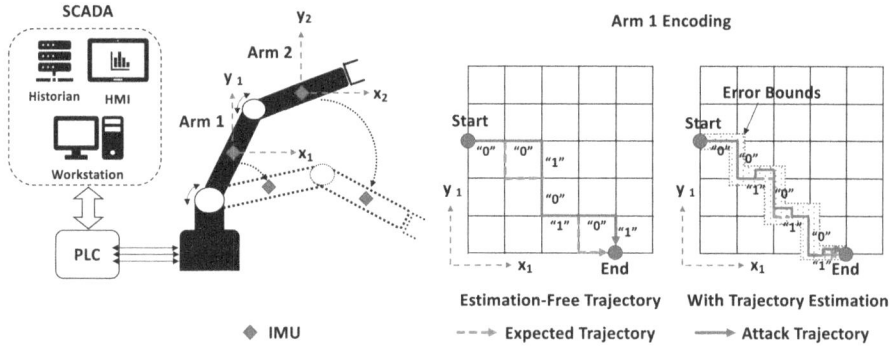

Fig. 2. Simplified CPS example to illustrate cyber-physical data exfiltration. An attacker may encode data into the movement of the robotic arm that is not being estimated or into the noise model associated with the estimated movement.

4.2 Motivating Example: Simplified Industrial Control System

Fig. 2 shows an example of an industrial control system (ICS) where a robotic arm is controlled by a programmable logic controller (PLC). We implement this industrial control system with a Dobot robotic arm controlled by a Siemens S7-1200 PLC. We emulate the SCADA components through an API that sends motion commands to the PLC, which calculates the appropriate actuation commands for the arm's stepper motors. The robot arm is composed of two segments independently controlled by stepper motors. Each segment is equipped with an inertial measurement unit (IMU) that is used to close the loop for the arm's controller. The robotic arm has an end attachment that can stand in for several example applications, such as 3D printing, laser etching, or a gripper for industrial automation. The PLC controls the arm in a closed feedback loop, reading in sensor data from the arm's two accelerometers and actuating its stepper motors to control the arm's movement.

For illustrative purposes, we restrict the arm from moving along the XY plane. In this case, we assume that the PLC takes an XY-coordinate setpoint from the SCADA entity and its internal control loop calculates the associated actuation commands necessary to arrive at the desired waypoint for arm segments. Given this system, we discuss the process of encoding data into physical channels and how this is affected by the presence and quality of the state estimation monitor.

4.3 Encoding Data into Physical Channels

As discussed, an attacker's encoding scheme into a particular physical channel will depend on the quality of the monitor's state estimation model – which is assumed to be known by the attacker. As such, we consider two different monitor models: (1) an attacker encoding data into a physical channel that is independent from any of the monitor's state estimator models, i.e., the control process for that state variable is locally autonomous; (2) an attacker encoding data into the "noise" of a channel that is being directly estimated by the system's state estimator. We detail these two cases in the following section.

Case 1: Local Autonomy and Estimation. If a state variable's associated control loop is locally autonomous to the edge device, i.e., there is no feedback control or estimation mechanism from a higher fidelity external entity, then an attacker essentially has little to no inhibitions with respect to stealthiness and can manipulate any aspect of the system as long as the system constraints and the utility of the application are maintained. For instance, in Fig. 2, the estimation-free trajectory scenario shows how an attacker may encode data into the path from a starting ("Start") XY-coordinate to an ending ("End") XY-coordinate. Such an attack would need to ensure that the *utility* of the function is maintained, e.g., that the encoding will have a mean noise of zero while ensuring that it reaches a distance within an error bound before the next sample. This also implies that the associated perturbations will not cause any collateral threshold violations for other states being estimated. And although this example shows the path trajectory between two sampled points as the physical channel of choice, any other cyber-physical channel that depends on the associated physical variables can also be utilized by an attacker, e.g., the acoustic noise of the stepper motors during the path trajectory. In any case, the attacker may engineer an encoding scheme that will transmit the data while ensuring the utility function's integrity is maintained. By specifying the utility function using STL, the attacker can test different encoding schemes that satisfy the constraints of the utility function. However, encoding data becomes more difficult for subsequent cases where the state variable is being estimated.

Case 2: Remote External Feedback Control. If a state variable's associated control loop relies on external state estimation and feedback control, the attacked state variable is being monitored with fine granularity – which complicates the design of the attacker's encoding scheme. However, it is infeasible for a monitor to have a perfect state estimator model for real-world systems due to sensing and process noise. Such an encoding mechanism requires an accurate noise model that is at least as granular as the noise model of the monitor. The plot on the right of Fig. 2 shows an attacker encoding bits into the trajectory of the end effector while staying within error bounds. In this case, the attacker is much more restricted in terms of how much noise can be introduced in the encoding scheme due to the fact that the bits are being encoded into an estimated variable. By specifying the state estimation model using STL, the attacker can ensure that the chosen encoding is sufficiently robust against detection by the monitor.

4.4 Decoding Cyber-Physical Encoded Data

In order to decode data that has been encoded with any of these encoding schemes, the attacker simply needs to mirror the modality and granularity of the encoding scheme. For instance, in the estimation-free trajectory attack of the ICS example, an attacker would need access to the finer-grained path trajectory between movement commands. Having access to either a faster sampling rate or even the IMU data would be ideal, but it is not realistic for a remote attacker– especially if we are assuming the monitor does not have access to these results. A more realistic approach is that an attacker may infer the cyber-physical encoding utilizing an air-gapped physical channel such as a microphone monitoring the noise of the device or by visually monitoring the movements of each component from a distance with a camera. For instance, we implemented a malicious motion command on the aforementioned PLC that encodes a bit string in the actuation of the arm's motors during a benign motion command. An attacker then tracks the arm's movement to covertly exfiltrate the encoded sensitive data. Next, we discuss design considerations for a communication protocol given an encoding and decoding scheme.

4.5 Communication Protocol

There are several domain-specific design parameters that need to be tuned for particular CPS. Ideally, the encoding should maximize both the rate of transmission as well as the signal-to-noise ratio (SNR), as well as the previously stated goal of stealthiness.

Channel Capacity and Bit Error Rate. There are several factors that determine the channel capacity of data exfiltration.

- **Physical system constraints.** The rate of encoding into a physical system is limited by the actuation speed of the system, which is determined by the system's kinematics. Faster encoding speeds require greater forces, which the system may not be able to support. Additionally, physical systems may have a minimum precision with which motions can be made consistently (e.g. a single motor step is $1.8°$ for the robotic arm).

- **The observer's frame rate and resolution.** The channel capacity is also limited by the capabilities of the observer. For a camera, the frame rate is analogous to the sampling rate, and we found that for the robotic arm, at least 3 frames were necessary to identify an encoded motion consistently. Additionally, the resolution of the observer is correlated with the encoding: a higher resolution means that smaller motions can be detected reliably, allowing a greater encoding rate within the constraints of the physical system.
- **Maintaining stealth.** In a scenario with a monitor performing state estimation on the system, a faster encoding produces more noticeable actuations, increasing the likelihood of revealing the exfiltration process to the monitor.

Table 1 shows the bit error rates for decoding in the robotic arm scenario. As we approached the limits of the encoding rate we found that the decoding accuracy decreases significantly due to increased system vibration coupled with fewer frames per encoded bit. We now briefly discuss design considerations for error checking.

Table 1. Bit error rates (BERs) for various encoding rates.

Bit Rate	FPS	Bit Error Rate
5 bit/sec	30	0%
10 bit/sec	30	0%
15 bit/sec	30	15.6%

Error Checking and Redundancy. Since the transmission channel is a one-way communication link, re-transmission can not be requested in case of a transmission error. Forward error correction such as cyclic-redundancy checks (CRC) [17], can be used to correct errors at the receiver at the cost of reducing transmission bandwidth for redundancy. Alternatively, if the same variables are being transmitted repeatedly (data values), then the values have a short "lifetime," and we can forgo error correction altogether, filtering out outliers at the receiver. The final design piece focuses on an attacker's means of maintaining *imperceptibility* from a monitor.

4.6 Maintaining Imperceptibility

We propose the following simple scheme for maintaining imperceptibility. For a given attacker strength, it is desirable to encode information at the lowest SNR so that the attacker can still decode reliably (e.g., with an acceptable bit error rate). This minimizes the differentiation between signal and noise for any observer and results in the least conspicuous encoding. Additionally, the frequency and choice of encoding should be chosen carefully to closely mirror normal operating characteristics. That being said, determining these parameters may be impractical in certain situations. We now evaluate each of the aforementioned attacker-monitor scenarios on a much more complex autonomous edge device.

5 Evaluation

We first evaluate the implementation of encoding and decoding schemes for two CPS applications: a robotic arm and a surveillance drone. We then evaluate both the attacker and monitor models presented in the previous section for the drone scenario. In particular, the drone is tasked to surveil an area, e.g., to search for particular objects of interest. Any inferences made by the drone will be reported back to a supervisory entity that is interacting with the drone. We describe our experimental setup in detail.

5.1 Experimental Setup

Robotic Arm Scenario. Our implemented decoding modality utilizes a camera on the arm to observe specific markers on the arm as shown in Fig. 3a. We applied color markers to the arm to simplify the implementation of a tracking algorithm, however a more sophisticated algorithm could perform position tracking without external markers specific to the CPS. For tracking the markers we utilized OpenCV, an open-source computer vision library. The resulting output of an encoded movement is shown in Fig. 3b.

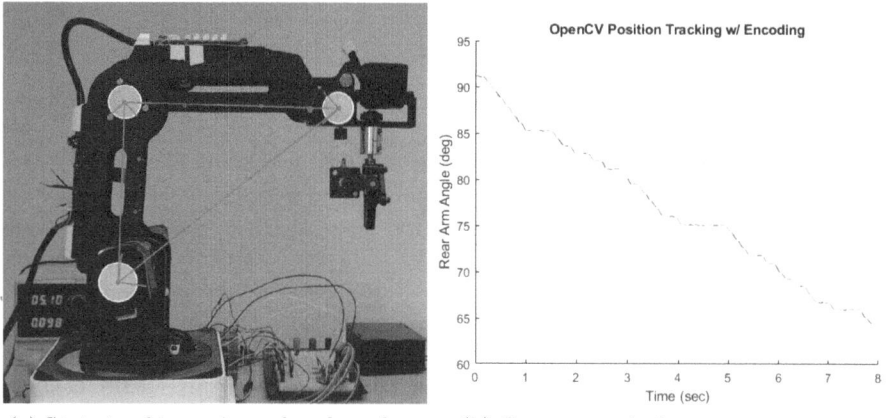

(a) State tracking using colored markers. (b) Camera-tracked system state trace.

Fig. 3. State tracking and encoding of robotic arms. On the right, each colored segment represents a single bit based on the perceived change in the angle of the rear arm segment.

Crazyflie Drone Scenario. As shown in Fig. 4, we use a Crazyflie quadcopter [22] and an Optitrack motion capture system [1] to provide the drone with external estimates of its 3D position and orientation. The Crazyflie quadcopter was chosen because it can be flown indoors, where the motion capture system needs to be calibrated. Our Optitrack setup uses 12 cameras to achieve sub-millimeter positioning accuracy, providing a more precise state estimator for the drone than outdoor localization schemes. We use the Robotic Operating System (ROS) to facilitate real-time communication between the drone, motion capture system, and host computer.

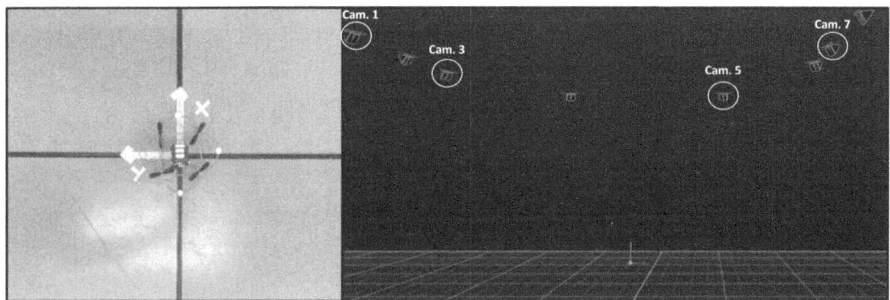

Fig. 4. Drone evaluation setup using the Optitrack motion capture system. The system consists of 12 Optitrack motion capture cameras. The circled cameras were the four different perspectives selected in our evaluation. For reference, the Crazyflie measures approximately 8×8 cm.

Similar to the robotic arm scenario, relative angles were established between the markers to track the system state for data exfiltration. To evaluate the feasibility of decoding the cyber-physical encoded data, we utilized video recordings of replays from the Optitrack system. Recordings were taken at 30 frames per second and can be zoomed in to simulate an enhanced observer tracking the drone to decode exfiltrated data. For simplicity, we demonstrate the impact a monitor can have when an attacker physically exfiltrates data through the drone's yaw.

5.2 Evaluating Across Different Defense Models

Flight Missions and Encoding Schemes. We consider two drone flight missions. In case 1, we start by considering a simple drone mission, in which the drone is tasked to execute a constant hover at 0.5 m. In case 2, we allow the monitor to check for both the position and yaw variable of the drone. Because the monitor would be able to easily detect a change in yaw in the stationary hovering case, we chose a slightly more complex drone task of flying in a one-meter radius circle 0.5 m from ground level. For both missions, we consider the following encoding schemes:

- **Encoded Bit 1** Attacker yaws drone approximately 5° counter-clockwise from the start. The attacker yaws back to the reference to complete the transmission.
- **Encoded Bit 0** Attacker yaws drone approximately 5° clockwise. The attacker yaws back to the reference to complete the transmission.

Baselines. Baselines with no attacker perturbation were first found for both the hover and circle scenarios. For the hover scenario, the observed ambient noise levels were low. For the circle (surveillance) scenario, the drone conducted circles with no attacker perturbation. Figure 5 shows the baseline position error and yaw of the drone as it completes ten circles with no attacker perturbation. Furthermore, Fig. 5 details five segments in each subplot, depicting the respective amount of position error or yaw as the drone navigates two full circle paths, resets, and starts again. We observe the XYZ position error variances are .00123, .00101, and .0001 m^2, for the case of no attacker, respectively.

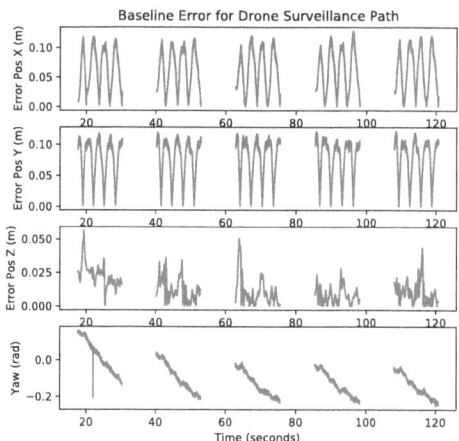

Fig. 5. Baseline errors for drone surveillance path.

Case 1: Local Autonomy and Estimation. In this case, the attacker is encoding into the yaw variable that is not being estimated by the monitor. The attacker exfiltrates using the first-bit encoding scheme. We repeat this process for encoding speeds of 1 bit/s, 2 bit/s and 5 bit/s. Figure 6 illustrates the hovering sequences of the drone. The data presented in Fig. 6 furthermore shows the yaw of the drone does not undergo significant drift as the attacker perturbs the system. Simple threshold values are sufficient for the monitor to detect an attacker in this experimental setup. Table 2 summarizes these results.

Table 2. Monitor Results for Case 1. The monitor has knowledge of attacker encoding strategy. Monitor uses threshold detection levels to detect attack.

Freq	Mean	Thresh Low	Thresh High	Accuracy
1 Hz	0	−.025	.025	100%
2 Hz	−.125	−.035	.035	93.75%
5 Hz	−0.14	−.030	0.30	93.75%

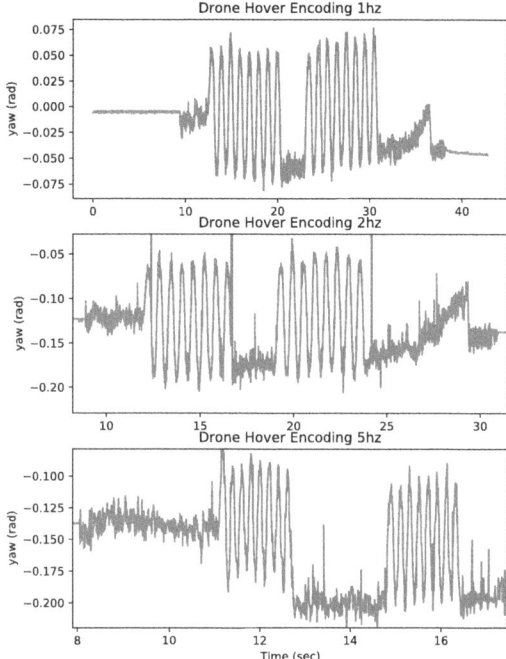

Fig. 6. Drone yaw (radians) vs flight time (seconds) for exfiltrating alternating 1's and 0's at 1 Hz, 2 Hz and 5 Hz frequencies

Isolating a single channel in Fig. 7 reveals a signal with acceptable signal-to-noise ratios (dB) for decoding (with sufficient signal processing). Of note in Fig. 7, as the encoding frequency approaches the channel capacity, physical system constraints become apparent, as the drone must either apply/experience greater forces or make smaller rotations (observed) to maintain the 5 Hz bitrate.

Case 2: Remote external feedback control. In this case, the attacker is encoding into the yaw variable as the drone flies along its circular surveillance path. Figure 8 shows the 3D position error of the CPS under encoding scheme 1. The attacker's encoding begins at each vertical red line, respectively. In this example, we first give the monitor the sub-task of monitoring the position of the drone. Figure 8 shows the position error traces as the drone follows the circular reference path. One key importance of Fig. 8 is the fact that the start of the attacker's encoding signal does not influence the baseline error of the 3D position (as seen in Fig. 5). This is further evident when comparing the associated XYZ position error variance levels from the attacker case (.00116, .00098, .000118) m^2, respectively. Thus, we observe the yaw attack has no effect on the position error variance levels, and the attack is unobservable.

Figure 9 depicts the yaw error about the reference path of the drone as the attacker performs encoding scheme one again. In this case, the monitor solely monitors the yaw variable, while the attacker perturbs this channel. The start

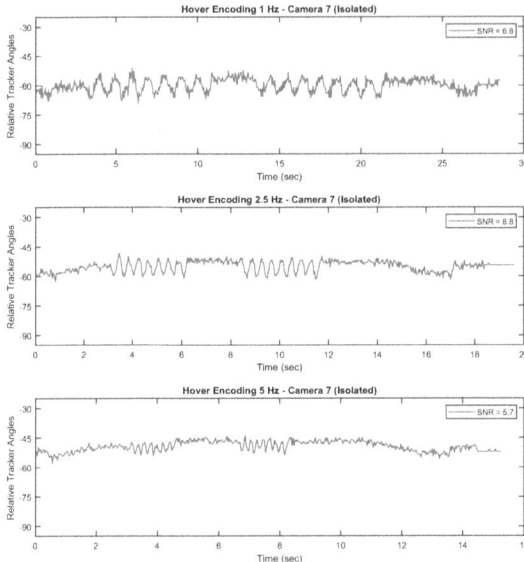

Fig. 7. The isolated, visually-reconstructed drone encoding trace for different encoding frequencies. Approximate peak-to-peak SNRs: 1 Hz = 6.8, 2.5 Hz = 8.8, 5 Hz = 5.7.

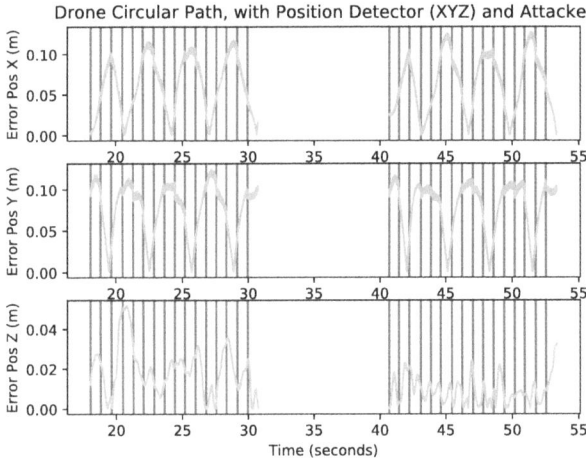

Fig. 8. Error in XYZ Position as attacker encodes bit sequence through yaw. Encoding is unobservable when monitoring just position. The start of the attacker's bit encoding is represented in red. (Color figure online)

of each attacker encoding is represented by a vertical red line. We see that maximums and minimums of yaw directly align with the attacker's encoding frequency over this channel.

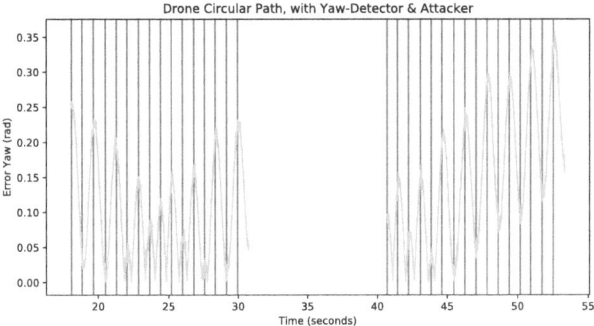

Fig. 9. Error in yaw (radians) vs flight time (seconds). Error in yaw is in sync with the start of each attacker perturbation command (red lines) (Color figure online)

As seen from Fig. 9, the drone's yaw slightly drifts as the attacker perturbs the drone's heading during its circular flight. We see implementing a simple thresholding technique here will not be as robust when compared to the hover case, since the flight data is clearly non-stationary in this example. Given that the monitor is monitoring the yaw state variable and knows the attacker's bit encoding scheme, a simple local min/max extrema search would detect the attacker's presence and encode the bit sequence. Table 3 displays the monitor's accuracy in correctly detecting the exfiltrated data through thresholding and local extrema finding.

Table 3. Monitor Results for Case 2: The monitor has knowledge of attacker encoding strategy. Comparison of threshold detection and local extrema accuracy.

Technique	Accuracy
Local Extrema	90.6%
Thresholding	53.1%

6 Discussion

We briefly discuss the practicality of such attacks as well as the practical design of defensive countermeasures. Additionally, we discuss the limitations of our work in the context of stronger defense models, and approaches for satisfying inherent issues relating to domain specificity and finding covert physical channels.

Practicality and Efficacy of Attacks. The attacks presented in this paper are significantly more complicated than the previous air-gapped attacks, e.g., encoding data into LEDs [36] is much easier than encoding into the movement of a drone while maintaining the utility of the application. However, such attacks may also be easier to mitigate as one can simply physically disable unnecessary actuators to harden the systems, e.g., by removing any LEDs from the system when possible. It is much less feasible to constrain particular movements of a CPS. Further, the attacks presented in this paper simple tasks and encodings. For more complex systems with more physical degrees of freedom, e.g., a swarm of drones or a factory automation floor with several robotic arms, more sophisticated encoding and decoding mechanisms can be instrumented to both increase the rate of data transmission as well as to further obfuscate the encoding scheme.

Practical Defensive Measures. The state estimators utilized by the monitors in this paper were idealistic as we used the Optitrack motion capture system that has sub-millimeter accuracy and typically requires significant calibration for a small and limited space. In reality, localization and state estimation of drones in the wild is much noisier and less predictable. In such cases, state estimation may not be reliable enough to detect an attacker and a monitor may need to rely on a means of attesting the software that is running on the CPS. Recent works have made strides towards attesting the behavioral integrity [3,44] as well as the integrity of controller software [21,43] in the context of industrial control systems. However, these have yet to be generalized to more complex CPSs such as drones.

Discovery of Physical Covert Channels and Encodings. In this work, we assume the attacker knows which physical channel to exploit. The search space for these channels can quickly explode with the complexity of the target CPS and the scenario. Future work can explore the formalization of the search space for covert physical channels through the use of simulators. Recent work [9,11] showed that given a CPS simulation model and desired attacker behavior against CPS security requirements in the form of signal temporal logic (STL) expressions along with security requirements, one can use Bayesian optimization to search for stealthy attack signals. While their approach focused on compromising the CPS, one can leverage these notions to discover stealthy CPS covert channels. A similar approach can be used for identifying encodings. While we discussed the use of STL to help with testing encodings, this approach could also be used to automate the generation and testing of different encodings against a physical system's specifications.

7 Related Work

In this section, we will discuss some of the related work on data exfiltration via covert channels as well as the generation of side channel attacks.

Air-gapped Covert Data Exfiltration. There is a large body of research on the topic of physical covert channel data exfiltration across air-gapped systems. Multiple works have shown that electromagnetic signals emitted from devices, e.g., signals from video displays, GSM frequencies emitted from workstations [24], or USB-generated electromagnetic emissions, can be picked up by mobile phones to establish a physical covert channel. Magnetic fields emitted from audio systems have been leveraged to inject malicious commands and compromise the voice interaction systems [35]. Power consumption has also been utilized as a transmitter of data by modulating the CPU utilization [26]. Similarly (but at a larger scale), it was shown that two PLCs in the context of an industrial power grid can communicate covertly with each other by modulating their associated actuators in a stealthy manner [19]. However, these systems are not necessarily air-gapped as they have direct access to the cyber-physical sensors. Thermal emissions between two PCs have also been utilized to establish bi-directional communication [25]. Acoustic covert channels have been utilized to exfiltrate data from physical hard drive noises, commodity desktop speakers, or desktop fans. It has even been shown that proprietary information of 3D printed models can be divulged from the noise of the motors [6]. There have also been several works that have shown a similar approach to optically encode and decode information by utilizing LEDs [36]. In all of these cases, these attacks were presented informally and, to the best of our knowledge, our work is the first to propose a control-theoretic model of covert physical channel exfiltration while maintaining *stealthiness* as well as the utility of the respective cyber-physical application. Despite the attempts to formalize the notion of *side-channels*, almost all of these related works do not implement these attacks in the context of cyber-physical applications and the associated proposed countermeasures discuss physical isolation or procedural security that are not applicable to cyber-physical edge devices in the wild.

Generating Cyber-physical Side Channel Attacks. The notion of an information theoretic model for side channels has been discussed to describe what an attacker may derive from other types of side channels. In these attacks, an attacker can query a system to observe its characteristics and infer characteristics about a secret key given a limited number of queries [32,41]. Note that the analysis of side channels is subsumed by our physical covert channel analysis as side channels are cyber-physical dependencies stemming from the memory-mapped I/O of the system. An attacker or a monitor can utilize our approach to analyze the cyber-physical dependencies of memory-mapped I/O and uncover possible side channels that may leak information. The major difference is that our model assumes an attacker can compromise the CPS binary to encode information and instrument side channels as covert channels.

8 Conclusion

In this paper, we characterized covert data exfiltration over air-gapped cyber-physical channels in the context of CPS applications on the edge. In particular, we formalized how an attacker may maintain the stealthiness and utility of a cyber-physical application while maximizing the rate of transmission. We detailed how to practically model attackers and monitors in this context using real-world examples of an industrial control system as well as an autonomous drone surveilling an area. We finally discuss the limitations of current defensive measures and discuss appropriate countermeasures.

Acknowledgements. The research reported in this paper was sponsored in part by the National Science Foundation (NSF) under awards FMitF-2425711, CNS-2231651, CNS-1705135, and CNS-2211301, Accenture, the Department of Energy (DOE) under award DE-CR0000056, and the DEVCOM ARL under Cooperative Agreement #W911NF-17-2-0196. The views and conclusions contained in this document are those of the authors and should not be interpreted as representing the official policies, either expressed or implied, of the funding agencies. Supported by the National Science Foundation (NSF) and Accenture. Mani Srivastava holds concurrent appointments as a Professor of ECE and CS (joint) at UCLA and as an Amazon Scholar. This paper describes work performed at UCLA and is not associated with Amazon.

References

1. OptiTrack - motion capture and and 3d tracking systems (2024). https://optitrack.com/applications/robotics/
2. Abdo, A., Malek, S.M.B., Zhao, X., Abu-Ghazaleh, N.: Avmon: securing autonomous vehicles by learning control invariants and residual prediction. In: Symposium on Vehicle Security and Privacy (VehicleSec) (2024)
3. Adepu, S., et al.: Control behavior integrity for distributed cyber-physical systems. In: 2020 ACM/IEEE 11th International Conference on Cyber-Physical Systems (ICCPS), pp. 30–40. IEEE (2020)
4. Ahmed, C.M., Mathur, A.P., Ochoa, M.: Noisense print: detecting data integrity attacks on sensor measurements using hardware-based fingerprints. ACM Trans. Priv. Secur. (TOPS) **24**(1), 1–35 (2020)
5. Akowuah, F., Kong, F.: Real-time adaptive sensor attack detection in autonomous cyber-physical systems. In: 2021 IEEE 27th Real-Time and Embedded Technology and Applications Symposium (RTAS), pp. 237–250. IEEE (2021)
6. Al Faruque, M.A., Chhetri, S.R., Canedo, A., Wan, J.: Acoustic side-channel attacks on additive manufacturing systems. In: 2016 ACM/IEEE 7th International Conference on Cyber-Physical Systems (ICCPS), pp. 1–10. IEEE (2016)
7. Bertsekas, D.P., Bertsekas, D.P., Bertsekas, D.P., Bertsekas, D.P.: Dynamic programming and optimal control. No. 2 in 1, Athena scientific Belmont, MA (1995)
8. Bonomi, F., Milito, R., Zhu, J., Addepalli, S.: Fog computing and its role in the internet of things. In: Proceedings of the First Edition of the MCC Workshop on Mobile Cloud Computing, pp. 13–16. ACM (2012)
9. Cairoli, F., Paoletti, N., Bortolussi, L.: Conformal quantitative predictive monitoring of stl requirements for stochastic processes. In: Proceedings of the 26th ACM International Conference on Hybrid Systems: Computation and Control, pp. 1–11 (2023)
10. Camurati, G., Francillon, A.: Noise-sdr: arbitrary modulation of electromagnetic noise from unprivileged software and its impact on emission security. In: 2022 IEEE Symposium on Security and Privacy (SP), pp. 1193–1210. IEEE (2022)
11. Chandratre, A., Hernandez Acosta, T., Khandait, T., Pedrielli, G., Fainekos, G.: Stealthy attacks formalized as stl formulas for falsification of cps security. In: Proceedings of the 26th ACM International Conference on Hybrid Systems: Computation and Control (2023)
12. Dash, P., Karimibiuki, M., Pattabiraman, K.: Out of control: stealthy attacks against robotic vehicles protected by control-based techniques. In: Proceedings of the 35th Annual Computer Security Applications Conference, pp. 660–672 (2019)

13. Ding, A., Chan, M., Hass, A., Tippenhauer, N.O., Ma, S., Zonouz, S.: Get your cyber-physical tests done! data-driven vulnerability assessment of robotic aerial vehicles. In: 2023 53rd Annual IEEE/IFIP International Conference on Dependable Systems and Networks (DSN), pp. 67–80. IEEE (2023)
14. Ding, A., Murthy, P., Garcia, L., Sun, P., Chan, M., Zonouz, S.: Mini-me, you complete me! data-driven drone security via dnn-based approximate computing. In: Proceedings of the 24th International Symposium on Research in Attacks, Intrusions and Defenses (RAID), pp. 428–441 (2021)
15. Donzé, A., Maler, O.: Robust satisfaction of temporal logic over real-valued signals. In: International Conference on Formal Modeling and Analysis of Timed Systems, pp. 92–106. Springer (2010)
16. Falliere, N., Murchu, L.O., Chien, E.: W32. stuxnet dossier. White paper, Symantec Corp., Secur. Resp. **5**(6), 29 (2011)
17. Freivald, M.P., Richards, M.S., Noble, A.C.: Change-detection tool indicating degree and location of change of internet documents by comparison of cyclic-redundancy-check (crc) signatures (Apr 27 1999), uS Patent 5,898,836
18. Garcia, L., Brasser, F., Cintuglu, M.H., Sadeghi, A.R., Mohammed, O.A., Zonouz, S.A.: Hey, my malware knows physics! attacking plcs with physical model aware rootkit. In: NDSS (2017)
19. Garcia, L., Senyondo, H., McLaughlin, S., Zonouz, S.: Covert channel communication through physical interdependencies in cyber-physical infrastructures. In: 2014 IEEE International Conference on Smart Grid Communications (SmartGridComm), IEEE (2014)
20. Gatlin, J., et al.: Encryption is futile: reconstructing 3d-printed models using the power side-channel. In: Proceedings of the 24th International Symposium on Research in Attacks, Intrusions and Defenses (RAID), pp. 135–147 (2021)
21. Ghaeini, H.R., et al.: Patt: physics-based attestation of control systems. In: International Symposium on Research in Attacks, Intrusions, and Defenses (RAID), Springer (2019)
22. Giernacki, W., Skwierczyński, M., Witwicki, W., Wroński, P., Kozierski, P.: Crazyflie 2.0 quadrotor as a platform for research and education in robotics and control engineering. In: 2017 22nd International Conference on Methods and Models in Automation and Robotics (MMAR), pp. 37–42. IEEE (2017)
23. de Gortari Briseno, J., Singh, A.D., Srivastava, M.: Inkfiltration: using inkjet printers for acoustic data exfiltration from air-gapped networks. ACM Trans. Priv. Secur. **25** (2022)
24. Guri, M., Kachlon, A., Hasson, O., Kedma, G., Mirsky, Y., Elovici, Y.: Gsmem: data exfiltration from air-gapped computers over gsm frequencies. In: USENIX Security Symposium, pp. 849–864 (2015)
25. Guri, M., Monitz, M., Mirski, Y., Elovici, Y.: Bitwhisper: covert signaling channel between air-gapped computers using thermal manipulations. In: Computer Security Foundations Symposium (CSF), 2015 IEEE 28th, pp. 276–289. IEEE (2015)
26. Guri, M., Zadov, B., Bykhovsky, D., Elovici, Y.: Powerhammer: Exfiltrating data from air-gapped computers through power lines. IEEE Trans. Inf. Forensics Secur. **15** (2019)
27. Han, Y., Chan, M., Aref, Z., Tippenhauer, N.O., Zonouz, S.: Hiding in plain sight? on the efficacy of power side {Channel-Based} control flow monitoring. In: 31st USENIX Security Symposium (USENIX Security 22), pp. 661–678 (2022)
28. Khan, H.A., et al.: Idea: Intrusion detection through electromagnetic-signal analysis for critical embedded and cyber-physical systems. IEEE Trans. Dependable Secure Comput. **18**(3), 1150–1163 (2019)

29. Kim, C.H., et al.: Securing real-time microcontroller systems through customized memory view switching. In: NDSS (2018)
30. Kim, H., et al.: A systematic study of physical sensor attack hardness. In: 2024 IEEE Symposium on Security and Privacy (SP), pp. 143–143. IEEE Computer Society (2024)
31. Kim, T., et al.: Reverse engineering and retrofitting robotic aerial vehicle control firmware using dispatch. In: Proceedings of the 20th Annual International Conference on Mobile Systems, Applications and Services, pp. 69–83 (2022)
32. Köpf, B., Basin, D.: An information-theoretic model for adaptive side-channel attacks. In: Proceedings of the 14th ACM Conference on Computer and Communications Security, pp. 286–296. ACM (2007)
33. Lee, C., Kim, D., Kim, G., Lee, S., Kim, T.: Lta: Control-driven uav testing and bug localization with flight record decomposition. In: Proceedings of the 22nd ACM Conference on Embedded Networked Sensor Systems, pp. 450–463 (2024)
34. Liu, C., DiValentin, L., Ding, A., Salem, M.B.: Build a computationally efficient strong defense against adversarial example attacks. In: ICISSP, pp. 358–365 (2024)
35. Liu, T., et al.: Magbackdoor: beware of your loudspeaker as a backdoor for magnetic injection attacks. In: 2023 IEEE Symposium on Security and Privacy (SP), pp. 3416–3431. IEEE Computer Society (2023)
36. Nassi, B., Pirutin, Y., Galor, T., Elovici, Y., Zadov, B.: Glowworm attack: optical tempest sound recovery via a device's power indicator led. In: Proceedings of the 2021 ACM SIGSAC Conference on Computer and Communications Security, pp. 1900–1914 (2021)
37. Papernot, N., McDaniel, P., Jha, S., Fredrikson, M., Celik, Z.B., Swami, A.: The limitations of deep learning in adversarial settings. In: 2016 IEEE European Symposium on Security and Privacy (EuroS&P), pp. 372–387. IEEE (2016)
38. Park, S., Kim, Y., Lee, D.H.: Scvmon: data-oriented attack recovery for rvs based on safety-critical variable monitoring. In: Proceedings of the 26th International Symposium on Research in Attacks, Intrusions and Defenses (RAID), pp. 547–563 (2023)
39. Sehatbakhsh, N., Yilmaz, B.B., Zajic, A., Prvulovic, M.: A new side-channel vulnerability on modern computers by exploiting electromagnetic emanations from the power management unit. In: 2020 IEEE International Symposium on High Performance Computer Architecture (HPCA), pp. 123–138. IEEE (2020)
40. Sun, P., Garcia, L., Zonouz, S.: Tell me more than just assembly! reversing cyber-physical execution semantics of embedded IoT controller software binaries. In: 2019 49th Annual IEEE/IFIP International Conference on Dependable Systems and Networks (DSN). IEEE (2019)
41. Tang, M., et al.: Modelguard: information-theoretic defense against model extraction attacks. In: 33rd USENIX Security Symposium (Security 2024) (2024)
42. Thibeault, Q., Anderson, J., Chandratre, A., Pedrielli, G., Fainekos, G.: Psy-taliro: a python toolbox for search-based test generation for cyber-physical systems. In: Formal Methods for Industrial Critical Systems: 26th International Conference, FMICS 2021, Paris, France, 24–26 August 2021, Proceedings 26, pp. 223–231. Springer (2021)
43. Tychalas, D., Benkraouda, H., Maniatakos, M.: {ICSFuzz}: Manipulating {I/Os} and repurposing binary code to enable instrumented fuzzing in {ICS} control applications. In: 30th USENIX Security Symposium (USENIX Security 21), pp. 2847–2862 (2021)

44. Zhou, X., Ahmed, B., Aylor, J.H., Asare, P., Alemzadeh, H.: Hybrid knowledge and data driven synthesis of runtime monitors for cyber-physical systems. IEEE Trans. Depend. Secure Comput. (2023)
45. Zhou, X., Kouzel, M., Alemzadeh, H.: Robustness testing of data and knowledge driven anomaly detection in cyber-physical systems. In: 2022 52nd Annual IEEE/IFIP International Conference on Dependable Systems and Networks Workshops (DSN-W), pp. 44–51. IEEE (2022)

AcousticScope: Understanding Biases in Voice Interaction via Automated Acoustic Testing

Shuhao Zhang[1(✉)], Mohammed Aldeen[1], Song Liao[2], Jeffery Young[1], and Long Cheng[1]

[1] Clemson University, Clemson, SC, USA
{shuhaoz,mshujaa,jay8,longcheng2}@clemson.edu
[2] Texas Tech University, Lubbock, TX, USA
song.liao@ttu.edu

Abstract. With advancements in AI and machine learning, Virtual Personal Assistants (VPAs) have become integral to daily life, enhancing efficiency by managing schedules, setting reminders, and sending emails. They also connect with smart home devices to control lights, temperature, and door locks, increasing convenience and comfort. VPAs enable natural human-computer interactions through voice commands, utilizing advanced speech recognition and natural language processing. Automatic Speech Recognition (ASR) technology, which is crucial for VPAs, allows them to recognize and transcribe human speech into text, facilitating seamless human interaction. However, as VPAs are increasingly used worldwide by people from diverse regions and ethnic backgrounds, it has been observed that VPAs exhibit varying responses based on users' demographic characteristics, such as race and gender. This discrepancy stems from the racial and gender biases inherent in the ASR technology employed by VPA systems. Although current researches have investigated the phenomenon of social bias in VPAs systems, these studies typically involve collecting voice data from volunteers and testing VPAs with the collected samples, which is time-consuming process that does not facilitate large-scale testing of VPAs systems. In this work, we designed and developed an automated voice testing tool called AcousticScope to explore social bias issues in VPAs systems. AcousticScope can generate voices representing different racial groups (White, Black, Indian, Chinese, and Kenyan) and genders (Male and Female) to interact with VPA systems, and automatically analyze the interaction content to assess social biases within VPAs systems. Our findings indicate that the Amazon Alexa Echo system exhibits racial and gender biases, with higher speech recognition accuracy for the White group (84.87%) compared to the Black group (70.42%) and the Indian group (75.21%). In terms of gender bias, Alexa demonstrates higher recognition accuracy for female voices at (74.55%) compared to (67.87%) for male voices. We also find that Alexa's recognition accuracy for adult voices (86.19%) is higher than that for children's voices (53.03%). Additionally, we tested speech-to-text tools from Google, Microsoft, and IBM, and the results indicate the existence of social biases in these ASR systems.

Keywords: Social bias · Virtual Personal Assistants · Acoustic Speech Recognition · Machine Learning · Tool Design

1 Introduction

Artificial intelligence and machine learning technologies have been widely used in commercial Virtual Personal Assistants (VPAs) systems, such as Apple Siri and Amazon Alexa, have become ubiquitous in daily life. These systems rely on advanced speech recognition technology to understand and respond to users' requests, making interactions more natural and seamless. However, when users interact with VPAs using voice commands, these systems may fail to respond correctly [21,23]. Since VPA systems rely on built-in automatic speech recognition (ASR) models to recognize human speech, and different companies use different audio datasets to train their ASR models, the quality of these models varies. A growing number of studies [11,19,32,35] have highlighted the prevalence of racial and gender biases in speech recognition systems, including state-of-the-art VPAs systems. Metz *et al.* [32] discussed the biases in speech recognition systems developed by major tech companies such as Apple, Amazon, and Google, finding that these systems are significantly less accurate in recognizing speech from black speakers compared to white speakers. Similarly, Robison *et al.* [35] highlighted that VPAs such as Siri, Alexa, Google Assistant, and Cortana also exhibit gender bias. Furthermore, Walker *et al.* [41] emphasized that attackers can exploit racial and gender biases in VPAs systems to interfere with and block their regular operation.

Although existing work [23,24,31] has revealed the presence of social bias in ASR systems, there are still some limitations. Typically, these studies collect existing voice datasets from the Internet or gather volunteers' voices to generate test datasets. This very time-consuming process hinders large-scale testing of VPAs systems. Additionally, due to the inherent limitations of these datasets or geographical constraints, the collected voice data often fails to represent races and groups comprehensively. To address these limitations, we developed an automated tool capable of testing the biases in VPAs systems named AcousticScope. This tool simulates interactions from users of diverse backgrounds with VPAs devices and automatically analyzes the accuracy of the VPAs devices' responses to various user commands.

To design AcousticScope, we faced two primary challenges: First, how to develop a method that can accurately capture the voice responses generated by VPAs. The next challenge was devising an effective method that can simulate human interactions with these devices. To address the above challenges, we designed and developed an automated testing tool called AcousticScope. It implements an automatic voice sampling algorithm for VPAs systems and uses ChatGPT to simulate human users' interaction with VPAs devices. Furthermore, it uses advanced AI speech synthesis technology to generate natural speech across various demographics, including different ages, ethnicities, and genders. This allow us to measure biases in the VPAs systems against different user groups.

In our project, we used our AcousticScope to generate ten voices, compared the spectrum of our tool synthesized speech and authentic human voices, and analyzed AcousticScope's detection results of VPAs systems bias. The results show that AcousticScope performs well in the naturalness and diversity of synthesized speech, and the spectrum analysis shows that synthesized speech is highly similar to the authentic human voice. What is more, the bias detection results further validated the effectiveness of AcousticScope in identifying racial and gender biases in VPA systems, providing a reliable tool and method for future large-scale testing of VPA systems. We summarize our contributions as follows:

- **Automated Testing Tool For Voice Personal Assistants (VPAs):** We designed and implemented an automatic bias testing tool for VPA systems named AcousticScope. To our knowledge, AcousticScope is the first automated testing solution explicitly designed to assess bias in VPA systems comprehensively. We plan to make our tool available to researchers and VPA system developers to facilitate future research.
- **Bias Measurement in ASR Systems:** With AcousticScope, we test voices from five racial groups (White, Black, Indian, Chinese, and Kenyan) and two gender accents (Male and Female). Our results reveal significant racial and gender biases in VPAs and speech-to-text tools. Specifically, Amazon's Alexa shows a preference for white and female voices. Additionally, Alexa has a higher recognition rate for adult voices than children's. We also tested three speech-to-text tools from Google, Microsoft, and IBM with AcousticScope, and the test results show that all three tools have lower word error rates (WERs) for white voices, indicating pervasive biases in these systems.

2 Background

2.1 Bias in VPAs Systems

Considering the diverse demographic distribution of VPA users across various countries, regions, races, and genders, there are notable disparities in the speech recognition performance of VPA devices across different groups. This has raised substantial concerns regarding social bias within VPA systems. The ASR technology in these systems is built on machine learning models trained using specific datasets and algorithms. However, these datasets often harbor inherent biases, and the optimization strategies employed in the algorithms may exacerbate the issue, leading to biased outcomes in the resulting models. In this study, we define ASR bias as the variations in speech recognition and transcription accuracy observed across different racial groups within ASR systems. It has been consistently shown that ASR systems tend to perform more accurately when recognizing the speech of white speakers, while their performance significantly declines for non-white groups. This disparity highlights the limitations of ASR systems in adequately serving diverse user populations.

Dataset Bias. Koenecke *et al.* [23] conducted an analysis of the ASR systems developed by five major tech companies: Apple, Google, Amazon, Microsoft, and IBM. They tested the systems using two datasets: the Corpus of Regional African American Language (CORAAL) and Voices of California (VOC). These datasets included 19.8 h of audio, gathered through interviews with approximately 115 volunteers from five different cities in the U.S., comprising 42 white and 73 Black speakers. The study revealed a significant disparity in the systems' performance: for Black speakers, the average Word Error Rate (WER) was 0.35, compared to 0.19 for white speakers. This means that, on average, the ASR systems failed to correctly transcribe 35 out of every 100 words spoken by Black speakers, whereas for white speakers, the systems only misinterpreted 19 out of 100 words. This difference is substantial and indicates a clear performance gap across all five ASR systems when processing speech from Black speakers.

The disparity is largely attributed to the lack of diversity in the training data used to develop these systems. Most ASR models are trained on datasets that predominantly feature speech from well-educated, middle-aged white males and females, often drawn from audiobook recordings. As a result, the systems struggle to recognize and transcribe speech patterns commonly used by Black speakers, particularly African American Vernacular English (AAVE). Since the acoustic models developed by these companies have not been sufficiently trained on audio data from AAVE speakers, they exhibit poor performance in recognizing Black speech. This underrepresentation in the training data highlights how biases can be embedded in ASR systems, resulting in unequal performance across different racial groups.

Algorithm Bias. In ASR systems, algorithmic bias usually stems from an imbalanced distribution of training data *et al.* [23], which directly affects the performance of the model. The core of the ASR system is a speech-to-text conversion model based on deep neural networks (such as convolutional neural networks, recurrent neural networks, or transformer models). The training of these models relies on a large amount of annotated speech data. When the speech data of certain groups accounts for too large a proportion of the training set, the model tends to optimize the speech recognition of these groups. Specifically, the ASR system recognizes different speech inputs by learning specific features. These features include acoustic features (such as spectral features, formants, etc.) and linguistic features (such as phonemes, lexical and syntactic structures, etc.). When there is a lot of data for a certain group, the model will be fully trained on these features, thereby improving the recognition accuracy of the group's speech. For groups with scarce data, due to insufficient speech samples, the model lacks sufficient diversity and generalization ability when extracting features of these groups, resulting in an increase in the recognition error rate Word Error Rate (WER).

In addition, the speech enhancement, feature extraction, and language modeling techniques commonly used in ASR systems may further exacerbate this bias [15,29,36]. For example, in the feature extraction stage, acoustic feature extraction algorithms such as MFCC (Mel-Frequency Cepstral Coefficients) or PLP (Perceptual Linear Prediction) may perform poorly for certain dialects or accents, especially when these dialects or accents are not fully included in the training data. Furthermore, in language models (such as those based on RNN or transformers), due to differences in language habits among different groups, the imbalanced distribution of languages in the training set may cause the language model to be more biased towards common language patterns during decoding, resulting in incorrect predictions for the speech of minority groups.

Algorithmic bias is also reflected in the post-processing stage [9], that is, through the joint decoding strategy of language model and acoustic model. Figure 1 shows the joint decoding strategy process. During decoding, the model combines acoustic features and language probabilities to predict the most likely word sequence. If the language usage pattern of a certain group is not fully represented, the decoder may mistakenly select words with high confidence but incorrect.

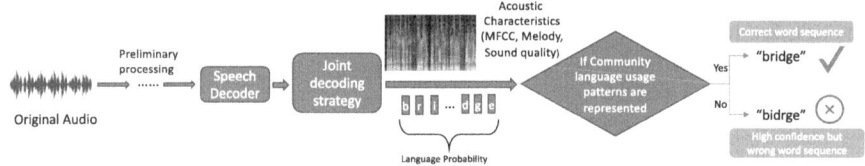

Fig. 1. Joint decoding strategy process

User Interaction Bias. In automatic speech recognition (ASR) systems, user interaction behaviors can be a significant source of bias. Variations in user speech, such as accent, speech rate, and volume, directly impact the accuracy of ASR systems. Since ASR models are often trained on speech samples from specific demographic groups, they tend to exhibit higher recognition accuracy for those groups, while showing increased error rates for others. Additionally, user language habits during interaction—such as vocabulary choice, grammatical structures, uncommon slang—and environmental factors like background noise, can exacerbate system bias. For instance, Koenecke *et al.* [23] demonstrates notable performance differences between ASR systems' recognition of Black and White speakers, linking racial bias to differences in pronunciation and prosody features, including rhythm, pitch, syllable stress, vowel duration, and tonal reduction. Further studies Chen *et al.* [13] examined speech characteristics across different racial and gender groups, revealing that while most acoustic feature differences among White, Black, Asian, and Hispanic subgroups were minimal, there were notable differences in specific metrics such as fundamental frequency (F0), formants (F1, F2), and entropy measures (PDF entropy, Perm entropy). These biases stemming from user interaction behaviors disproportionately affect minority groups, non-standard speakers, and non-native users, leading to higher error rates and a suboptimal user experience when interacting with ASR systems.

Furthermore, biased speech recognition systems are vulnerable to exploitation due to their discriminatory nature. Researchers [41] have demonstrated that attackers can exploit gender and racial biases in ASR systems to disrupt regular use. For instance, attackers might use gender and ethnic voices more easily recognized by ASR systems to interfere with normal user commands, disrupting or manipulating system responses. Such attacks pose significant threats to the integrity of speech recognition systems and the safety of their users.

Despite extensive research into bias in VPAs and continuous improvements by companies, addressing these issues remains challenging due to the complexity of natural language processing and the diversity of accents and dialects in human speech. To the best of our knowledge, there is currently no tool capable of conducting large-scale automated testing of VPAs devices to detect bias issues.

2.2 Voice and Interaction in VPAs

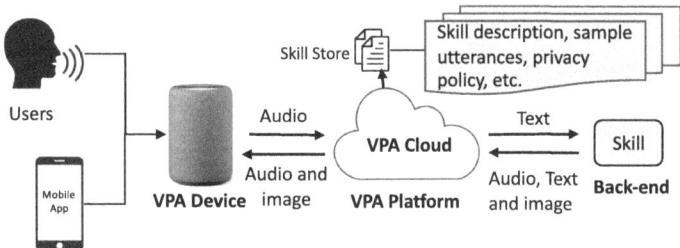

Fig. 2. VPA user interaction

Figure 2 illustrates the working principle and skill response process of VPAs devices. VPAs systems operate through software agents that provide services by receiving and executing human voice commands. Skills, akin to mobile apps on Android and iOS platforms, are primarily developed by third-party developers and are available through platforms like the Amazon Alexa Skills Store. Each skill has a dedicated web page displaying essential information, including the skill name, developer details, and sample utterances, which are a set of phrases provided by skill developers. Users can initiate interactions using a range of supported phrases and invocation names offered by Alexa. For example, saying "Alexa, open weather channel" automatically triggers the interaction.

The response process for each skill involves both front-end and back-end code. The front-end code primarily handles receiving and responding to user requests, while the back-end code analyzes the acquired user voice stream and generates the corresponding response. When a user issues a voice command to the VPA device, the device first sends the received audio stream to a cloud server (e.g., Amazon Web Services, AWS) for analysis. The server processes the audio, converts speech to text, identifies the most appropriate skill to handle the request, and returns the result to the VPA device, ultimately presenting the response to the user.

The response of a skill is also influenced by its algorithmic mechanism. Amazon designed a dynamic response mechanism to ensure diversity in Alexa's responses, allowing Alexa to generate varied responses to the same voice command. For instance, consider a skill named "Open Escape The School." Although the output follows a standard template ("Welcome to Escape the School, the game started..."), the specific content changes dynamically based on the user's actions each time. This dynamic response behavior enables users to experience slightly different interactions each time they use the same skill, making interactions more engaging and fresh. However, this leads to an issue: due to the dynamic response mechanism, Alexa generates various contents, making it difficult to compare two responses given the same input simply. How do we determine if VPAs correctly recognize a user input? We discuss analyzing Alexa's responses in Sect. 3.2 (Bias Assessment).

2.3 Skill Behavior

Skill behavior refers to the handlers and responses executed by Alexa skills when invoked by the user in accordance with the user's intention. These behaviors are dictated by the skill's design and functionality, which determine its interactions with users. The behavior of each skill varies according to the design specifications set by different developers. These behaviors encompass the functionalities and interactions that constitute an Alexa skill, enabling it to provide users with a wide range of services and experiences.

Skill behaviors can be categorized based on their intent. For instance, some skills are designed for information retrieval, providing users with updates on news, weather, or lifestyle information. Entertainment skills engage users through games, music, and storytelling. Trivia skills facilitate various types of question-and-answer dialogues and provide continuous feedback. When a skill accurately recognizes a user's command, it calls the corresponding handlers and performs the appropriate operation to provide feedback. If the skill fails to recognize the user's command, it generates responses like "Sorry, I don't understand" or "Sorry, I didn't get that, please say it again" to indicate that it did not recognize the command and prompt the user to re-enter it.

3 System Design

3.1 System Overview

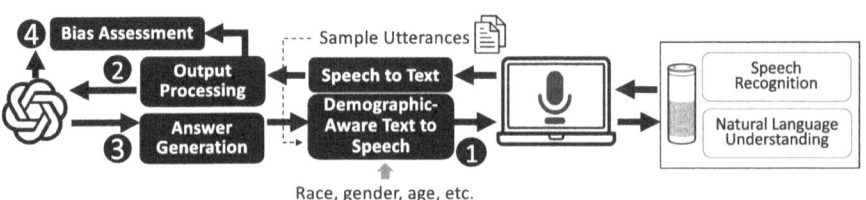

Fig. 3. System overview

Traditional testing methods require collecting large amounts of voice data, which is time-consuming and makes it difficult to fully cover accents of different races and genders. To address this issue, we designed and developed an automated bias speech testing tool called AcousticScope. Figure 3 presents the system overview of AcousticScope. First, we use the skill sample utterances from the Amazon Alexa Skill Store as input text for the initial interaction with Alexa (❶). The input can be any form of voice command; we utilize the skill sample utterances for batch-testing purposes. These utterances are then fed into the model, and an AI-driven text-to-speech tool converts them into various accents representing different races, genders, and ages from diverse regions and countries. By adjusting the parameters of the social-aware speech-to-text conversion module's parameters in AcousticScope, AcousticScope generates different accents to simulate daily interactions between users from various regions, races, and genders with Alexa. After generating speech with multiple accents, the VPAs systems are acoustically tested. First, the speech stream generated by the social-aware speech-to-text conversion module is transmitted to the Amazon Alexa Echo device through the speaker. When Alexa receives a user's voice command and generates a response, the microphone captures Alexa's voice response. The speech-to-text module then converts the captured audio into text format and stores text in the output processing module (❷). The output Processing module saves all the conversation content and then sends the converted text to ChatGPT. ChatGPT generates a text response by simulating human-like thinking and reasoning processes (❸). Subsequently, the generated text response is converted into speech via the social-aware speech-to-text conversion module and broadcast to the Amazon Alexa Echo device, simulating the interaction cycle between the user and the VPAs systems. Finally, all content generated during the interaction between Alexa and ChatGPT (saved in the output processing module) will undergo preprocessing before being sent to the bias analysis module in AcousticScope (❹). The bias analysis module will then analyze these interactions to evaluate the presence of social bias in the Amazon Alexa Echo device based on the analysis results.

3.2 System Design and Implementation

Demographic-Aware Speech Generation. Our goal is to leverage AI technology to synthesize the voices of diverse demographic groups to investigate social bias in VPAs systems. We utilized the generative AI speech platform Lovo AI [2], renowned for its consistently high-quality audio generation to mimic natural user speech closely. The cloned speech produced by Lovo AI closely resembles natural human voices. The Lovo AI platform also offers nearly a hundred voice samples encompassing various races, genders, and age groups, each equipped with an identity ID for quick reference. By employing the voice cloning technology provided by Lovo AI or modifying the identity parameters of the platform's voice samples, we can rapidly generate speech representing different countries, regions, races, and genders.

Output Processing. After generating accents representing different races and genders across various countries and regions, we broadcast the generated speech through a speaker to an Alexa Echo device and use a microphone to capture Alexa's voice response. To simulate the interactions of real users with Alexa, we utilize ChatGPT [34] to mimic human cognitive and behavioral processes during interactions with the Alexa Echo device. ChatGPT is an artificial intelligence model based on advanced natural language processing technology with the following features: (1) ChatGPT can understand complex language structures and semantics, accurately parsing the intent and content of user input to generate coherent and contextual responses, thereby imitating human communication and conducting natural conversations. (2) ChatGPT can maintain contextual memory across multiple rounds of conversations, ensuring that responses are relevant to the preceding text. It provides a coherent interactive experience and can adjust the language style and content according to the conversation's context, making communication more aligned with real-life scenarios. (3) ChatGPT possesses a broad knowledge base encompassing various fields, can provide information and insights on different topics, and can simulate human thinking logic for reasoning and problem-solving, aiding in answering questions or offering suggestions. These features demonstrate that ChatGPT can achieve nearly human-like interactive capabilities. Therefore, we input the converted text into ChatGPT to generate responses for interaction with the Alexa Echo device.

While capturing audio streaming from the Amazon Alexa Echo device, we encountered two main challenges: (1) Each Alexa skill has an unpredictable startup delay when awakened. This delay is caused by network latency, as the skill's response is generated by back-end code and sent back to the VPA through the cloud, rendering it uncertain when to initiate the listening process. If the listening process starts too early, the program will automatically exit because it has not captured any sound. Conversely, the captured voice will be incomplete if the listening process starts too late. (2) The response length of each Alexa skill varies, making it impractical to use a fixed duration to capture all responses generated by the Alexa skill. For instance, if the response content is short but the listening time is extended, Alexa will exit and interrupt the conversation. Conversely, recording all the voice content will be impossible if the response content is long but the listening time is insufficient. To solve this problem, we designed and implemented a dynamic monitoring mechanism based on existing speech-to-text technology. Our objective is to record complete voice data generated by each skill while precisely controlling the start and end time of the listening process. Algorithm 1 details the specific design of this dynamic monitoring mechanism. We first set four parameters for audio signal sampling: *Timeout (O)*, *Maximum Duration Time (T)*, *Silence Threshold (α)*, and *Sampling Rate (S)*. Next, we initialize variables related to recording audio. Timeout (O) controls the initial listening duration. Recording starts if an audio signal is detected within the timeout period; otherwise, it continues listening until the timeout. We then set the maximum duration time T to 120 s, ensuring that all voice stream data generated by the Amazon Echo device can be captured (Alexa's long response

Algorithm 1: VPAs Systems Speech Sampling Algorithm

1 **Parameters:** $Timeout(O)$; $Max\_duration(T)$; $Silence\_threshold(\alpha)$; $Sample\_rate(S)$
2 **Result:** *Processed audio output*
3 **initialize various parameters**
4 **while** $current\_time - start\_time < timeout$ **do**
5 $\quad chunk \leftarrow record\_duration$;
6 $\quad$ **if** $\max(abs(chunk)) \geq Silence\_threshold(\alpha)$ **then**
7 $\quad\quad audio\_chunks \leftarrow chunk$;
8 $\quad\quad$ **break**;
9 $\quad$ **end**
10 **end**
11 **while** $True$ **do**
12 $\quad chunk \leftarrow record\_duration$;
13 $\quad audio\_chunks \leftarrow chunk$;
14 $\quad$ **if** $\max(abs(chunk)) < Silence\_threshold$ **then**
15 $\quad\quad silent\_chunk\_count +1$;
16 $\quad\quad$ **if** $silent\_chunk\_count \geq max\_silent\_chunks$ **then**
17 $\quad\quad\quad$ **break**;
18 $\quad\quad$ **end**
19 $\quad$ **end**
20 $\quad$ **if** $audio\_chunks \geq Max\_duration(T)$ **then**
21 $\quad\quad$ **break**;
22 $\quad$ **end**
23 **end**
24 $output \leftarrow$ concatenate $audio\_chunks$;
25 **return** $output$;

content is typically around 30–60 s). We precisely control the start and end of the program by setting the silence threshold α. The silence threshold α is responsible for detecting voice activity. Within the timeout period (O), if the detected sound amplitude exceeds the set silence threshold, it indicates voice activity, and recording begins. Conversely, if the detected sound amplitude is lower than the silence threshold, it indicates the absence of voice activity, and the program continues monitoring until the timeout is reached.

During recording, we set the sampling rate (S) to 44.1 kHz, a common standard for high-quality audio recording, and collected the audio in fixed-length segments. Each segment is evaluated against the silence threshold (α). If multiple consecutive segments are silent, the recording stops; if it exceeds the maximum duration time (T), it also stops. After sampling, all collected audio fragments ("*audio chunks*") are combined into a continuous signal ("*audio*") and resampled to 16 kHz to meet the input requirements of the speech-to-text model. Next, we used the processor in the Wav2Vec2 model [8] to perform feature extraction and format the resampled audio data and utilized the Wav2Vec2 speech-to-text model to generate the transcribed text content. By setting the maximum duration time (T) and the silence threshold (α), we successfully captured the Amazon Echo device's voice dynamically, ensuring precise control over when the listening process starts and ends and the complete voice data stream was recorded.

Answer Generation. We aim to make AcousticScope think and respond like a natural human, so we employ ChatGPT to generate answers. After converting the captured Alexa speech to text, the transcribed text is subsequently fed into ChatGPT, which processes the query and generates responses tailored to the text request. The ChatGPT-generated text response is then input into Lovo AI's text-to-speech model, which converts the text into natural speech for transmission to Amazon Echo devices. It is noteworthy that we employed ChatGPT to emulate human-like cognitive processes. Through prompt engineering, we directed ChatGPT to generate responses for each skill interaction. We set two characters, A and B, and had ChatGPT play character B to generate answers based on input from character A. To illustrate, we employed few-shots prompting, *e.g.*, *"Focus on providing a direct and concise response to the input from 'A.' Your reply should be limited to 1–3 words, embodying the persona 'B.' Ensure your answer is a straightforward continuation of the conversation without additional explanations or unnecessary words. If the question cannot be directly answered, generate a fictional response within the same word limit, and look at the previous answers to understand the context"*. Additionally, we set the model temperature to 0.2 to make the model produce more focused and deterministic responses and explicitly requested concise answers due to Alexa's limitations in processing lengthy responses. We also instructed ChatGPT to avoid generating similar responses that have been included in the given array to explore diverse interaction branches. During the testing phase, each interaction with the skill was logged in an array and subsequently fed into ChatGPT. This approach enabled the model to grasp the complete context from previous interactions, thereby producing more accurate responses.

Bias Assessment. We collected all content generated during interactions. Each pair of responses is tagged with a corresponding skill ID and skill name to distinguish content generated by different skills. After collecting sufficient interactive content, we use an NLP-based method to analyze the data. First, we utilized two advanced natural language processing libraries, SpaCySpaCy [5] and LanguageTool [4], to perform data preprocessing on the collected text. The SpaCy library eliminates data noise like blank lines and converts all text to lowercase, while LanguageTool is employed to check and correct grammatical and spelling errors. After preprocessing the text data, we extract the interaction content between the simulated user and Alexa from the processed results and input it into ChatGPT for analysis. This analysis aims to determine whether Alexa has effectively responded to the user's instructions.

As shown in Fig. 4, the analysis begins by extracting Alexa's response and the corresponding simulated user's response from the preprocessed data. First, the user-generated response is input into ChatGPT so that it understands the instructions the user sends to Alexa. Then, the response generated by Alexa is input into ChatGPT to determine whether Alexa has made an accurate and effective response based on the user's instructions. Each analysis includes a user response and the corresponding Alexa response. If the reaction generated by Alexa is logically relevant to the user's command, the conversation is marked as "Understand"; otherwise, it is marked as "Not Understand." After the analysis is completed, the system returns documents with these tags.

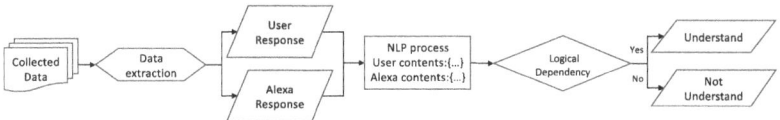

Fig. 4. Data flow of bias Assessment

To ensure the reliability and accuracy of our analysis, we employ prompt engineering to configure ChatGPT's behavior for improved results. ChatGPT, an advanced natural language processing model, can understand complex language structures and contextual semantic relationships. We frame the scenario as a conversation between two individuals and have ChatGPT determine whether the content of their conversation is logically related. By clearly defining logical relationships, providing contextual information, and using heuristic analysis techniques, ChatGPT is guided to analyze and reason about given sentences. This allows us to evaluate whether Alexa's responses are appropriate. For example, consider the following prompt: *"Imagine a conversation between two individuals, A and B. Analyze the interaction between A and B to assess whether B has correctly responded to A's command. If B's response logically relates to A's content, mark B as 'Understand.' Conversely, if B's response lacks logical relevance to A's content or includes replies such as 'Sorry, I did not understand,' 'I am not sure,' or other similarly negative meanings, mark B as 'Not Understand.'"*

We provide the following two examples to further illustrate how we use ChatGPT to analyze responses. As shown in Fig. 5(a), when Alexa begins interacting with the user, assume AcousticScope initiates with the sample command, *"Hi, Alexa, let's play a word game"* to wake up Alexa. If Alexa accurately recognizes the user's voice, it will generate a reply such as, *"Welcome to the word game. Let's play some word games. What food name can you make from the letters O, A, O, T, P, T?"*. Upon receiving Alexa's response, the user will generate the answer *"Potato"* based on the question and provide it to Alexa. If Alexa correctly recognizes the user's voice again, it will generate a reply such as, *"Correct! Let's play the next game."* Such a coherent and logical conversation will be treated as a valid response and marked as "Understand." Conversely, for an invalid response, as shown in Fig. 5(b), if Alexa does not recognize the user's voice upon initiation, it will generate replies such as *"Sorry, I can't understand," "Sorry, I didn't get that,"* or *"I am not sure, please try again later,"* indicating that the user command was not understood. Alternatively, Alexa may randomly generate a response entirely unrelated to the user command. This situation may arise from errors in Alexa's speech-to-text transcription, leading to a "misunder-

38 S. Zhang et al.

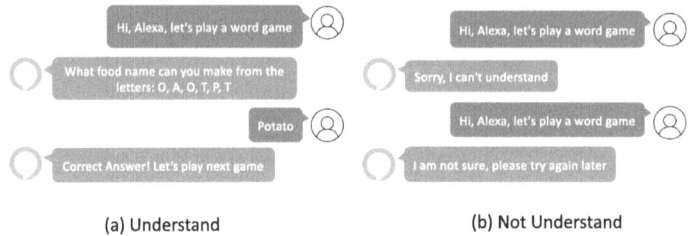

Fig. 5. Different responses generated by Alexa

standing" of the user command. In these cases, the response will be considered invalid and marked as "Not Understand."

4 Prototype Implementation

4.1 Experiment Setup

To determine the social bias in the Alexa system, we analyzed the content generated by Alexa for accents of different races and genders. Our initial step involved selecting test skills from the Amazon Alexa Skills Store. Each skill is identified by a unique skill ID number, skill name, and invocation example. As illustrated in Fig. 6, the skill name and invocation example can be obtained directly from the skill's web page, while the skill ID number is derived from the skill's web URL link. This skill ID uniquely identifies each skill. By default, each skill is initially disabled. Given that multiple skills with the same name exist in the Amazon Alexa Skills Marketplace, we used a single Amazon account to locate each skill by its unique ID number and then enabled the skills required for testing. This approach ensured consistency in the skills tested across all trials. We randomly selected 100 trivia skills from the Amazon Alexa Skills Marketplace that had been verified to perform their functions properly and generate meaningful output. The trivia category was chosen because these skills necessitate numerous

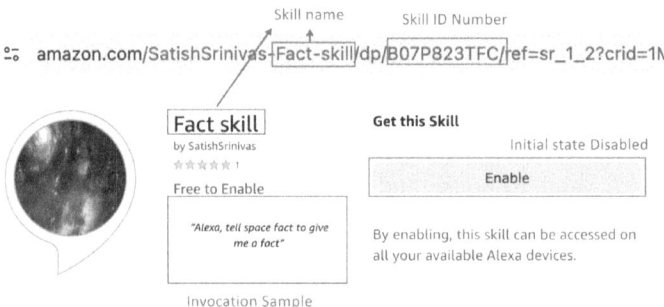

Fig. 6. Alexa skill sample utterance

user interactions and can produce lengthy content. Thus, we utilized trivia skills with multiple interactions to test for social bias in the Alexa Echo device.

The experiment was conducted in a soundproof and quiet environment to exclude external noise. We leveraged the Lovo AI tool to generate natural speech samples representing ten different demographics: White male and female, Black male and female, Chinese male and female, Indian male and female, and Kenyan male and female. And use selected 100 trivia skills to test all voice types. To minimize the impact of Alexa's dynamic reply mechanism on the experimental results, we used every kind of voice to test each skill 15 times to enrich the sample data as much as possible. During the entire experiment, we collected a total of 15,000 data samples.

4.2 Data Analysis

As mentioned in Sect. 3.2, our work evaluates potential social bias in Amazon's voice assistant, Alexa, by examining its effectiveness in responding to user commands. To this end, we established two classification labels: "Understand" and "Not Understand," and used ChatGPT to analyze the logic of the conversations. Additionally, to verify the reliability of the ChatGPT analysis results, we performed an manual analysis. In the manual analysis phase, we validated the analysis results generated by ChatGPT to ensure their accuracy and reliability. The detailed steps of the manual analysis process are as follows: (1) We randomly sampled 20% of the conversations from the entire dataset, ensuring that the samples were diverse in terms of content, user command complexity, accent, gender, and race, to better assess Alexa's performance across different scenarios. (2) Independent evaluators were selected as manual annotators, all of whom possessed relevant knowledge of speech recognition systems and had the ability to comprehend English conversation content. We ensured that these annotators were unbiased toward the experimental design and ChatGPT's analysis prior to their participation in order to avoid any preconceived notions. (3) The manual annotators employed the same classification labels as ChatGPT—"Understand" and "Not Understand"—to evaluate whether Alexa accurately understood each command. During the analysis, the annotators focused on whether Alexa's responses were appropriate, whether it correctly interpreted the user's intent, and whether any misunderstandings or biases were present. (4) The results of the manual analysis were compared to those generated by ChatGPT, identifying any potential discrepancies. In cases where discrepancies occurred, the research team conducted further in-depth analysis to determine whether adjustments to the ChatGPT model were necessary or if the context of the conversation needed to be reevaluated. Through these steps, we ensured the rigor of the analysis and the reliability of the results.

4.3 AcousticScope Extension

To expand the application of AcousticScope and investigate bias in various ASR systems, we evaluated AI-powered speech-to-text tools from Google AI [6],

Microsoft Azure [7], and IBM Watson AI [3]. Our primary objective was to identify potential biases in these systems using AcousticScope. Initially, we successfully accessed the speech-to-text services provided by Google AI, Microsoft Azure, and IBM Watson AI via their respective APIs. Then we randomly selected ten pieces of text content, each approximately one minute long, from online sources to serve as our test corpus. For this test, we focused on the voices of the white and black groups, as these two groups exhibit significant social and linguistic differences [23]. Racial bias in VPA systems can be more easily detected through comparative analysis. Using AcousticScope, we used four speeches, including white male, white female, black male, and black female voices, to read collected texts separately. This approach ensured that ten distinct speech samples were created for each accent, thereby mitigating errors from randomness and chance in speech recognition. Subsequently, we transcribed these speech samples using the speech-to-text tools from Google, Microsoft, and IBM and saved the transcribed results. In total, 40 speech samples were generated and tested throughout the experiment.

After testing and transcribing all speech samples, we utilized the Word Error Rate (WER), the most widely adopted metric for evaluating the performance of ASR systems. WER quantifies the percentage of words with errors (including substitutions, deletions, and insertions) in the transcribed text compared to the original text. Variations in WER can intuitively reflect the accuracy of different speech-to-text tools in recognizing speech from various races and genders. This analysis aids in identifying and understanding potential social biases in these ASR systems.

4.4 Experiment Device

Figure 7 illustrates our study's experimental setup and equipment. In this experiment, we used a single microphone to simulate the auditory function of the human ear and a speaker to simulate human vocal expression. Our voice model runs on an NVIDIA RTX 2060 GPU to enhance computational efficiency and provide the processing power needed for real-time audio analysis and synthesis. Additionally, for virtual personal assistants (VPAs) devices, we selected the Amazon Echo Dot (second generation) as the test subject. For the speech-to-text tools of Google, Microsoft, and IBM, we run their API on the Apple Macbook.

5 Evaluation

In this section, we evaluate AcousticScope by answering the following research questions:

RQ1: Whether there exists a social bias in Amazon Alexa?
(Sect. 5.1, Sect. 5.2, and Sect. 5.3)
RQ2: Whether there exists a social bias in the ASR system of Google, Microsoft, and IBM? (Sect. 5.4)
RQ3: How effective is AcousticScope in generating different accents? (Sect. 5.5)

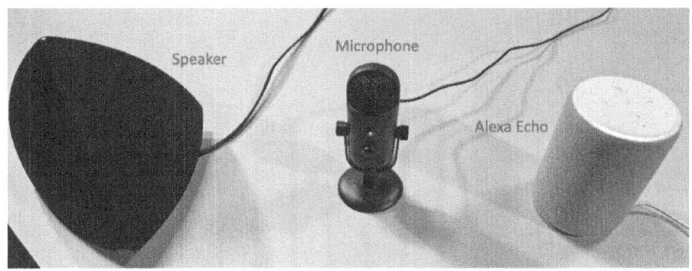

(a) Experiment setup for Alexa

(b) Experiment setup for Speech-to-text tool

Fig. 7. Experiment setup

5.1 Racial Bias

Our experimental results clearly reveal issues with racial bias in Amazon Alexa. Figure 8 shows Alexa's speech recognition accuracy for the five test races. Among them, the white group has the highest recognition accuracy, reaching 84.87%, followed by the Indian group, with a recognition accuracy of 75.21%; and the black group, with a recognition accuracy of 70.42%, which are 9.66% and 14.45% lower than white voices respectively. In addition, Alexa performed the worst when recognizing Chinese voices, with an accuracy of only 58.87%. These results show significant differences in Alexa's speech recognition accuracy for different races, revealing clear racial bias in Amazon's Alexa system.

5.2 Gender Bias

After analyzing racial bias, we also examined gender bias in the Alexa voice assistant. As shown in Fig. 9, Alexa's recognition accuracy for female voices is 74.55%, whereas its recognition accuracy for male voices is 67.87%. This 6.68%

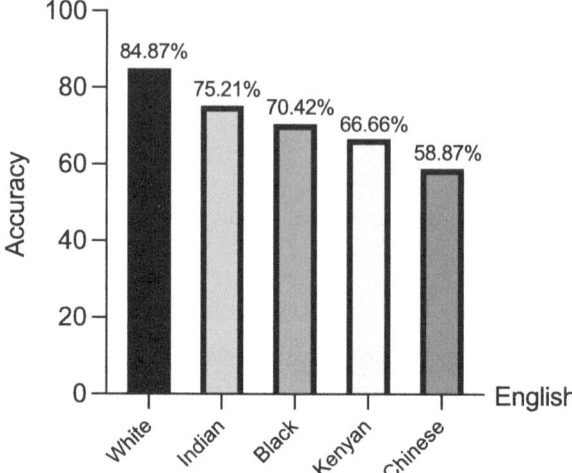

Fig. 8. Comparative results of Alexa's English accent recognition accuracy for different races

difference indicates that Alexa performs better at recognizing female voices than male voices across different genders.

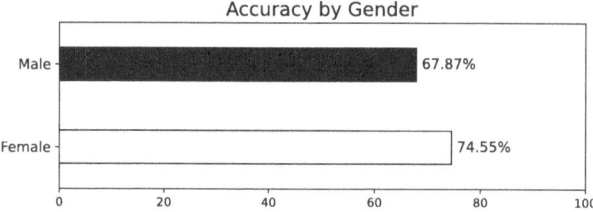

Fig. 9. Comparative results of Alexa's English accent recognition accuracy for different genders

Figure 10 presents detailed speech recognition accuracy for males and females across five different races. The results demonstrate that among white individuals, Alexa's accuracy in recognizing female voices is 86.19%, higher than the 83.54% accuracy for male voices. Among black individuals, Alexa's recognition accuracy for female voices is 75.0%, surpassing 65.83% for male voices. In the Kenyan group, Alexa's recognition rate for female voices is 68.99%, compared to 64.36% for male voices. The most pronounced gender difference is observed in the Chinese group, where Alexa's recognition rate for female voices is 68.93%, significantly higher than the 48.81% rate for male voices. This analysis of male and female voices across various races further confirms the gender differences present in the Amazon Alexa system.

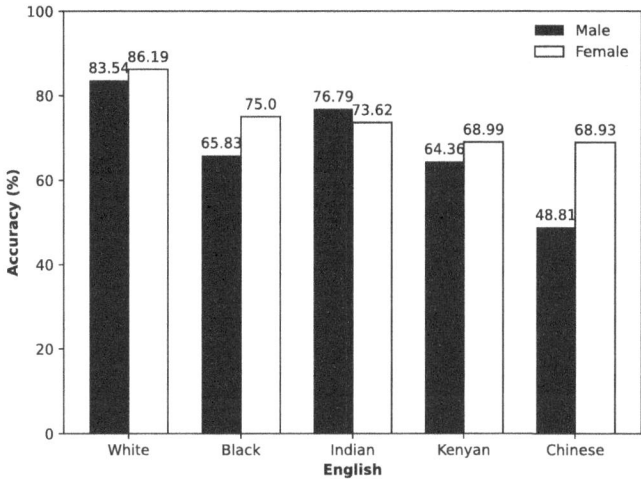

Fig. 10. Comparative results of Alexa's English accent recognition accuracy for different races and genders

5.3 Age Bias

We also utilized AcousticScope to examine the differences in how Alexa responds to children's voices versus adult voices. Since Lovo AI provides only the voices of white female children, We compared only white adult female voices and white female children's voices to ensure the validity of the comparison results. As shown in Table 1, there is a stark contrast in Alexa's recognition rates: 86.19% for adult voices compared to a significantly lower 53.03% for children's voices. This substantial disparity in recognition efficiency between adults and children has profound implications. It indicates that Alexa's voice recognition system is more adept at understanding and processing adult voices. Consequently, children may experience frustration and reduced functionality when using Alexa, impacting their overall user experience and limiting their ability to interact effectively with the system.

Table 1. Accuracy of Alexa's recognition between adult and children

Type	Accuracy
Adult	86.19%
Children	53.03%

5.4 Bias in Different ASR Systems

After analyzing social bias in Amazon Alexa, we utilized AcousticScope to evaluate the speech-to-text systems developed by Google AI, Microsoft Azure, and

IBM Watson AI. Figure 11 presents the transcription performance for White and Black speakers across these systems. The comparison reveals significant racial disparities, particularly in the recognition results from IBM Watson AI. The data indicate a clear disparity in WER between the two racial groups across all platforms. IBM Watson AI reports the highest overall WER, particularly for Black speakers, with a WER of 0.135, compared to 0.095 for White speakers. In contrast, Google AI shows a WER of 0.035 for Black speakers and 0.025 for White speakers, while Microsoft Azure reports a WER of 0.045 for Black speakers and 0.026 for White speakers. Despite the relatively lower overall WERs in Google AI and Microsoft Azure, both systems still show significantly higher error rates for Black speakers. These results demonstrate that all three ASR platforms exhibit systemic bias, favoring White speakers, as reflected by their consistently lower WERs. This finding underscores the widespread racial bias present in contemporary speech-to-text tools.

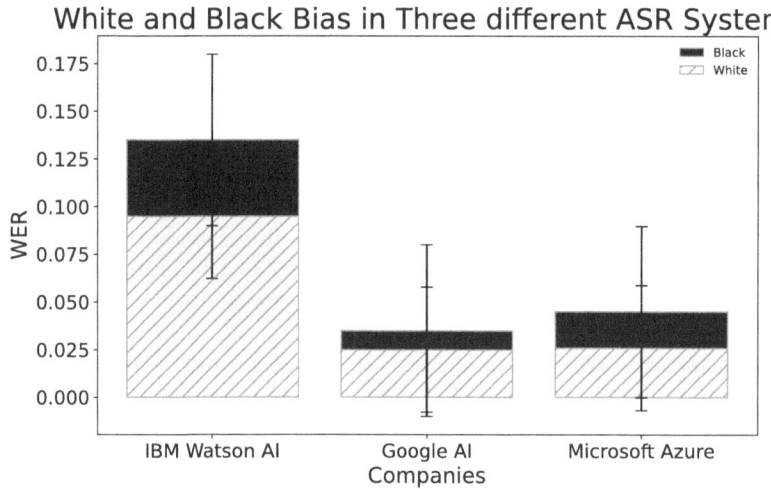

Fig. 11. Bias of White and Black in Three different ASR System

To further investigate the ASR system's performance across different racial groups, we expanded the test groups to include Indian, Kenyan, and Chinese speakers. Figure 11 presents the average Word Error Rate (WER) for these racial groups across three major ASR systems:IBM Watson AI, Google AI, and Microsoft Azure. Among them, IBM Watson AI shows the highest WER for Indian and Kenyan speakers, at 0.205 and 0.185, respectively. The WER for Black speakers is 0.135, while it is relatively lower for White and Chinese speakers, at 0.095 and 0.130, respectively. In contrast, Google AI demonstrates overall lower WERs, but still performs best for White speakers (0.025), with slightly higher WERs for Indian and Chinese speakers at 0.045 and 0.040, respectively. Microsoft Azure exhibits a similar trend, with the lowest WER for White speakers at 0.026, while Black and Indian speakers experience higher WERs, at 0.045 and 0.040, respectively. Overall, these three ASR systems exhibit significant differences in their performance across racial groups, particularly with IBM Watson AI showing notably higher WERs for non-White speakers, suggesting potential racial bias in the training data. Although Google AI and Microsoft Azure perform relatively more evenly, they still display slight bias. These findings reveal inconsistencies in ASR performance across different racial groups, indicating the presence of potential racial bias in the systems.

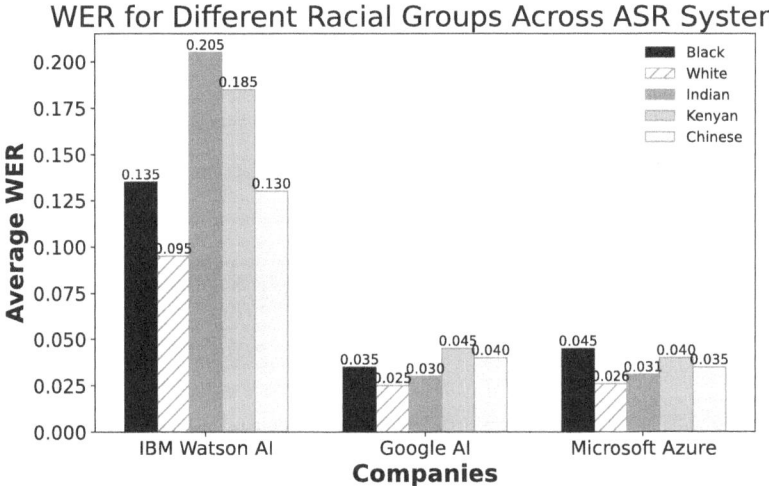

Fig. 12. Average WER in IBM, Google, and Microsoft for different races

5.5 Performance Evaluation

To ensure the reliability of the AI-generated sound by AcousticScope, we evaluate the similarity between AI-generated speech and natural human speech by analyzing their spectral features. Initially, we randomly selected clear, noise-free speech samples from YouTube for four categories: White male, White female, Black male, and Black female. Each sample was about one minute in duration. Using YouTube's subtitle feature to transcribe the speech content of these natural human voices. Subsequently, we utilized Lovo AI to clone these four types of voices, ensuring that each cloned voice matched its corresponding natural human voice in content, thereby maintaining consistency in word pronunciation and intonation.

After successfully collecting the four types of natural human voices and AI-generated cloned voices, we performed spectral analysis using the librosa library [1]. Firstly, we resampled the natural and cloned voices to ensure consistent sampling intervals, thereby enhancing the reliability of subsequent feature extraction and analysis. Next, the resampled audio files were converted to mono to reduce data dimensionality and highlight the primary information of the audio. We then extracted the MFCC (Mel-frequency cepstral coefficients) features of natural and cloned human voices and applied t-SNE (t-distributed stochastic neighbor embedding) for dimensionality reduction and visualization of the extracted MFCC features.

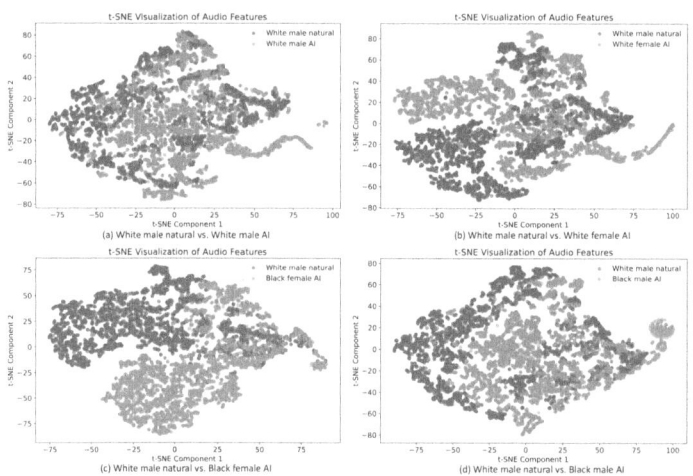

Fig. 13. Spectrum analysis of authentic voice and AI clone voice

Figure 13 illustrates the comparison of MFCC features between the natural human voice of a White male and the AI-cloned voices of different categories. Figure 13 (a) shows a high overlap in MFCC features between the natural human voice of the White male and the AI-cloned voice of the White male; (b), (c), and (d) respectively show significant differences in MFCC features between the natural human voice of the White male and the AI-cloned voices of the other three categories.

Furthermore, as detailed in Table 2, we utilize natural white male speech as a baseline and conduct controlled tests with voices representing four demographic categories generated by AcousticScope: AI white male, AI white female, AI black male, and AI black female. The results show that the AI-generated white male voice has the highest resemblance to the natural white male voice. This conclusion is supported by both mean difference and statistical significance analyses. First, the mean difference between the natural white male voice and its AI-generated counterpart is 67.03, the smallest among all voice combinations. In comparison, the mean differences between the natural white male voice and the AI-generated white female, black male, and black female voices are 70.59, 73.79, and 78.31, respectively, indicating a progressively larger gap as the demographic characteristics differ. This comparative analysis confirms that the AI effectively reproduces voice characteristics across different racial and gender groups. Second, in terms of the 95% confidence interval (CI), the CI for the natural white male voice and its AI clone is (67.01, 67.05), showing that the mean difference lies within a narrow range of 67.01% to 67.05%. This CI is tighter compared to the other groups, indicating a smaller deviation between the natural and AI-generated white male voices. The narrow CI further suggests high precision in these estimates. Additionally, the p-values across all comparisons are significantly below 0.05, confirming that these differences are statistically significant. Thus, the results convincingly demonstrate AcousticScope's ability to accurately simulate natural human speech.

Table 2. Spectral similarity analysis between authentic voice and AI clone voice

Natural voice	AI clone voice	Mean Difference	STD	95% CI	p-value
White male	White male	67.03	31.46	(67.01, 67.05)	3.66e-08
White male	White female	70.59	31.52	(70.57, 70.61)	2.16e-88
White male	Black male	73.79	34.42	(73.77, 73.81)	2.33e-36
White male	Black female	78.31	34.84	(78.29, 78.32)	1.28e-53

6 Discussion

6.1 Implication

This study uses AcousticScope to evaluate two ASR applications, including VPAs devices and speech-to-text tools. Unlike traditional manual collection methods, our tool leverages AI speech generation technology to create synthetic voices with diverse demographic characteristics. This approach allows for the rapid generation of voices representing different genders and races, overcoming geographical limitations and enabling efficient testing of multiple ASR devices.

The racial and gender biases observed in ASR systems not only harm different user groups but also impact social fairness [18,19,23]. Our research results indicate that the Alexa voice assistant performs better when recognizing and responding to the voices of white people and females. Specifically, speech recognition accuracy is significantly higher for white users than for other racial groups. Additionally, female users experience a higher probability of their voice commands being correctly recognized and executed than male users. Furthermore, the results from testing speech-to-text tools from Google, Microsoft, and IBM illustrate the prevalence of racial bias in ASR systems—white individuals consistently receive the lowest WER. These biases hinder non-White groups from fully benefiting from voice recognition technology. Such disparities could significantly impact non-White groups when employers use speech recognition software to evaluate interview candidates or when criminal justice agencies use it to record court proceedings.

Our findings also show that speakers of non-standard English, including regional dialects and non-native English accents, experience poor performance when using VPA systems. Due to social biases within VPAs, Alexa can more accurately recognize standard White English, resulting in a better user experience for white users. In contrast, African Americans and non-native English speakers encounter significant barriers when using Alexa, leading to a poor user experience. Therefore, addressing social equity issues in VPAs systems is critical.

6.2 Limitation

Although we used the AI voice generated by AcousticScope to simulate real human interactions with VPA devices and verified the reliability of the generated voice through spectrum analysis, and explored the social biases in VPA devices such as Amazon Alexa and three speech-to-text tools from Google, Microsoft, and IBM, there is still room for improvement. First, we only selected 100 skills from the trivia skill category of the Amazon Skills Marketplace for testing, although Amazon offers over 100,000 skills across various functions and types. We focused on the trivia category because it supports continuous long-term conversations. Additionally, the rapid updates in the Amazon Skills Market policy render many skills outdated or dysfunctional, and some skills fail to provide high-quality, meaningful responses. These limitations constrained our sample size. In future work, we plan to collect a broader range of skills that generate high-quality conversation content to enhance our experimental data and verify the generalizability of our results. Second, the child voice category in the Lovo AI speech synthesis tool only includes the voices of white children, lacking representation of children of other races and genders. This limits the comprehensiveness of our analysis of Alexa's processing of adult and child voices. Future work will aim to create more diverse categories of children's voices to enrich our sample and thoroughly explore these differences. Although we used spectral analysis for AI-synthesized speech to verify its effectiveness and ensure the voice sounds authentic and natural, it still has limitations in accents, dialects, and emotional pattern changes. This restricts its ability to fully simulate the dynamic interaction capabilities of real humans with VPAs devices.

6.3 Mitigation

Our research reveals significant social bias in ASR systems, including Amazon Alexa and speech-to-text tools from Google, Microsoft, and IBM. To mitigate this bias, it is essential to diversify training data to include various ethnicities, genders, and age groups. Employing a diverse team for data annotation can also reduce subjective bias. Algorithmic improvements should focus not just on overall WER but also on the performance of underrepresented user groups, making targeted optimizations by analyzing WER for different groups separately. Adjusting model weights to prioritize data from these groups can ensure a meaningful impact. Effective user feedback mechanisms are crucial for identifying and addressing bias, with timely responses and model adjustments enhancing user experience. ASR developers should regularly evaluate and publicly report their progress in mitigating bias to promote transparency and accountability. Implementing these measures will reduce social bias in ASR systems, improving service quality for all users.

7 Related Work

VPA devices have seen widespread adoption globally. With their increasing usage, concerns about the security implications of VPA devices have garnered significant attention, making this a prominent topic in related research. For instance, studies on voice injection attacks aim to manipulate user commands without their knowledge [12,37,42,43]. At the same time, other research focuses on the security and privacy of VPA devices, examining how these devices collect and use users' private information [14,16,26]. The issue of social bias in speech recognition systems has garnered significant attention [11,21,22,39]. Numerous studies have demonstrated the existence of substantial social biases in speech-related systems [20,27,30,40]. Blodgett *et al.* [10] investigated the variations in African American English (AAE) on social media platforms and analyzed how demographic factors such as age, gender, and location influence dialect differences in AAE. Utilizing a large dataset from Twitter, their research highlights the linguistic diversity within the African American community and its impact on natural language processing (NLP) models. Markl and Nina [29] explored language variation and algorithmic bias in British English automatic speech recognition systems. Their findings indicate significant biases in the system's handling of different language variations, which affects the system's fairness and accuracy. Tatman [38] examined gender and dialect bias in YouTube's automatic subtitle system. Analyzing a substantial amount of YouTube video data, Tatman found that the subtitle system had a higher error rate when processing female voices and performed worse with non-standard English dialects, such as Southern American accents and African American English. Feng *et al.* [17] analyzed performance differences among various user groups in ASR systems through a series of studies and experiments; they collected speech data from diverse genders, races, and dialects and tested the performance of current mainstream ASR systems. Zhang *et al.* [44] demonstrated using RNN-Transducer with language bias to enhance speech recognition performance in a Chinese-English mixed speech recognition task. Meyer *et al.* [33] introduced the Artie Bias Corpus, an open dataset designed to detect demographic bias in speech applications. This dataset, created by Josh Meyer, Lindy Rauchenstein, and Joshua D. Eisenberg, sheds light on the issue of demographic bias in speech recognition systems. Lai and Holliday [25] investigated the sources of racial bias in ASR systems, focusing on prosodic changes. They detailed how prosodic variations—features such as rhythm, stress, and intonation—differ among users from various ethnic and linguistic backgrounds, contributing to bias in ASR systems. Liu and Garcia [28] discussed enhancing the performance of transcoding speech recognition (i.e., switching languages within the same paragraph) through interactive language bias. They introduced a dynamic language bias mechanism to adjust the ASR system's language conversion in real time, thereby improving recognition accuracy. Finally, Sudo et al. [36] developed a method to improve contextualized ASR systems through attention-based biased phrase-enhanced beam search. Their experimental results show that this model significantly improves recognition accuracy across various scenarios, outperforming traditional ASR models when processing speech with specific phrases and contexts. These studies provide valuable insights into understanding and mitigating social bias in ASR systems.

8 Conclusion

In this work, we investigated social bias in the VPAs ecosystem and introduced an automated interactive testing tool called AcousticScope to evaluate bias in different ASR systems. We tested Amazon Alexa voice assistant devices and speech-to-text tools from Google, Microsoft, and IBM using ten accents from various demographic groups generated by AcousticScope. 15,000 utterances were collected for the Amazon Alexa test, while 40 were collected for the speech-to-text tools. The results indicate that Amazon Alexa prefers white and female groups, while speech-to-text tools from Google, Microsoft, and IBM all exhibit lower WERs for white groups. Additionally, we evaluated the performance of AcousticScope by comparing the spectral similarity of the speech generated by AcousticScope with that of natural human speech, demonstrating its ability to produce realistic, natural, and fluent speech. Through this study, we aim to raise awareness of social bias in VPAs systems, promote fairness and inclusiveness in technology, and thus have a positive and far-reaching impact on the entire VPAs ecosystem.

Acknowledgment. This work is supported by the National Science Foundation (NSF) under Grant No. 2239605, 2228616, the South Carolina Research Authority, and partially based upon the work supported by the National Center for Transportation Cybersecurity and Resiliency (TraCR) (a U.S. Department of Transportation National University Transportation Center) headquartered at Clemson University, Clemson, South Carolina, USA. Any opinions, findings, conclusions, and recommendations expressed in this material are those of the author(s) and do not necessarily reflect the views of TraCR, and the U.S. Government assumes no liability for the contents or use thereof.

References

1. librosa: v0.10.0. https://librosa.org/doc/latest/index.html. Accessed: 23 Jun 2024
2. Lovo ai: Ai voice generator. https://lovo.ai/. Accessed 23 May 2024
3. Ibm speech to text, 2024. Accessed 01 Jul 2024
4. Languagetool: An open source proofreading software. https://github.com/languagetool-org/languagetool (2024). Accessed 25 May 2024

5. spacy: Industrial-strength natural language processing in python. https://github.com/explosion/spaCy (2024). Accessed 25 May 2024
6. Speech-to-text: Automatic speech recognition — google cloud (2024). Accessed 01 Jul 2024
7. Speech to text: Convert speech to text — microsoft azure (2024). Accessed 01 Jul 2024
8. Facebook AI. wav2vec 2.0: A framework for self-supervised learning of speech representations, 2021. Accessed 01 Jul 2024
9. Bircan, T., Ceylan, D.: Machine discriminating: Automated speech recognition biases in refugee interviews. J. Immigrant Refugee Stud. 1–16 (2024)
10. Blodgett, S.L., Green, L., O'Connor, B.: Demographic dialectal variation in social media: a case study of African-American English. arXiv preprint arXiv:1608.08868 (2016)
11. Buolamwini, J., Gebru, T.: Gender shades: intersectional accuracy disparities in commercial gender classification. In: Conference on Fairness, Accountability and Transparency, pp. 77–91. PMLR (2018)
12. Carlini, N., et al.: Hidden voice commands. In: 25th USENIX Security Symposium (2016)
13. Chen, X., Li, Z., Setlur, S., Wenyao, X.: Exploring racial and gender disparities in voice biometrics. Sci. Rep. **12**(1), 3723 (2022)
14. Cheng, P., Roedig, U.: Personal voice assistant security and privacy–a survey. Proc. IEEE **110**(4), 476–507 (2022)
15. Diverse Education. Algorithmic bias continues to impact minoritized students (2024). Accessed 10 Sep 2024
16. Edu, J.S., Such, J.M., Suarez-Tangil, G.: Smart home personal assistants: a security and privacy review. ACM Comput. Surv. (CSUR) **53**(6), 1–36 (2020)
17. Feng, S., Halpern, B.M., Kudina, O., Scharenborg, O.: Towards inclusive automatic speech recognition. Comput. Speech Lang. **84**, 101567 (2024)
18. Fenu, G., Lafhouli, H., Marras, M.: Exploring algorithmic fairness in deep speaker verification. In: Gervasi, O., et al. (eds.) ICCSA 2020. LNCS, vol. 12252, pp. 77–93. Springer, Cham (2020). https://doi.org/10.1007/978-3-030-58811-3_6
19. Fenu, G., Medda, G., Marras, M., Meloni, G.: Improving fairness in speaker recognition. In: Proceedings of the 2020 European Symposium on Software Engineering, pp. 129–136 (2020)
20. Gorisch, J., Gref, M., Schmidt, T.: Using automatic speech recognition in spoken corpus curation. In: Proceedings of the Twelfth Language Resources and Evaluation Conference, pp. 6423–6428 (2020)
21. Harwell, D., Mayes, B., Walls, M., Hashemi, S.: The accent gap. The Washington Post, 19 2018
22. Kitashov, F., Svitanko, E., Dutta, D.: Foreign English accent adjustment by learning phonetic patterns. arXiv preprint arXiv:1807.03625 (2018)
23. Koenecke, A., et al.: Racial disparities in automated speech recognition. In: Proceedings of the National Academy of Sciences, vol. 117, no. 14, pp. 7684–7689 (2020)
24. Kulkarni, A., Tokareva, A., Qureshi, R., Coucelro, M.: The balancing act: unmasking and alleviating asr biases in portuguese. arXiv preprint arXiv:2402.07513 (2024)
25. Lai, L.F., Holliday, N.: Exploring sources of racial bias in automatic speech recognition through the lens of rhythmic variation. In: Proceedings of the IEEE International Conference on Acoustics, Speech, and Signal Processing, pp. 1284–1288 (2023)

26. Liao, S., Cheng, L., Cai, H., Guo, L., Hu, H.: Skillscanner: detecting policy-violating voice applications through static analysis at the development phase. In: Proceedings of the 2023 ACM SIGSAC Conference on Computer and Communications Security, pp. 2321–2335 (2023)
27. Liu, C., et al.: Towards measuring fairness in speech recognition: casual conversations dataset transcriptions. In: ICASSP 2022-2022 IEEE International Conference on Acoustics, Speech and Signal Processing (ICASSP), pp. 6162–6166. IEEE (2022)
28. Liu, H., Garcia, L.P., Zhang, X., Khong, A.W., Khudanpur, S.: Enhancing code-switching speech recognition with interactive language biases. In: ICASSP 2024-2024 IEEE International Conference on Acoustics, Speech and Signal Processing (ICASSP), pp. 10886–10890. IEEE (2024)
29. Markl, N.: Language variation and algorithmic bias: understanding algorithmic bias in British English automatic speech recognition. In: Proceedings of the 2022 ACM Conference on Fairness, Accountability, and Transparency, pp. 521–534 (2022)
30. Martin, J.L., et al.: Spoken corpora data, automatic speech recognition, and bias against African American language: the case of habitual'be. In: Proceedings of the 2021 ACM Conference on Fairness, Accountability, and Transparency, pp. 284–284 (2021)
31. Martin, J.L., Tang, K.: Understanding racial disparities in automatic speech recognition: the case of habitual be. In: Interspeech, pp. 626–630 (2020)
32. Metz, C.: There is a racial divide in speech-recognition systems, researchers say. The New York Times (2020)
33. Meyer, J., Rauchenstein, L., Eisenberg, J.D., Howell, N.: Artie bias corpus: an open dataset for detecting demographic bias in speech applications. In: Proceedings of the Twelfth Language Resources and Evaluation Conference, pp. 6462–6468 (2020)
34. OpenAI. Openai api documentation (2024)
35. Robison, M.: Voice assistants have a gender bias problem. what can we do about it? (2020)
36. Sudo, Y., Shakeel, M., Fukumoto, Y., Peng, Y., Watanabe, S.: Contextualized automatic speech recognition with attention-based bias phrase boosted beam search. arXiv preprint arXiv:2401.10449 (2024)
37. Sugawara, T., Cyr, B., Rampazzi, S., Genkin, D., Fu, K.: Light commands:{Laser-Based} audio injection attacks on {Voice-Controllable} systems. In: 29th USENIX Security Symposium (USENIX Security 20), pp. 2631–2648 (2020)
38. Tatman, R., et al.: Gender and dialect bias in youtube's automatic captions. In: Proceedings of the first ACL Workshop on Ethics in Natural Language Processing, pp. 53–59 (2017)
39. Tatman, R., Kasten, C.: Effects of talker dialect, gender & race on accuracy of bing speech and youtube automatic captions. In: Interspeech, pp. 934–938 (2017)
40. Vanmassenhove, E., Hardmeier, C., Way, A.: Getting gender right in neural machine translation. arXiv preprint arXiv:1909.05088 (2019)
41. Walker, P., McClaran, N., Zheng, Z., Saxena, N., Gu, G.: Biashacker: voice command disruption by exploiting speaker biases in automatic speech recognition. In: Proceedings of the 15th ACM Conference on Security and Privacy in Wireless and Mobile Networks, pp. 119–124 (2022)
42. Yan, Q., Liu, K., Zhou, Q., Guo, H., Zhang, N.: Surfingattack: interactive hidden attack on voice assistants using ultrasonic guided waves. In: Network and Distributed Systems Security (NDSS) Symposium (2020)

43. Zhang, G., et al.: Dolphinattack: inaudible voice commands. In: Proceedings of the 2017 ACM SIGSAC Conference on Computer and Communications Security, pp. 103–117 (2017)
44. Zhang, S., Yi, J., Tian, Z., Tao, J., Bai, Y.: Rnn-transducer with language bias for end-to-end mandarin-English code-switching speech recognition. In: 2021 12th International Symposium on Chinese Spoken Language Processing (ISCSLP), pp. 1–5. IEEE (2021)

Practitioner Paper: Decoding Intellectual Property: Acoustic and Magnetic Side-Channel Attack on a 3D Printer

Amirhossein Jamarani[1(✉)], Yazhou Tu[2], and Xiali Hei[1]

[1] University of Louisiana at Lafayette, Lafayette, LA 70504, USA
{C00550518,c00404592}@louisiana.edu
[2] Auburn University, Auburn, AL 36849, USA
yzt0065@auburn.edu

Abstract. The widespread accessibility and ease of use of additive manufacturing (AM), widely recognized as 3D printing, has put Intellectual Property (IP) at great risk of theft. As 3D printers emit acoustic and magnetic signals while printing, the signals can be captured and analyzed using a smartphone for the purpose of IP attack. This is an instance of physical-to-cyber exploitation, as there is no direct contact with the 3D printer. Although cyber vulnerabilities in 3D printers are becoming more apparent, the methods for protecting IPs are yet to be fully investigated. The threat scenarios in previous works have mainly rested on advanced recording devices for data collection and entailed placing the device very close to the 3D printer. However, our work demonstrates the feasibility of reconstructing G-codes by performing side-channel attacks on a 3D printer using a smartphone from greater distances. By training models using Gradient Boosted Decision Trees, our prediction results for each axial movement, stepper, nozzle, and rotor speed achieve high accuracy, with a mean of 98.80%, without any intrusiveness. We effectively deploy the model in a real-world examination, achieving a Mean Tendency Error (MTE) of 4.47% on a plain G-code design.

Keywords: 3D Printer · Side-channel Attack · G-code Reconstruction · Physical-to-cyber Attack · Intellectual Property

1 Introduction

The emergence of 3D printers dates back to the 1980s. The foundational concept laid the groundwork for the development of modern 3D printers. Today, as additive manufacturing (AM) systems grow and become prevalent globally for various purposes ranging from industrial usage to healthcare [4,14], biomedical [19], aviation [18], energy, and consumer products [8,18,19] the importance of keeping the Intellectual Property (IP) of these systems safe and secure has significantly increased. The ease of working with 3D printers and their efficiency have led industries to produce both high-tech and regular goods using G-code because it is cost-effective, flexible, accessible, and reliable. As predicted in [30], the global revenue of 3D printers reached over 20.2 billion dollars in 2021. It is calculated to produce revenue of 162.7 billion dollars by 2030 at a Compound Annual Growth Rate (CAGR) of 23.6% [1,24]. The concept of cyber-physical attacks on 3D printers has been explored since 2014, with research monitoring and analyzing these attacks [23,31].

3D printers emit acoustic signals and generate magnetic fields, raising the question: *could these emissions be recorded by smartphones and used to reconstruct the G-code?* Since 3D printers have digital acoustic signatures [5] for each movement and also generate magnetic fields, attackers could utilize a smartphone's built-in sensors including the microphone to capture these data without physically contacting the printer. By analyzing these recordings, attackers can potentially reconstruct the original G-code and commit successful IP theft [17]. Moreover, advancements in smartphone sensors have made it increasingly easier for attackers to accurately and discreetly collect data, allowing them to access IPs without being physically close to the 3D printer.

In this paper, we conducted a real test-bed attack on a 3D printer by analyzing multiple side channels emitted during the printing process using a smartphone for data collection. We examined the relationship between G-code commands and IP through distinct movements of the 3D printer: vertical or horizontal movements (left and right or up and down), header, and strata movements. Using the side-channel data, we trained Gradient Boosted Decision Trees on each movement. Subsequently, we applied the Side-Channel Reconstruction of the G-code (SCReG) technique, which utilizes acoustic and magnetic emissions generated by a 3D printer to infer and reconstruct the original G-code instructions. By collecting data through a smartphone's sensors, this method employs machine learning models to analyze the side-channel information and predict the printer's movements. The reconstructed G-code can then be used to replicate the printer's operations, potentially bypass the security measures, and access the IP coded in the printing process. This approach demonstrates the feasibility of reverse-engineering 3D printer instructions by only monitoring side channels.

The Mean Tendency Error (MTE) of our research attained the lowest percentage of 4.47%, highlighting that the reconstructed G-code and the reverse-engineered printed object were very similar to the initial object printed by the user. The accuracy of our models varied for each movement of the 3D printer discussed in Sect. 5. Our study introduces an approach to reconstructing G-code commands using machine learning algorithms with minimal MTE and inaccuracy. This research illuminates previously unexplored areas of side-channel analysis, such as setting up the smartphone for data collection at further distances of the 3D printer and the non-intrusiveness nature of the attack. Also, it utilizes feature extraction from acoustic and magnetic data to achieve more accuracy. The key contributions of our study are outlined below:

- We fully analyze the side channels produced by the 3D printer in different axes of nozzle movements and train a model to predict the movements.

- We reconstruct the G-code commands with the usage of a machine learning algorithm, Gradient Boosted Decision Trees, with high accuracy and low MTE. We illuminate the procedure of how to use a smartphone to collect data from 3D printer in an effective way.
- This study provides a technical and comprehensive taxonomy of the attack model. We discuss/identify open challenges and future trends in side-channel attacks on 3D printers.

The remaining parts of this study are organized as follows: we survey the related literature in Sect. 2. We provide a background and operational mechanism of AMs and 3D printers with the commonly used third-party tools to interpret the design to the 3D printer in Sect. 3. Then, we introduce a threat model and examine the side channels on the 3D printer in Sect. 4. We present the trained acoustic and magnetic models to predict movements in Sect. 5. In the following section, Sect. 6, we showcase our model and results in a real-world test-bed. Finally, Sect. 7 concludes the study.

2 Related Work

Extensive research has been conducted in the domain of either acoustic or magnetic side channels on 3D printers. However, less focus has been on the area where both side channels are employed. This section reviews the related studies in the field. The comparison of the works that are most related to ours is discussed in Sect. 6.1.

In [30], the authors' approach to reconstruct the G-code reached an MTE score of 5.87% by implementing a five-layer operational analysis consisting of Layer movement, which modeled to diagnose if the 3D printer has Z-axis movement (changing to another layer) or if it is in X-Y plane. Then, header movement was examined to detect if the nozzle was printing or if it was just changing the position to align with no material extrusion. The subsequent layer was axial movement to discern if the nozzle was moving in X-axis or Y-axis. The last two layers were designed to spot if the nozzle is in X-axis movement or if it is moving in X-left or X-right. The same was designed for the Y-axis movement to perceive if the nozzle is moving in Y-up or Y-down. Our study aligns with the work done in [30]; however, the main differences are the distance the smartphone was placed to the 3D printer, the algorithm used, and applying different feature extractions to have more robust and clean data.

Authors in [9] completed a thesis on cyber-physical attacks in additive manufacturing systems. They updated that physical-to-cyber attacks exploit manifestations of cyber-domain information through physical actions like motion and temperature changes, leaking confidential data via side-channels such as acoustic, thermal, and power. Their thesis investigated how acoustic side-channels can be used to gain the confidentiality of AMs, such as 3D printers, by reconstructing the G-code commands and IPs. Their attack model, including digital signal processing and machine learning algorithms, restored test objects with 78.35% axis prediction accuracy and 17.82% length prediction error.

In cyber-physical domains [10], the integration of side channels makes the systems vulnerable to attacks, exploiting information, like thermal, acoustic, and power conduits, to extract data without any disturbance to the functional system. As a case study, the authors implemented an FDM-based model on 3D printers, depicting the fact that how acoustic data can represent information about what is being printed on the 3D printer. With a model of an attack and the usage of machine learning methods, they reconstructed G-code to access IPs stored in the cyber domain. Their method gained an average axis prediction accuracy of 86% and an average length prediction error of 11.11% on different simple objects.

Authors in [15] only deployed acoustic side-channel attack on a 3D printer. The concept was faster motor rotation (higher speed), which results in higher amplitude and frequency sounds, so by collecting only acoustic data and training a model (regression model), they could access the G-code. Their attack methodology was comprised of two phases: training and attack itself. During the training phase, they recorded audio signals, pre-processed, and examined feature extractions in time and frequency domains. Then, features were mapped to corresponding G-codes. Finally, regression and classification models were trained. For the attack phase, they collected audio frame data and pre-processed it as similar to what was done in the training phase. Features were passed to the trained models, and the Predicted data was used to reconstruct the G-code.

For their classification model, the authors [15] introduced four sections, naming phi 1 to phi 4, each classifying in the Z-axis (Z) and no movement in the Z-axis (-Z), one-dimensional (1D), and two-dimensional (2D) movement. If the movement is 1D, this classifier determines whether it is along the X-axis or Y-axis, and: If the movement is 2D and along the XY axes, this classifier determines whether the X and Y motors are moving at the same speed or different speeds. The attack model reconstructed a square (simple shape) with classification accuracy reaching 98.55% and a Mean Absolute Percentage Error (MAPE) of 3.13% after post-processing. The authors in [34] looked at 3D printers as a weapon. They explored potential risks aligned with the malicious intent of using 3D printers. They also pointed out that by only having minor modifications on the IP and the physical properties of the 3D printers, the overall object at the end can be turned into dangerous items. The paper offered different taxonomies covering possible types of attack scenarios and discussed the weaponizing scenarios based on each attack model.

In another aspect of securing G-codes and preventing any malicious modifications, the two studies in [27, 28] proposed an approach for encrypting the G-code at its initial level once it is out of Stereo-Lithography (STL) file for being sliced layer-by-layer. In [28], the main purpose was to implement an approach to protect sensor data in cyber-enabled advanced manufacturing systems from cyber-physical attacks, in specific terms of unauthorized access and malicious tampering. However, in [27], the authors mainly focused on blockchain-based G-Code storage and asymmetry encryption of the row by row of the G-code to provide a secure path for sender and receiver. They came up with two case scenarios of the attack: First, unintended design modifications, and second on intellectual property theft where unauthorized access to the G-code allows the attacker to reproduce the product without the owner's permission.

The attack vectors on infrared (IR) thermography, used for quality control in metal additive manufacturing, was studied in [29]. The research diagnosed each possible attack scenario, such as manipulating calibration data and compromising thermal cameras, which can lead to defects like increased porosity or lack of fusion in the final product. The research highlighted the differences between open-loop and closed-loop systems to demonstrate the vulnerability of the 3D printer that while both can be achieved, the effects on part quality may change depending on the system's configuration. In [7], the aim was to secure AMs from cyber-physical attacks by identifying a physical hash method. This method uses a quick read code that encodes a hash of process parameters and toolpaths, ensuring that any deviations during manufacturing are detected in a synchronized manner of the time. The research contributes to enhancing the security and quality assurance of AM processes by integrating this physical hash with side-channel monitoring systems.

IP protection in additive layer manufacturing (ALM) was explored in [33]. The authors introduced an outsourcing model for ALM that aimed to address the limitations of traditional outsourcing methods for the purpose of securing IPs. Authors in [31] also did similar work to the research done in ALM. They explored potential cyber-physical attack vectors within the AMs process chain. The research showcased that the current detection methods, such as machine operators, virus-checking tools, and STL validation software, are insufficient for identifying sophisticated cyber attacks.

3 Background and Operational Mechanism

As depicted in Fig. 1, the typical life cycle of AM involves several stages. Users begin by designing a 3D object using design tools such as Fusion, Blender, or Sketchup [11]. After designing, a STL file is generated through Computer-Aided Design (CAD). Subsequently, Computer-Aided Manufacturing (CAM) tools are used to produce slicing instructions and layer description files, such as G-code.

IP protection in AM is critical due to the risk of theft and unauthorized reproduction. Different implementations have been proposed to secure IPs, including digital watermarking, encryption [28], and blockchain-based solutions [27]. In recent years, block chaining the G-codes has emerged as an effective resolution for securing the entire lifecycle of AM products by providing a tamper-proof record of all modifications and operations performed on the STL and G-code.

For production purposes, the components of the 3D printer, such as motors, steppers, nozzle, fan, and extruder, are commanded to run the lines specified to them by the G-code or the STL file. In this step, the printing process of an object happens. As known, the 3D printers emit acoustic sound and generate magnetic field while printing [30]; here, the attacker can take advantage of these data by recording them. Having analyzed and trained the data by machine learning or

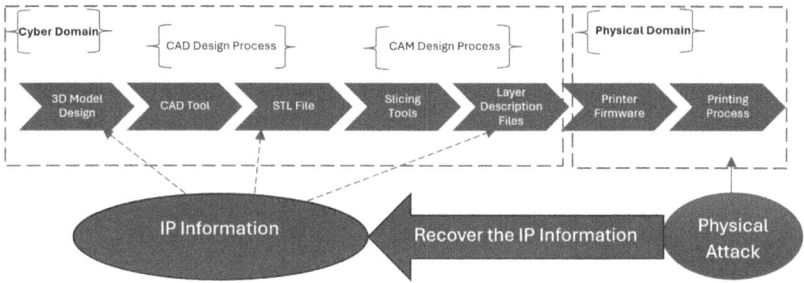

Fig. 1. Lifecycle of AMs under physical attack

other applicable methods, the attacker recovers the IP data from the physical attack, which will lead him to gain full access to the IP information that was initially generated by the designer of the 3D object.

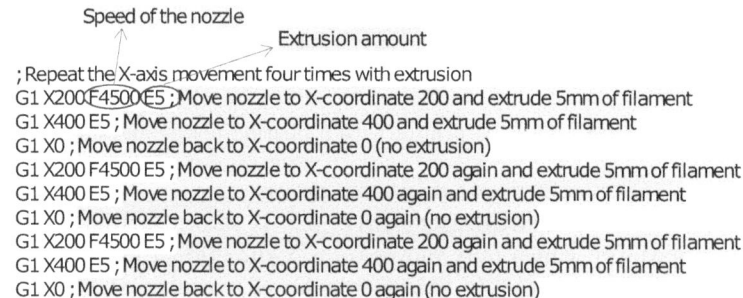

Fig. 2. Specific G-code commands for only moving the nozzle on the x-axis

The major Segments of a 3D printer are actuators and motors. These motors control the nuance movements of the printer's parts, including the extrusion nozzle and the print bed, along the X, Y, and Z axes. The main responsibility of an extruder is to feed the filament into a nozzle, where it is melted by the pre-heated nozzle and released onto the pre-heated bed. The extruder contains a stepper motor that drives the filament through the nozzle. As the filament is extruded, the stepper motors move the nozzle according to the G-code instructions, constructing the object layer by layer. The pulse Width Modulation (PWM) technique is applied in each 3D printer to balance the power and current rushed to the motors and stators. The bed is the surface on which the object is printed. The stepper motors also move the build platform up and down along the Z-axis as new layers are added. Some 3D printers include auto-leveling sensors to ensure the print bed is level before printing begins, improving the quality and accuracy of the final object [25].

Figure 2 represents a G-code for only moving the nozzle in the X-axis from right to left and reverse. Each row specifies the nozzle speed and movement coordinates. Also, the amount of filament per spot is controlled and can be modified based on each object's design. These commands are considered the IPs of a 3D printer, each of which can be reconciled with a distinct emission of sound and magnetic field.

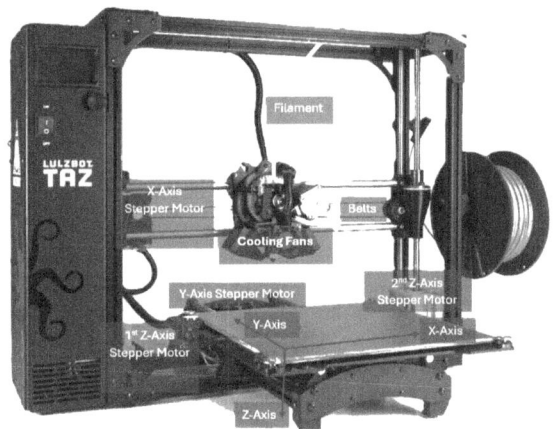

Fig. 3. LULZBOT TAZ 3D printer with a heating nozzle and platform.

A common 3D printer, LULZBOT TAZ [22], is shown in Fig. 3. It consists of a heating platform, which only moves on a Y-directional axis with the help of the Y-Axis stepper motor. In this particular 3D printer, there are two Z-axis stepper motors, each at the very edge of the 3D printer's platform. These only move top and bottom to adjust the nozzle for different printing process layers. The nozzle, the cooling fans, and the X-axis stepper motor are connected to the z-axis mover. This gives the 3D printer flexibility to align and print at each single spot on the platform. The X-axis stepper motor moves the nozzle only horizontally to either left or right, with the stepper connected to the transitional belts in parallel. The Y-axis stepper motor sole positions the platform and moves vertically, either backward or forward. The printing process generally involves heating the nozzle to change the material from a solid to a semi-solid stage. Temperature regulation is gained through cooling fans and heaters connected to the nozzle and the platform. The motors and their actuation systems manage the entire printing operation as those run the commands on the G-code or STL files.

4 Threat Model and Side-Channels

As 3D printers produce acoustic sound and generate magnetic field while printing [6,30], attackers can reach the initial G-code by recording the acoustic and magnetic data either separately or concurrently using a smartphone application.

For the simultaneous recording of magnetic field and acoustic sound, there are applications, such as [2], where the users can record acoustic data and magnetic data at the same time. The output will be in a comma-separated value file, where can easily be analyzed and labeled for training a model.

A potential attack model has been depicted in Fig. 4. In this case scenario, the attacker places a smartphone near the 3D printer to record emitted data, both acoustic and magnetic. Then, he uses pre-trained learning algorithms to reconstruct the G-code. This allows the attacker to acquire the IP of the object being printed without even touching the 3D printer or disrupting the printing process.

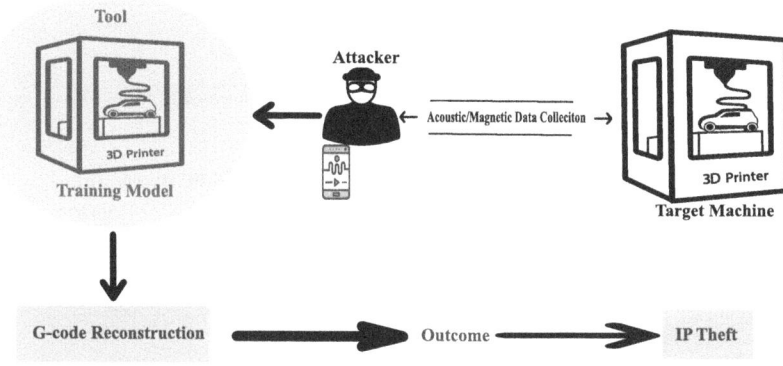

Fig. 4. Threat model

4.1 Acoustic Side-Channel

By recording the acoustic data emitted from the 3D printer's nozzle, extrusion, layer, header, axial, and directional movements, the IP reconstruction will be feasible by an attacker. As depicted in Fig. 5, the acoustic data of a 3D printer [22] was collected with a microphone of a smartphone, Samsung Galaxy S22 plus [26]. As it is apparent in Fig. 5, there are trends at which level of Decibel (dB) the sound is emitted. Passing over time (in seconds), a continuous cycle of dBs ranges at certain points. In such case, to have better and more accurate data, feature extraction techniques [13,20] were applied to the dataset. This helps to remove any background or unwanted noise, which leads to precise data.

Applying Formula (1), Zero-Crossing Rate (ZCR), the rate at which a signal changes from positive to negative or back is monitored. Here, $x[n]$ is the signal at sample n and N is the total number of samples.

$$\text{ZCR} = \frac{1}{N-1} \sum_{n=1}^{N-1} \mathbb{1}\{x[n] \cdot x[n-1] < 0\} \qquad (1)$$

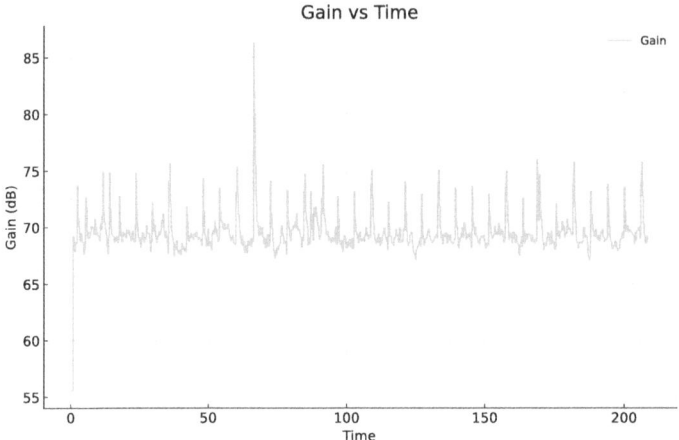

Fig. 5. Gain vs. time for x-axis movements

We applied the short-time energy Formula (2), which filters out low-energy segments that are not relevant to the overall acoustic data collection of the 3D printer. This improved the total performance of processing stages by spotting only the significant parts of the signal.

$$\text{STE}[n] = \sum_{m=0}^{N-1} x^2[n-m] \tag{2}$$

where $x[n]$ is the audio signal and N is the window length. We applied Formula (3), Root Mean Square (RMS), as it is less sensitive to short-term signal fluctuations. This makes RMS a more robust metric for evaluating the total level of our noisy signals.

$$\text{RMS} = \sqrt{\frac{1}{N} \sum_{n=0}^{N-1} x^2(n)} \tag{3}$$

As we implemented a machine learning algorithm to train the data, it was necessary to use the spectral centroid, Formula (4), as it classifies different spectrums of sounds. With the help of spectral centroid, our model could detect the background noise and the noise emitted out of the 3D printer.

$$\text{Spectral Centroid} = \frac{\sum_{k=0}^{N-1} f(k) \cdot |X(k)|}{\sum_{k=0}^{N-1} |X(k)|} \tag{4}$$

where $f(k)$ is the frequency bin and $X(k)$ is the magnitude of the Fourier Transform.

Spectral Bandwidth in Formula (5) was used to assist the model in understanding the spectral characteristics of a signal and applying necessary modifications.

$$\text{Spectral Bandwidth} = \sqrt{\frac{\sum_{k=0}^{N-1}(f(k)-C)^2 \cdot |X(k)|}{\sum_{k=0}^{N-1}|X(k)|}} \quad (5)$$

where C is the spectral centroid.

Lastly, we used Gaussian filter to smooth our data in Formula (6).

$$G(x) = \frac{1}{\sqrt{2\pi\sigma^2}} \exp\left(-\frac{x^2}{2\sigma^2}\right) \quad (6)$$

where x is the distance from the center of the filter, and σ is the standard deviation.

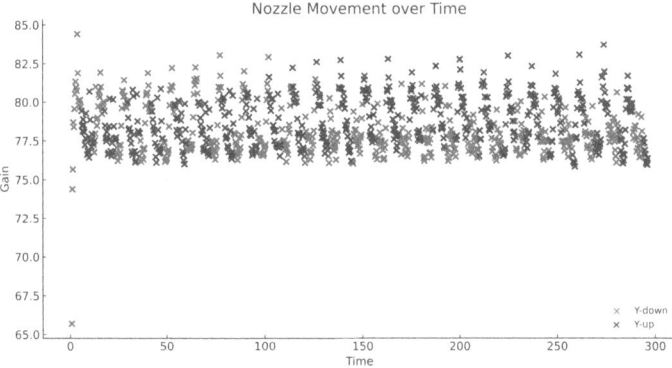

Fig. 6. Trained model while the nozzle moving on y-axis backward and onward without extrusion

We mainly applied Mel-Frequency Cepstral Coefficients (MFCCs), Formula (7), to provide a relatively tight Portrayal of the spectral data of an acoustic signal. By analyzing the key information in a small number of coefficients, MFCCs greatly reduce the dimensionality of the data, making the machine-learning models to train the model easier and more classified. MFCCs are derived as follows:

$$\text{MFCC}(n) = \sum_{m=1}^{M} \log(S(m)) \cos\left[n(m-0.5)\frac{\pi}{M}\right] \quad (7)$$

Mathematically, if $S(m)$ is the mel spectrum.

By applying all the feature extractions above to our acoustic data, we successfully trained a model to detect the movement of the nozzle in the y-axis, as illustrated in Fig. 6. As it is apparent, the model predicted the movement with a high rate of accuracy with only some points missing due to the initial sound emitted by the 3D while it starts to initiate the G-code commands. Blue crosses are highlighted when the nozzle moves in Y-up, and red ones specify the Y-down movement of the nozzle.

4.2 Magnetic Side-Channel

The electromagnets are initiated as the stepper motors receive current by PWM technology. With the implementation of the rotor, the electromagnet power is transmitted to the turning movement of the bearings and spacers. Connecting this rotation movement to the 3D printer by belts and long threaded screws, the nozzle will gain the power to move in a three-dimensional platform. As depicted in Fig. 7, the stepper consists of a stator, which generates a magnetic field by the usage of coils, which send currents to the rotor. This helps with accurate positioning and alignment of each movement.

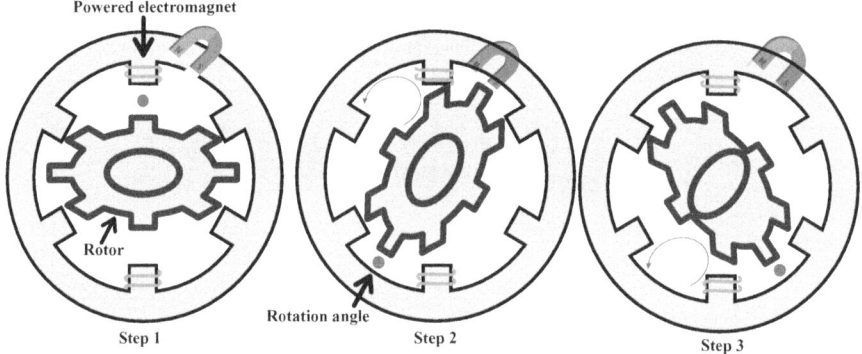

Fig. 7. Exploded view of a common stepper motor used in 3D printers

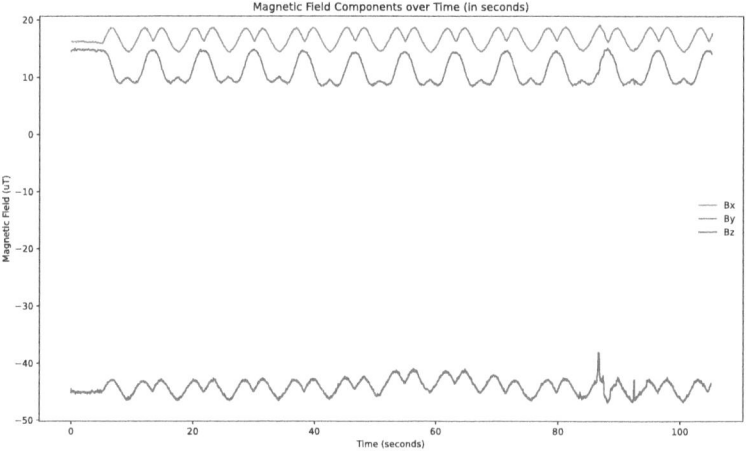

Fig. 8. Magnetic data while the nozzle moves in x-axis

Knowing how steppers function, we used [26] to record magnetic data while the 3D printer only printed on the x-axis. We designed and sent a specific G-code through the slicing tool [12]. The data was collected at 100 Hz in the unit of micro-Tesla (μT). As is apparent in Fig. 8, the pattern of the nozzle moving in the x-axis for X and Y magnetic fields was interestingly similar. This shows how the magnetometer received magnetic fields while the nozzle was in a left-right movement. As the nozzle gets closer to the left, where the smartphone was positioned, the magnetic field grows and picks at close to 20 μT, but as it moves to the right, further to the smartphone's built-in sensors, the magnetic field will become weaker. The blue wave also shows the z-axis movement, which is negative as obvious since we did not have any z-axis movement.

To make the data effective and classified, we use feature extraction as we did for acoustic data. This is considered as a pre-processing step for further analysis of the dataset. As to help the model to learn the central tendency better, we used Formula (8). It assist a single value that represents the center of the dataset.

$$\text{Mean} = \frac{1}{N} \sum_{i=1}^{N} x_i \tag{8}$$

Standard deviation (Std Dev) is applied because it detains how the spread of the data is. We harnessed Formula (9), which classifies different movements of a 3D printer.

$$\text{Std Dev} = \sqrt{\frac{1}{N} \sum_{i=1}^{N} (x_i - \text{Mean})^2} \tag{9}$$

Skewness highlights whether data points are more concentrated on one side of the mean. A positive skewness indicates a right-tailed distribution (more values are concentrated on the left), and a negative skewness indicates a left-tailed distribution (more values are concentrated on the right).

$$\text{Skewness} = \frac{\frac{1}{N} \sum_{i=1}^{N} (x_i - \text{Mean})^3}{\left(\frac{1}{N} \sum_{i=1}^{N} (x_i - \text{Mean})^2\right)^{3/2}} \tag{10}$$

The kurtosis Formula (11) function calculates the tailedness or rocketedness of a distribution analogized to the normal distribution.

$$\text{Kurtosis} = \frac{\frac{1}{N} \sum_{i=1}^{N} (x_i - \text{Mean})^4}{\left(\frac{1}{N} \sum_{i=1}^{N} (x_i - \text{Mean})^2\right)^2} - 3 \tag{11}$$

5 Acoustic and Magnetic Models

Upon inspecting the 3D printer movements while printing an object, we found three key movements, as illustrated in Fig. 9. By collecting distinct acoustic and magnetic data during each key stage and doing feature extraction, the 3D printer's initial IP will be at risk of being accessible. First, the nozzle moves either vertically or horizontally. If it moves vertically, then it is on the X-axis, moving left and right. However, if it moves horizontally, the nozzle moves up and down. The next step, header movement, determines if the nozzle is printing or aligning to position at the right spot and then starts printing. The speed of the motor can detect the major difference in printing or aligning as it aligns with maximum speed but slows down while printing to prevent any string act or bad quality printed shape.

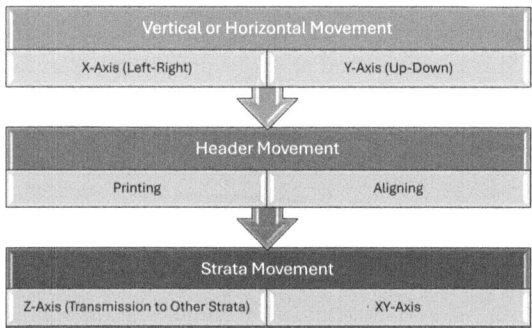

Fig. 9. Taxonomy of the chronological order in 3D printers

5.1 Analysis and Results

This section outlines our approach to identifying precise movement patterns in 3D printers. We employed Gradient Boosted Decision Trees trained on both acoustic and magnetic data. Using a Samsung smartphone equipped with sensors and strategically placed microphones, we captured magnetic data (100 μT) and acoustic data (dB) for thorough analysis. Data pre-processing included Gaussian filtering and segmenting signals into 100 ms frames, which were then organized into distinct training and testing datasets.

5.2 Test Setup

Having detected each specific movement mechanism in 3D printers, we formulated different models using Gradient Boosted Decision Trees to train the acoustic and magnetic data. The data collection setup involved positioning a smartphone [26], equipped with built-in sensors, including an accelerometer, gyroscope, and magnetometer, alongside primary and secondary microphones capable of features like audio zoom and directional recording. Placed 15 cm away from the 3D printer at a 45-degree angle, the smartphone captured data optimally as shown in Fig. 10.

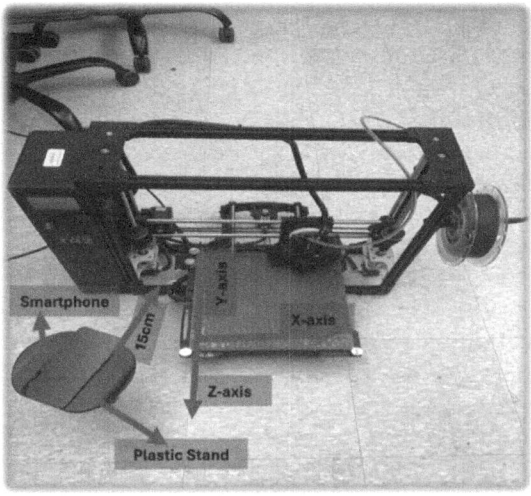

Fig. 10. Experiment setup

The magnetic data, measured in 100 μT, and the acoustic data, recorded in dB, were both collected for detailed analysis of the printer's movements and performance. Initially, the side-channel data underwent smoothing using a Gaussian filter. Subsequently, the signal was segmented into frames, each lasting 100 ms. These segments were then categorized into training and testing sets tailored to the requirements of various models, ensuring robust analysis and prediction capabilities.

5.3 Results

Applying feature extractions to the collected data mentioned in Sects. 4.1 and 4.2 and smoothing the data with the Gaussian filter, we were able to test our models and reach high accuracy in predicting which direction the nozzle is moving, as shown in Fig. 11.

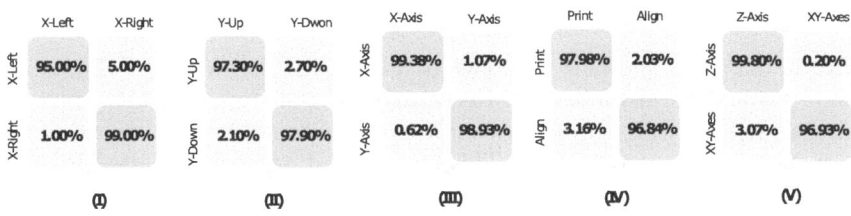

Fig. 11. Model's accuracy per each movement

Figures 11 (I) and 11 (II) show the vertical and horizontal models, which determine whether the printer operates in the X-left, X-right, or Y-up, Y-down plane. The training set comprises 1000 magnetic frames for each step, and the testing set includes a total of 3000 magnetic frames. This model can distinguish between the nozzle moving on either the X or Y axis and then detect if it is moving on the left and right axis or up and down. As apparent, the model successfully predicted 99.00% of the movement while the nozzle was moving and printing in the X-left direction.

Figure 11 (III) depicts the model's accuracy on diagnosing if the nozzle is moving in X or Y planes. This model is critical as it provides the foundation for knowing the movements in Figs. 11 (I) and 11 (II). The training set comprises 1000 magnetic frames for each category, and the testing set includes a total of 3000 magnetic frames. This model effectively reached an average accuracy of 99.49%. The high accuracy of this model stems from the fact that the magnetic data emitted while moving the nozzle in the X or Y plane was quite strong, so more precise data could be recorded by the smartphone magnetometer.

To firmly determine if the nozzle is actually printing or just positioning on the platform, we developed a model in Fig. 11 (IV). Since the term speed plays a critical role in detecting the movements while the header is adjusting or printing, we used only acoustic data here and removed the magnetic data for the purpose of having tailored data with low deviation rate. The average accuracy for this model hit 97.25%. Achieving high accuracy in nozzle detection, whether it is positioning or printing, is a complex scenario due to a combination of mechanical, physical, and software-related factors. Even small errors in any of these areas can add up and lead to noticeable inaccuracies in the final print.

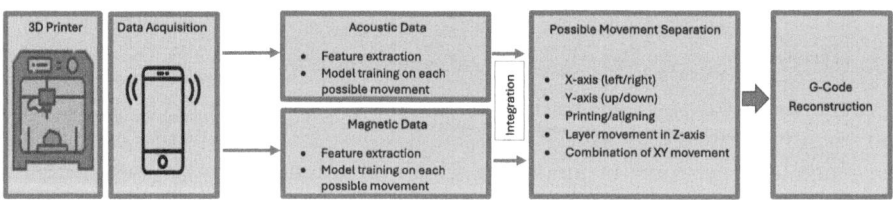

Fig. 12. Overview of the stages to reconstruct G-code

We developed a specialized model to distinguish between the Z-axis and XY-axis movements of the nozzle. Recognizing the distinct acoustic signatures associated with vertical versus horizontal movements, we focused inclusively on acoustic data and magnetic data. This streamlined approach resulted in a tailored dataset with minimal deviation, contributing to an average accuracy of 97.44% as showing in Fig. 11 (V).

6 Real Environment Implementation

To evaluate our models in a real-world examination, we decided to print a shape that involves every movement of the 3D printer for 3 layers: a square with dimensions of 1 cm * 1 cm. We followed steps in Fig. 12. We positioned the smartphone near (within 15 cm) the 3D printer and started to collect acoustic and magnetic data simultaneously, as shown in Fig. 10, while the printer was executing our specified G-code for the shape of the square. Once the printing process was finished, we did the feature extraction steps on the data and used a Gaussian filter to smooth the dataset. We inserted this dataset as an input to the model we trained. With the implementation of data simulation from G-code instructions on Python, we were able to reconstruct the initial G-code of the printer with high accuracy in the final shape. As depicted in Fig. 13, the overall square commands are successfully reconstructed with only some modifications in the axis and speed of the nozzle. Also, as it is apparent in Fig. 14, there are some edge lay-offs and or increases in length for the reconstructed G-code.

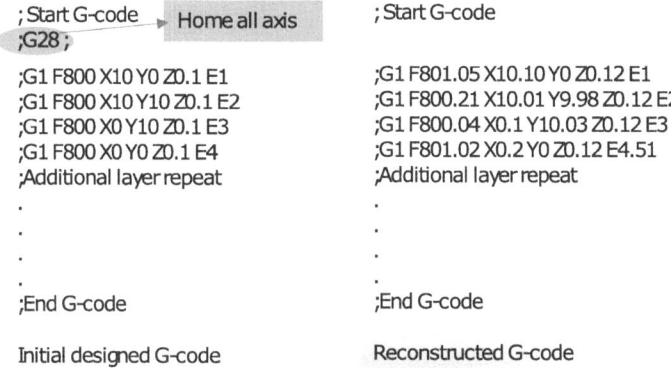

Fig. 13. Comparison of the initial G-code with the reconstructed one by the use of side-channel attacks

MTE rate on this square shape reconstruction was 4.47% which typically relates to a statistical measure used to assess the accuracy or bias of a forecasting or prediction model. In this case, the lower the MTE is, the higher the precision is on the reconstructed shape. It would generally involve calculating the average difference between predicted values and actual values over a dataset [30], reflecting whether the model tends to underpredict or overpredict systematically. This metric helps in understanding the overall directional bias of the model's predictions.

6.1 Discussion

In this section, we discuss various limitations, including the distance between the smartphone and the 3D printer, differences in speed, equal loads on step-

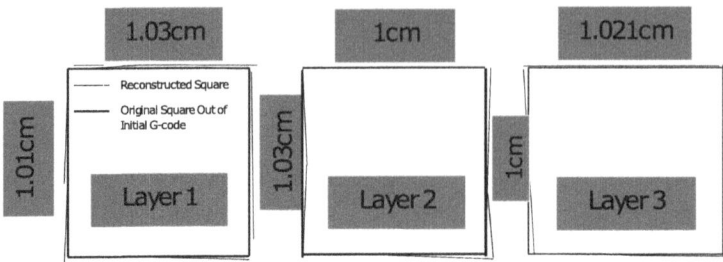

Fig. 14. The visual comparison of the reconstructed and original square shape out of magnetic and acoustic data

per motors, and background noise. Additionally, we compare our overall results with the related works that deployed relatively the same methods as ours to reconstruct the G-code.

Distance has posed a significant constraint in the realm of side-channels [16]. As demonstrated in Table 1, distance profoundly influences both the MTE and the G-code reconstruction, which is at the final stage of IP access. In the same way, there is a direct relationship between the complexity of the object that is designed to be printed with the overall MTE rate; The more complicated the object is, the lower the MTE rate would be as intricate object demands high speed of nozzle movement and nuance extrusion on the nozzle, which leads to the lower accuracy of collecting data. This is primarily due to the diminishing signal strength received by smartphones at greater distances. Consequently, the training of models is hindered by increased data variability, posing challenges for accurate movement prediction. Considering this limitation, our experiment and methods used reached high accuracy in reconstructing the G-code with low rates of MTE in collecting data with the smartphone within three different distances of 15 cm, 20 cm, and 30 cm.

Table 1. MTE rates at different distances

Distance (cm)	MTE Rate (percentage)
20 cm	5.10 (%)
30 cm	6.09 (%)

However, one of the key points of the attack scenario tested in this study is the non-intrusive nature of the attack. This adds to the stealthiness of the experiment as, in opposition to traditional attack overviews, there were apparent modifications and intrusive actions on the printing process to collect the data. Although the distance between the smartphone and the 3D printer directly affects the accuracy and overall performance of the attack, still the same object can be reproduced by reconstructing the G-code with minor differences in the quality and specifications of the final object. It is noted that if the magnetic field is beyond the coverage range, smartphones will be unable to detect magnetic data. This limitation can lead to significant inaccuracies in object reconstruction.

Future work could explore using multiple smartphones and sensor fusion to further extend the attack distance. Additionally, it might be possible to investigate conducting such attacks without a direct line of sight to the victim device. Since acoustic and magnetic side channels do not always require a clear line-of-sight transmission between the sensor and the target object or victim [21,32], adversaries could better conceal their devices in certain non-line-of-sight scenarios in real-world attack settings.

In addition, speed and consistent loads on stepper motors present additional limitations. The rapid alignment speeds typical in 3D printers result in unreliable magnetic and acoustic data collection during these fast operations. This rapid nozzle movement, driven by the high-speed actions of the steppers, complicates the accurate capture of data within short time frames. Similarly, when stepper motors operate under similar loads, distinguishing between them becomes challenging due to their production of magnetic and acoustic data that lacks significant distinctions. Training models under these conditions become particularly difficult due to the absence of clear differentiation between motor behaviors.

Table 2. Comparison of accuracy metrics between different studies

Study	Metric	Value	Metric	Value
Our Study	Avg. Accuracy	98.80%	MTE	4.47%
[30]	Avg. Accuracy	94.97%	MTE	5.87%
[3]	Avg. Accuracy	78.35%	MTE	Not Mentioned
[10]	Avg. Accuracy	86%	MTE	Not Mentioned
[15]	Avg. Accuracy	98.55%	MTE	Not Mentioned

Background noise presents another challenge in real-world IP attacks. While magnetic data remains largely unaffected, background noise significantly impacts the accuracy of acoustic data used to record stepper speed and movement. Despite efforts to mitigate this interference through feature extraction and Gaussian filters to reduce and smooth unwanted noise, the overall accuracy of data analysis and model predictions is inevitably influenced.

As shown in Table 2, the average accuracy and MTE rate of each related study have been summarized. The nearest study which had a high accuracy to ours was in [15]. Some of the studies did not mention the MTE rate once the G-code was reconstructed; however, [15] highlighted a MAPE rate of 3.13% instead of MTE.

7 Conclusion and Future Work

AMs, specifically 3D printers, have become pervasive throughout the globe and are being used in different sectors. The importance of 3D printers in this world has caused researchers to protect the 3D printer's IPs and the G-code. However, since stepper motors produce magnetic field and acoustic sound, by recording and analyzing the data close to the printer, attackers can reconstruct the G-code and modify it as their intent. In this study, we explained how stepper motors generate data and outlined different movements in 3D printers. We trained a model using Gradient Boosted Decision Trees to diagnose the nozzle movement on the platform. Reaching high accuracy in predicting the movements, we implemented data simulation from G-code instructions on Python to gain access to the initial G-code. Having tested this method on a real-world object, we successfully reconstructed the G-code with an MTE of 4.47%. In our future works, we plan to collect data from a greater distance using two smartphones positioned at different angles. This will help us investigate whether attackers can record data and regenerate the G-code from further away. Also, we will test our models on more complicated objects with more layers and nuance movements of the nozzle to check if the accuracy remains high. We hope the outcomes in this paper help to secure the valuable IPs of 3D printers.

Acknowledgment. This work was supported in part by the U.S. National Science Foundation under grants OIA-1946231, CNS-2117785, and CNS-2231682.

References

1. https://www.verifiedmarketresearch.com/product/global-3d-printing-market-size-and-forecast/
2. https://play.google.com/store/apps/details?id=com.chrystianvieyra.physicstoolboxsuite&hl=en_US&pli=1
3. Al Faruque, M.A., Chhetri, S.R., Canedo, A., Wan, J.: Acoustic side-channel attacks on additive manufacturing systems. In: 2016 ACM/IEEE 7th International Conference on Cyber-Physical Systems (ICCPS), pp. 1–10. IEEE (2016)
4. Awad, A., Trenfield, S.J., Gaisford, S., Basit, A.W.: 3D printed medicines: a new branch of digital healthcare. Int. J. Pharm. **548**(1), 586–596 (2018)
5. Belikovetsky, S., Solewicz, Y.A., Yampolskiy, M., Toh, J., Elovici, Y.: Digital audio signature for 3D printing integrity. IEEE Trans. Inf. Forensics Secur. **14**(5), 1127–1141 (2018)
6. Bilal, M.: A review of internet of things architecture, technologies and analysis smartphone-based attacks against 3D printers. arXiv preprint arXiv:1708.04560 (2017)
7. Brandman, J., Sturm, L., White, J., Williams, C.: A physical hash for preventing and detecting cyber-physical attacks in additive manufacturing systems. J. Manuf. Syst. **56**, 202–212 (2020). https://doi.org/10.1016/j.jmsy.2020.05.014, https://www.sciencedirect.com/science/article/pii/S0278612520300789
8. Brooks, G., Kinsley, K., Owens, T.: 3D printing as a consumer technology business model. Int. J. Manage. Inf. Syst. (Online) **18**(4), 271–280 (2014)

9. Chhetri, S.R.: Novel side-channel attack model for cyber-physical additive manufacturing systems. University of California, Irvine (2016)
10. Chhetri, S.R., Canedo, A., Faruque, M.: Confidentiality breach through acoustic side-channel in cyber-physical additive manufacturing systems. ACM Trans. Cyber-Phys. Syst. **2**(1), 1–25 (2017)
11. Chopra, A.: Introduction to Google Sketchup. Wiley (2012)
12. Cura: Ultimaker. https://ultimaker.com/software/ultimaker-cura/
13. Dhanalakshmi, P., Palanivel, S., Ramalingam, V.: Classification of audio signals using SVM and RBFNN. Expert Syst. Appl. **36**(3), 6069–6075 (2009)
14. Dodziuk, H.: Applications of 3D printing in healthcare. Kardiochirurgia i Torakochirurgia Polska/Po. J. Thorac. Cardiovasc. Surg. **13**(3), 283–293 (2016)
15. Faruque, M.A.A., Chhetri, S.R., Canedo, A., Wan, J.: Acoustic side-channel attacks on additive manufacturing systems. In: 2016 ACM/IEEE 7th International Conference on Cyber-Physical Systems (ICCPS), pp. 1–10 (2016). https://api.semanticscholar.org/CorpusID:19155564
16. Heyszl, J., Merli, D., Heinz, B., De Santis, F., Sigl, G.: Strengths and limitations of high-resolution electromagnetic field measurements for side-channel analysis. In: Smart Card Research and Advanced Applications: 11th International Conference, CARDIS 2012, Graz, Austria, 28–30 November 2012, Revised Selected Papers 11, pp. 248–262. Springer (2013)
17. Holland, M., Stjepandić, J., Nigischer, C.: Intellectual property protection of 3D print supply chain with blockchain technology. In: 2018 IEEE International Conference on Engineering, Technology and Innovation (ICE/ITMC), pp. 1–8. IEEE (2018)
18. Kalender, M., Kılıç, S.E., Ersoy, S., Bozkurt, Y., Salman, S.: Additive manufacturing and 3D printer technology in aerospace industry. In: 2019 9th International Conference on Recent Advances in Space Technologies (RAST), pp. 689–694. IEEE (2019)
19. Lai, J., Wang, C., Wang, M.: 3D printing in biomedical engineering: processes, materials, and applications. Appl. Phys. Rev. **8**(2) (2021)
20. Lambrou, T., Kudumakis, P., Speller, R., Sandler, M., Linney, A.: Classification of audio signals using statistical features on time and wavelet transform domains. In: Proceedings of the 1998 IEEE International Conference on Acoustics, Speech and Signal Processing, ICASSP 1998 (Cat. No. 98CH36181), vol. 6, pp. 3621–3624. IEEE (1998)
21. Liu, Y., Huang, K., Song, X., Yang, B., Gao, W.: MagHacker: eavesdropping on stylus pen writing via magnetic sensing from commodity mobile devices. In: Proceedings of the 18th International Conference on Mobile Systems, Applications, and Services, pp. 148–160 (2020)
22. LULZBOT: 3D printer (1999). https://lulzbot.com/store/taz-6
23. Moore, S., Armstrong, P., McDonald, T., Yampolskiy, M.: Vulnerability analysis of desktop 3D printer software. In: 2016 Resilience Week (RWS), pp. 46–51. IEEE (2016)
24. Ree, B.J.: Critical review and perspectives on recent progresses in 3D printing processes, materials, and applications. Polymer 127384 (2024)
25. Saggiomo, V.: A 3D printer in the lab: not only a toy. Adv. Sci. **9**(27), 2202610 (2022)
26. Samsung: S22+ (2023). https://www.samsung.com/us/business/mobile/phones/galaxy-s/galaxy-s22-128gb-unlocked-sm-s901uzgaxaa/

27. Shi, Z., Kan, C., Tian, W., Liu, C.: A blockchain-based G-code protection approach for cyber-physical security in additive manufacturing. J. Comput. Inf. Sci. Eng. **21**(4), 041007 (2021). https://doi.org/10.1115/1.4048966
28. Shi, Z., Oskolkov, B., Tian, W., Kan, C., Liu, C.: Sensor data protection through integration of blockchain and camouflaged encryption in cyber-physical manufacturing systems. J. Comput. Inf. Sci. Eng. **24**(7), 071004 (2024). https://doi.org/10.1115/1.4063859
29. Slaughter, A., Yampolskiy, M., Matthews, M., King, W.E., Guss, G., Elovici, Y.: How to ensure bad quality in metal additive manufacturing: in-situ infrared thermography from the security perspective. In: Proceedings of the 12th International Conference on Availability, Reliability and Security, ARES 2017. Association for Computing Machinery, New York (2017). https://doi.org/10.1145/3098954.3107011
30. Song, C., Lin, F., Ba, Z., Ren, K., Zhou, C., Xu, W.: My smartphone knows what you print: exploring smartphone-based side-channel attacks against 3D printers. In: Proceedings of the 2016 ACM SIGSAC Conference on Computer and Communications Security, CCS 2016, pp. 895–907. Association for Computing Machinery, New York (2016). https://doi.org/10.1145/2976749.2978300
31. Sturm, L.D., Williams, C.B., Camelio, J.A., White, J., Parker, R.: Cyber-physical vulnerabilities in additive manufacturing systems: a case study attack on the .stl file with human subjects. J. Manuf. Syst. **44**, 154–164 (2017). https://doi.org/10.1016/j.jmsy.2017.05.007, https://www.sciencedirect.com/science/article/pii/S0278612517300961
32. Tu, Y., Shan, L., Hossen, M.I., Rampazzi, S., Butler, K., Hei, X.: Auditory eyesight: demystifying {μs-Precision} keystroke tracking attacks on unconstrained keyboard inputs. In: 32nd USENIX Security Symposium (USENIX Security 23), pp. 175–192 (2023)
33. Yampolskiy, M., Andel, T.R., McDonald, J.T., Glisson, W.B., Yasinsac, A.: Intellectual property protection in additive layer manufacturing: requirements for secure outsourcing. In: Proceedings of the 4th Program Protection and Reverse Engineering Workshop, PPREW-4. Association for Computing Machinery, New York (2014). https://doi.org/10.1145/2689702.2689709
34. Yampolskiy, M., Skjellum, A., Kretzschmar, M., Overfelt, R.A., Sloan, K.R., Yasinsac, A.: Using 3D printers as weapons. Int. J. Crit. Infrastruct. Prot. **14**, 58–71 (2016). https://doi.org/10.1016/j.ijcip.2015.12.004, https://www.sciencedirect.com/science/article/pii/S1874548215300330

Security Techniques for Cyber-Physical Systems

An Efficient and Applicable Physical Fingerprinting Framework for the Controller Area Network Utilizing Deep Learning Algorithm Trained on Recurrence Plots

Rafi Ud Daula Refat[✉][iD], Alireza Mohammadi[iD], and Hafiz Malik[iD]

University of Michigan - Dearborn, Dearborn, MI 48128, USA
{rerafi,amohmmad,hafiz}@umich.edu

Abstract. The Controller Area Network (CAN) is widely used in the automotive industry for its ability to create inexpensive and fast networks. However, it lacks an authentication scheme, making vehicles vulnerable to spoofing attacks. Evidence shows that attackers can remotely control vehicles, posing serious risks to passengers and pedestrians. Several strategies have been proposed to ensure CAN data integrity by identifying senders based on physical layer characteristics, but high computational costs limit their practical use. This paper presents a framework to efficiently identify CAN bus system senders by fingerprinting them. By modeling the CAN sender identification problem as an image classification task, the need for expensive handcrafted feature engineering is eliminated, improving accuracy using deep neural networks. Experimental results show the proposed methodology achieves a maximum identification accuracy of 98.34%, surpassing the state-of-the-art method's 97.13%. The approach also significantly reduces computational costs, cutting data processing time by a factor of 27, making it feasible for real-time application in vehicles. When tested on an actual vehicle, the proposed methodology achieved a no-attack detection rate of 97.78% and an attack detection rate of 100%, resulting in a combined accuracy of 98.89%. These results highlight the framework's potential to enhance vehicle cybersecurity by reliably and efficiently identifying CAN bus senders.

Keywords: CAN · Deep leaning · Transfer learning · MobileNetV2 · EfficientNet

1 Introduction

One of the most popular in-vehicle networking protocol is called the controller area network (CAN) through which vehicle computing devices communicate with each other. The famous protocol was first introduced by Robert GmBH in 1983 and became a defacto for in-vehicle communication due to two specific reasons. 1) by design the protocol is applicable for hard real-time environments that guarantees communication with minimal time latency. 2) it reduced the wiring problem of a vehicle and was able to reduce the cost of vehicle manufacturing [1]. That is why the CAN bus protocol is used in all modern vehicles as the backbone of in-vehicle network communication.

By default, the CAN protocol is broadcasting in nature which means messages that are sent to the bus are accessible by all the entities connected to the network. It brings simplicity in terms of design but on the other hand the simplistic design can be leveraged by hackers [2,3], as it lacks a basic security feature i.e. implementation of a message authentication mechanism which makes it vulnerable to a variety of spoofing attacks [4,5]. In a single CAN message packet, a field that contains information of the source is absent. Because of the absence of the sender information, any electronic control unit (ECU) on the network can impersonate other ECUs in the network. An adversary can leverage that vulnerability of this protocol to launch various attacks leading to malfunctioning of the vehicle.

For example, in 2015 Charlie Miller and Chris Valasek remotely took control of a vehicle by injecting CAN data in the network. Surprisingly, the vehicle could not differentiate the impersonating CAN message and moved into a ditch [2]. Another demonstration was shown by the Keen Security Lab of Tencent team in 2016 where researchers remotely controlled a Tesla Model S. The researchers have gained entrance remotely by using Wi-Fi/Cellular as backdoor and was able to compromise many in-vehicle systems like IC, CID, and Gateway. Moreover, the team injected malicious CAN message into the network [3]. In December 2019, a gray-hat hacker created an android application that used an arduino microcontroller in order to inject CAN message to a Mercedes vehicle. The basic functionality of the application was to add features such as locking and unlocking doors, display custom text in instrument cluster, control hazard light etc. [4]. These evidences clearly indicate that the researchers took advantage of a known weakness of CAN protocol to spoof the network, i.e. the absence of source identification field.

To solve the above-mentioned security vulnerability, different approaches have been implemented by the security researchers [6,7]. These solutions can be broadly categorized into two categories. (1) cryptography based solutions [8–10], (2) intrusion detection system based solution [11–13]. The traditional cryptography based solutions can provide some degree of security but they are computationally expensive and uses the network bandwidth which is critical for CAN based vehicle networks [14]. Moreover, these cryptography based solutions are vulnerable to replay attack [11]. Recently, researchers have proposed intrusion detection system based solutions for detecting CAN cyberattacks by implementing the famous physical layer identification [15] techniques [5,14,16]. The fundamental idea of this approach is, the analog signal behaviors of data transmitters has slight variations which are introduced in the design, fabrication and manufacturing process. Researchers show that even manufactured in the same production lot, two same digital devices has unique artifacts in their signaling behavior which is difficult to control and duplicate [17]. Avatefipour et al. was able to extract those unique artifacts and proposed a framework based on neural network for CAN sender identification by utilizing the extracted distortions [5]. Likewise, in the last 5 years researchers [14,18,45] have proposed a lot of frameworks that are effective in CAN transmitter identification. While, the frameworks offer high percentage of accuracy, but the core architecture of these methods depend on handcrafted feature engineering. As the approaches rely on neural network based methods, the feature engineering remains an essential step in testing phase of the framework. In some cases, the feature engineering becomes computationally so expensive, that the real time sender identification remains a challenge. So, here is the research question in this paper, "Is it possible to identify the source of CAN message sender using an in-expensive approach that leverages deep neural network"?

In order to integrate a transmitter identification strategy to the existing CAN protocol, this paper proposes a framework that is based on the intersection between physical layer identification [15] and computer vision technique. To fingerprint, the proposed framework first extracts distortions from the analog signals sent by the ECUs. Then the distortions are converted into visual representation (images) by using recurrence plot technique [36] which are distinctive in human eyes. To automate the process of ECU identification, the images are fed into deep neural networks (Mobilenetv2 [23] and EfficientNet [39]) to learn patterns of the signals from those generated images. Finally, the trained model is tested to evaluate the performance of the proposed framework. According to the evaluation, it achieves better accuracy in identifying the CAN senders and is lightweight in terms of computational cost.

The main contribution of the paper is as follows:

- To the best of our knowledge, this is the first CAN sender identification framework that utilizes the concept of deep learning based computer vision task and transfer learning.
- The framework takes advantage of recurrence plot to visualize the dynamics of the senders to fingerprint ECUs.
- Based on the experimental settings with 8 ECUs, the framework identifies senders with an accuracy of 98.34% where 0.05 ms is needed to process a feature of a single observation for identification.
- The framework does not change the underlying architecture of basic CAN protocol, thus making it applicable to all CAN protocol based vehicles.

The rest of the paper is organized as follows. Section 2 provides the background of CAN, then the state-of-the-art of CAN cybersecurity is presented in Sect. 3. The methodology of our framework is presented in Sect. 4. Section 5 describes the experimental result of the framework and finally, the paper is concluded with the conclusion/future work section that is followed by acknowledgement section.

2 Background

2.1 Controller Area Network (CAN)

In this subsection, the background of controller area network (CAN) is presented. To highlight the overview of CAN protocol, the protocol characteristics and its representation in terms of OSI model [22] is described here. Moreover, the security issues originated from the basic architectural design of the protocol is described at the end of the subsection.

By design the controller area network (CAN) is a broadcasting protocol where ECUs communicate with each other using a single wire. This enables the system manufacturer to reduce complex wiring design of many point to point connections between ECUs and make the system easily maintainable [20]. While connected to a standard CAN network, an ECU can send 0–8 bytes of data with an eleven bit identifier. The identifier maintains the priority scheme of CAN protocol which is, message with lower arbitration ID has high priority while going through the bus [17]. On the other hand, any entity connected to the bus can listen to all the traffic in the network for its broadcasting nature.

The CAN protocol is specified in ISO 11898 and is defined in the physical layer and data link layer of Open Systems Interconnection model (OSI model) [22]. In the CAN physical layer, the data is handled as binary bits and the core functionality of this layer is to ensure bit encoding/decoding, bit synchronization and indicate physical wire orientation and on the other hand, CAN data link layer handles CAN data as frames and performs complex tasks like data encapsulation, frame encoding, frame error detection [22]. Physically, the CAN bus is actually a twisted pair wire, terminated with $120\,\Omega$. The twisted pair is called the CAN high (CAN-H) and the CAN low (CAN-L) and provides protection against electromagnetic interference. In terms of physical layer orientation of OSI model, CAN protocol follows differential signaling (shown in Fig. 1) where the final voltage of a single bit data is extracted by subtraction between CAN-H and CAN-L. When there is a 0 bit in the bus (dominant bit), CAN-H pulls $3.5\,V$ where CAN-L contains $1.5\,V$. In terms of a bit with value 1 (recessive), CAN-H and CAN-L both set the voltage to 2.5.

In data link layer, a CAN protocol handles data as frames. By default a standard CAN packet has 108 bits in total as shown in Table 1. It starts with a single bit of data called start of frame (SOF) field. Then it is followed by 11 bit arbitration ID (AID), 1 bit remote transmission request (RTR), 6 bit control field, 0–64 bits of data field, 16 bits cyclic redundancy check (CRC), 2 bits acknowledgment (ACK) field, 7 bits of end of frame (EOF) field [1]. While connected in a network, an ECU can send CAN packet to the traffic by sending a CAN data frame by putting dominant bit in the RTR field and an ECU can request data from another ECU by sending a CAN remote frame with a recessive bit in the RTR field. Although there is an AID field presented in a CAN packet, but there is not a single field available that indicates the source address. There is CRC field in a CAN packet which only protects the data field. So, the absence of source field and the broadcasting nature of the protocol clearly

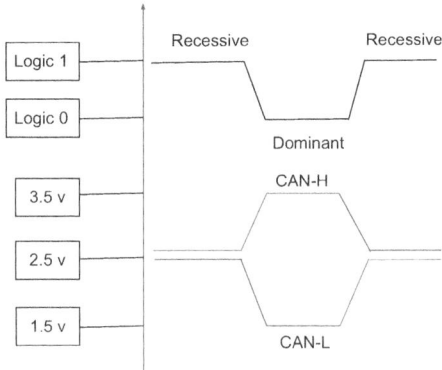

Fig. 1. CAN differential signaling

indicates that the CAN protocol lacks one of the concepts of the famous CIA triad (confidentiality, integrity and availability) i.e. integrity. The work proposes a framework to identify senders thus ensuring integrity to make CAN network security proven.

Table 1. A standard CAN data packet

Field name	Number of bits
Start of frame	1
Arbitration ID	11
Remote transmission request	1
Control fields	6
Data field	0–64
Cyclic redundancy check	16
Acknowledgement	2
End of frame	7
Total	108

3 Sender Identification: State-of-the-Art

To ensure integrity in the CAN bus one approach is to implement message authentication scheme [28] by including a message authentication code (MAC) inside CAN frame. While it makes the CAN bus secure but according to the standards, the least size of the MAC is 64 bit to prevent collisions [14]. So, the challenge of implementing the MAC based approaches is to add 64 bit MAC

Table 2. Computational complexity of common state-of-the-art statistical features

Feature name	Equation	Time complexity
Minimum	$min = min(x_i)$	$\Theta(n)$
Maximum	$max = max(x_i)$	$\Theta(n)$
Mean	$\bar{x} = \frac{\sum_{i=1}^{n} x_i}{n} = \frac{x_1 + x_2 + \ldots + x_n}{n}$	$\Theta(n)$
Variance	$s^2 = \frac{\sum_{i=1}^{n}(x_i - \bar{x})^2}{n-1} = \frac{\sum_{i=1}^{n} x_i^2 - n\bar{x}^2}{n-1}$	$\Theta(n^2)$
Skewness	$skewness = \frac{\sum_{i=1}^{n}(x_i - \bar{x})^3}{(n-1) * \sigma^3}$	$\Theta(n^2)$
Kurtosis	$kurt = \frac{\mu_4}{\sigma^4}$	$\Theta(n^2)$

along with the data that needs to be transported to the network where the data field can only hold up to 64 bits of data 1. To overcome the approach, researchers proposed two kind of MAC implementations. one is instead of using 64 bit MAC, they were using a truncated MAC to include integrity to CAN protocol [26–28] and the other approach is to use CAN+ protocol, an improvement of the existing CAN [29,30] where additional data can be sent in time intervals to authenticate CAN messages. For example, researchers in crafted a 4 byte MAC and put it into the data field of the CAN packet to authenticate CAN message. The disadvantage of truncating CAN data field to include MAC [26,27] is, it limits the size of data payload to be transmitted in a CAN packet and restrict the CAN protocol to transmit 8 bytes data payload. The proposed works in [29] sends two CAN messages where one contains the data payload the other one contains the MAC address. The approach resolves the issues originated by the truncated MAC approaches but it uses the limited traffic bandwidth of CAN network (1 Mbit/s) [5] as it needs to send two packets of data to securely send a single CAN data payload.

Apart from the CAN message authentication techniques, researchers have considered to fingerprint CAN senders by using physical unclonable characteristics such as clock skews [31] and voltage [5,14,18]. The main idea of this approach is to identify the source of CAN transmitters. The concept is adopted from the famous physical layer identification (PLI) [15] technique where the unique characteristics of transmitters are extracted to link the physical signals to the senders. The techniques for CAN PLI can be classified into two categories.

Clock Skew Based Fingerprinting. The quartz crystal clock determines the different clock frequencies on an ECU, resulting in random clock drifts which can be used to uniquely identify an ECU. Cho and Shin proposed a Clock-based IDS (CIDS) [31] which exploits the intervals of periodic message to estimate the clock skews as the fingerprint of the transmitter ECU. The idea was used to estimate clock behaviors of ECUs to detect the intrusion and identify the source of the message. However, this method is effective in a temperature-stable environment [32].

Voltage Based Fingerprinting. Authenticating the CAN message transmitter based on the unique and immutable physical characteristics such as the voltage, is termed as physical fingerprinting. This area of research has gained popularity now a days where utilizing the voltage characteristics is the core idea. For example, Kneib et al. [32] used voltages for fingerprinting ECUs, utilizing rising edge, falling edge of the dominant bits. The framework achieved an accuracy of 99.85% in identifying ECUs by using statistical features like mean, standard deviation, variance, skewness, kurtosis, root mean square, maximum and energy etc. Researchers in [5] extracted time domain and frequency domain statistical features using voltages captured from the ECUs and proposed a neural network based ECU classifier. They achieved an accuracy of 98.3% on an experimental setup using microcontrollers. Authors in [14] proposed an edge based identification method using voltage collected using picoscope (software defined oscilloscope) and a naive bayes classifier. As a feature they used statistical time domain features such as mean, variance, skewness, kurtosis, radio max plateau, plateau, overshoot height, irregularity, centroid, flatness, power and maximum. Similar work has been proposed in [33] that uses 10 time domain features and 10 frequency domain features and achieved an accuracy of 98.94% accuracy at maximum while voltage data is collected using an oscilloscope at a sampling rate of 2 GS/s. Bellaire et al. [18] proposed a machine learning based ECU fingerprinting framework by handcrafting signal processing features on voltage data such as transient response length, maximum transient voltage, energy of the transient period, average dominant bit steady-state value, peak noise frequency and average noise. Similar kind of approaches are also proposed in [34,35].

The research works described above achieved high accuracy in identifying CAN signal senders, but the feature extraction is highly expensive in terms of computational complexity. Table 2 represents the common statistical features and their corresponding computational cost. To overcome this, this paper proposed a novel framework that eliminates the necessity of extracting highly computational statistical features described above by utilizing images generated from the uniqueness presented in the voltage data to identify CAN signal transmitter. The image is generated using recurrence plot method whose computational complexity is $\Theta(n^2)$ whereas the computational complexity of any framework that uses feature shown in Table 2, is $3 * (\Theta(n^2) + \Theta(n))$. Experimental result shows that the proposed framework processes features to identify ECUs with a lower computational time than the state-of-the-art work.

4 Methodology

In this section the proposed framework for identifying CAN message senders is described. The phases of this methodology is described in a bottom up fashion, where the core idea of physical layer identification in the subsection A is presented first. Then the technique of image generation using the physical characteristics is describes in subsection B and finally in subsection C, the entire proposed framework is explained.

4.1 Linking CAN Signal to the Transmitter

Physical layer identification is a popular concept for identification of senders in connected networks for so many years [15,37]. The fundamental idea of this approach is, the behavior of senders in terms of analog signal has slight variations. The differences are introduced in the design, fabrication and manufacturing process, even two identical digital devices that are manufactured in the same production lot, have unique artifacts which is difficult to control and duplicate [5,14,32]. In a practical world, although it can be reproduced by reverse-engineering, but the process is difficult if not impossible for a determined attacker. Figure 2 illustrates the amount of inherent variation between two different CAN transmitters.

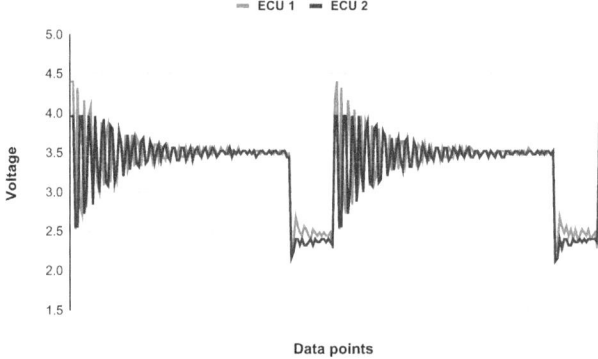

Fig. 2. Analog signal difference of two ECUs

The paper uses the above mentioned inherent variation of the CAN transmitter and uses them to fingerprint the transmitter as it is unique. The Fig. 3 shows how a CAN signal stays in an idea condition and how it distorts in practical world. The spikes from the idea line is considered as the impurity of each CAN transmitter which is identical to it. The proposed work uses it to create a unique signal characteristics profiling for transmitters.

Again the question remains how to extract the tiny variations? Which is called distortions of the analog voltage. Lets assume, V is a collection of analog voltage signal captured from the CAN-H wire where,

$$V = (V_1, V_2, V_3....V_n) \qquad (1)$$

Ideally, V_i should be 3.5 when it is a dominant bit and 2.5 when it is a recessive bit. In real world, the unique artifacts add noise to the ideal value and creates spikes (see 3). In order to extract the unique variations, the spiking points needs to be subtracted from 3.5 or 2.5 depending on it is a dominant or recessive bit. So, the unique artifacts (Distortions, D_i) of an ECU is,

$$D_i = (V_i - T_j) \qquad (2)$$

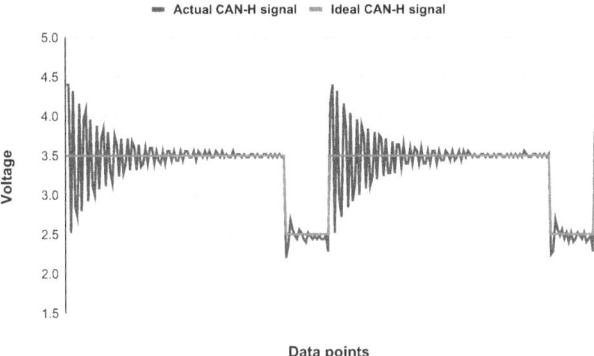

Fig. 3. CAN-H signals with unique artifacts

where T_j is either 3.5 or 2.5 depending on if the bit is dominant or recessive.

4.2 Representing Signal Profiling by Recurrence Plots

In this phase of the proposed method, the extracted unique slight variations of CAN senders are used to create images for each transmitter. The variations are turned into images via the recurrence plot (RP) [36] technique where each image represents the pattern of a sender. The initial purpose of recurrence plots was to create a visualisation of the recurrences of a system's states in a phase-space (with dimension n) within a small deviation ϵ. That means, a recurrence of a state at time i and at a different time j is marked within a two-dimensional squared matrix with ones and zeros (black and white points in a plot), where both axis represent time. The RP can be formally expressed by the following matrix in Eq. 3 [36].

$$R_{ij} = \Theta(\epsilon - |\vec{D_i} - \vec{D_j}|) \qquad i,j = 1....n \tag{3}$$

where D_i and D_j is the distortions extracted in Eq. 2. The matrix can be used to create an image which is actually a correlation plot. The image is the representation of a CAN transmitter as it is created form the unique physical characteristics. The proposed work uses the images in identifying the source of CAN signals by considering it as an image classification problem.

4.3 Source Identification of ECUs

Finally, the paper proposes a framework that uses the above mentioned steps to identify the source of CAN ECUs. Figure 4 shows overall architecture of our source identification framework. The proposed architecture first extracts the unique artifacts of the CAN transmitters. Then the recurrence plots are created from the extracted distortions which are a representation of uniqueness of CAN transmitters. Finally, the recurrence plots are used to train and test a deep learning model. In the Fig. 4 the green line shows the training phase and the red line represents the testing phase of the system. The data processing needed for the framework is to extract the unique artifacts and the creation of recurrence plot, which is same for both the training and testing phase. For the selection of deep learning architecture, we have chosen two popular network MobileNetV2 [38] and EfficientNet [39] for the experiment.

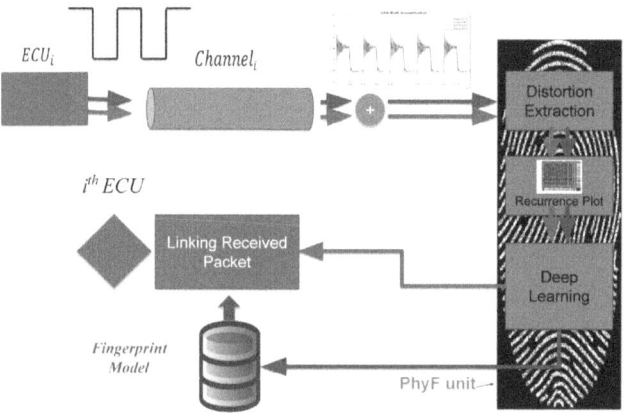

Fig. 4. Source identification of CAN message senders

5 Experimental Result

In this section the effectiveness of the proposed methodology is evaluated by conducting experiments in the laboratory where Subsect. 5.1 presents the experimental setup. It is followed by Subsect. 5.2, 5.3 i.e. the generation of recurrence plot of CAN senders and the performance of the proposed framework consecutively. Then the effect of environmental factors over the proposed framework is discussed in Subsect. 5.4 and effect of information aware downsampling is presented in Subsect. 5.5. Finally, the performance of spoof detection in a vehicle test bench and the comparison with the state-of-the-art is presented in Subsect. 5.6 and 5.7 respectively to conclude the section.

5.1 Experimental Setup

To verify the effectiveness of the proposed framework, an experiment is designed (Fig. 5) on a CAN protocol based test bed that has total 8 ECU built by Arduino Uno microcontrollers connected to CAN transceivers where each ECU has 1 m channel length. Physical signal for each ECU is captured using a DSO1012A oscilloscope with a sampling rate of 2 GS/s, 100 MHz bandwidth, and 8-bit vertical resolution. The data was collected in a laboratory environment from the CAN-H pin which ideally ranges from 3.5 v to 2.5 v. Multiple programming languages are used in this experiment as the microcontrollers are programmed using C programming language and Python is used for training & testing deep learning models and result analysis. The experimental testbed is set up in a plug and play mode, because some of the experiments were done using 4 ECUs and some of the experiments were conducted using 8 ECUs. To check the performance of the proposed framework on a real vehicle, an experiment is designed on a vehicle test bench that is based on the GM Sierra 2020 model for spoof detection also (elaborated extensively in Subsect. 5.6).

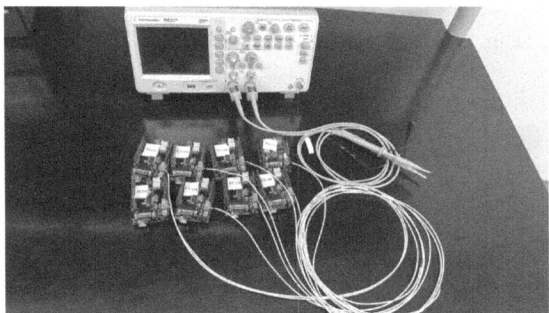

Fig. 5. Experimental settings

5.2 Generation of Recurrence Plot of CAN Senders

The goal of this subsection was to create recurrence plot by using distortions captured from the CAN transmitters and visualize them in human eyes. To perform that experiment we collected analog signals from 4 ECUs using the testbed described in Subsect. 5.1 and extracted the distortions of the ECUs. The distortions are mapped to create recurrence plot and saved as images for visualization. Each image was generated from 96 voltage data points (length of CAN-H dominant bit) started from the peak of the voltage signals. Figure 6 shows the generated plots of 4 ECUs where each row has images for each ECU. It indicates that, the images has their own patterns and they are different to each other visually.

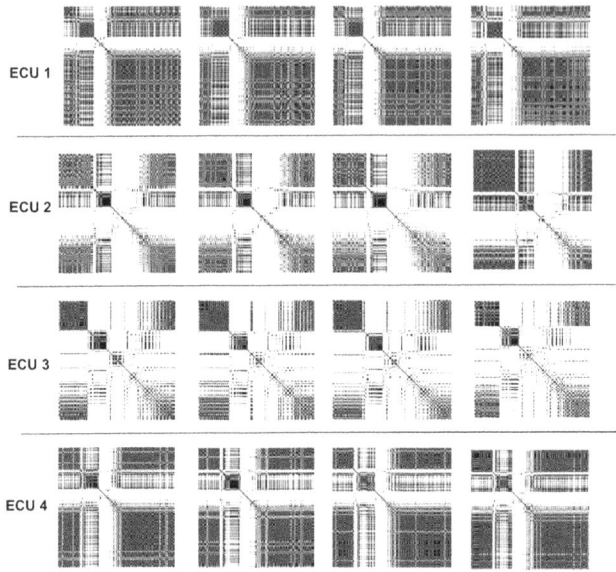

Fig. 6. Recurrence plot representing 4 ECUs

Visually RPs provide some useful insights about the sender ECUs. But the question arises how much information the RPs contain to distinguish the CAN transmitters. In order to do so, an experiment was deigned to quantify the RPs generated from the CAN ECU signals. To achieve that, a recurrence quantification analysis was performed to quantify the RPs by extracting recurrence properties from the generated RPs. We used recurrence parameters [44] such as recurrence rate, entropy diagonal lines, longest diagonal line length, laminarity, divergence, additional diagonal line length to perform data analysis. To do so, a Python program is written to generate RPs from CAN high analog voltages collected from 8 ECUs and then the images are used to perform recurrence quantification analysis (RQA) in an Apple M1 chip computer with 8 GB RAM. For RQA, the images are fed into python library PyRQA [42] and the parameters are extracted for rigorous analysis. To see the feature differences of the ECUs in terms of RQ parameters the data is plotted in a box plot shown in Fig. 7. The figure shows that the 8 ECUs have notable variations when compared against the 6 recurrence parameters.

5.3 Performance of the Proposed Framework

This experiment evaluates the accuracy of sender identification in a CAN network using the proposed framework. The main goal of this subsection is to demonstrate the applicability of image classification via deep learning models in CAN physical fingerprinting. To achieve this, the proposed framework is validated against deep learning networks using (Mobilenetv2 [23] and EfficientNet

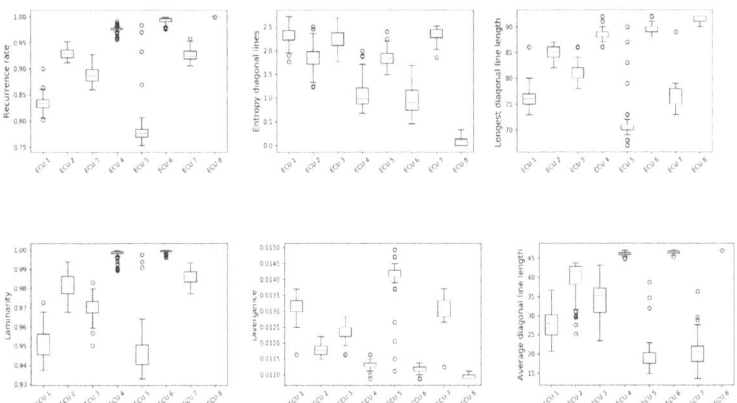

Fig. 7. Feature differences of recurrence quantification parameters

[39]) architecture. First the data from 8 ECUs are collected from the testbed described in subsection A, then data processing which involves extraction of distortion & image generation is done in an Apple M1 computer with 8 GB RAM and finally, the training and testing of deep learning architecture is performed in a Google-Colab environment. The code for the data processing, model training and model testing is written in Python programming language.

Table 3. CAN sender identification using Deep learning models

Algorithm	Accuracy(%)	Feature processing time (ms)
efficientnet	95.04	0.07
Mobilenetv2	97.52	0.07

To conduct the experiment 131,760 analog voltage data points are gathered in total and 1098 images were created as described in subsection B where each ECU has 250 images and each image has a dimension of 192X192 pixels. To handle the smaller number of images we used transfer learning [41], where a trained model is tuned to solve a problem which is unknown to the trained model. In order to do so, a pre-trained MobileNetV2, trained with a public dataset (imagenet [40]) with 1,700,505 parameters is selected and it's weights are used to retrain the model to solve the CAN sender identification problem. For retraining the model, 70% data was used, while the retrained model is tested with 15% and validated with remaining 15% data. Table 3 shows the result of the simulation. It indicates, Mobilenetv2 architecture achieves a maximum validation accuracy of 97.52%. To check the performance of the proposed methodology while using a different deep learning algorithm, the same experiment was repeated using an EfficientNet model pre-trained on imagenet dataset [40] with 5,338,572 parameters. It was re-trained using 70% data and tested using 15% data where the image

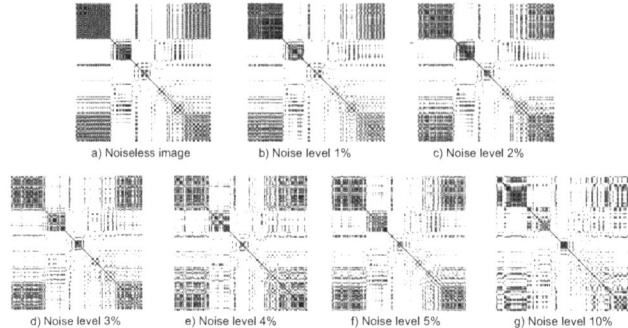

Fig. 8. Images generated from voltages under different environmental conditions

dimension was 224 × 224 pixels. Finally, the performance of the methodology was measured by validating the trained model with 15% remaining data. The experimental result shows that the EfficientNet achieves a validation accuracy of 95.04%. While using both MobileNetv2 and EfficientNet it takes 0.07 ms per image on average for data processing task which involves noise extraction and image creation.

5.4 Effect of Environmental Factors on the Proposed IDS

This subsection represents the analysis of the effect of environmental conditions on the performance of the proposed framework. It is important because, the foundation of the proposed methodology is image classification where the images are created from the distortions present in electrical signals and the signal characteristics are sensitive to environmental factors like temperature, amount of moisture contamination, aging, etc. [21]. These factors, if not accounted for, could lead to incorrect identification of the senders in real-world scenarios. In order to verify their effect, data is collected from a setup testbed with 8 ECUs and then analysis is performed to measure performance of the proposed framework under the presence of noise that may be produced by environmental factors. To add the noise, a simulation is created by adding Additive White Gaussian Noise (AWGN) [43] of different percentage of the voltage distortions (1%, 2%, 3%, 4%, 5% & 10%) to the voltage signals. The overall experiment is divided into two different steps, first one is adding AWGN to the electrical signals and creating the images by using the noisy distortions. Figure 8 shows the distorted images that are generated from different level of noisy voltages. In the subfigure (a) represents an image generated without AWGN, while subfigure b, c, d, e, f, g represents images with 1%, 2%, 3%, 4%, 5% & 10% added AWGN. The noiseless and noisy figures clearly indicates that the noises caused by environmental factors has significant effect on the images generated by the voltage distortions while there is deviation from the original image increases with the addition of level of noises to the voltages.

In the second step of the experiment, a deep learning model is trained using the noiseless original noise, while the images with noise is tested against the trained model and the model performance is evaluated in terms of sender identification accuracy (shown in Table 4). While introducing 1% noise in the testing data the performance of the proposed framework degrades by a 30.23% so the proposed model is sensitive to environmental noise. To check the performance of the proposed approach when the model is retrained, again the trained model is retrained by adding images with 1% AWGN and tested against noisy images. Later the trained model was retrained with 2% noise and model performance against noisy images was evaluated again. The experimental result is summarized and shown in Table 4. According to it, when the generated images with 1% AWGN are introduced during the model training with noiseless images, testing accuracy improves significantly for noisy images (1%, 2%, 3%, 4%, 5% and 10% GN). Although the training data had only noisy images with low AWGN (1%), the trained model was able to classify noisy images with an improvement of maximum 34.47% and minimum 19.37%. Again, when the model was again retrained by introducing noisy images with 2% AWGN and the model can identify senders with an maximum upgrade of 9.2% in terms of accuracy. So, it can be concluded that, the proposed framework is performs better if the model is retrained with noisy images.

5.5 Effect of Selective (Information Aware Down Sampling) Sampling on the Proposed Framework

Since, the amount of data to be processed for generating each image has a larger influence on the required computing power, a major goal is to reduce the required amount of sampling points. To reduce the sampling points considered to create the image, an experiment with rigorous analysis is conducted. If we look carefully, the backbone of the methodology is the images which are created from the distortions of the ECUs. Again, the distortions are created from analog voltage signal of the CAN signals. Figure 9 shows the plot of analog signal captured form an ECU and it is clearly visible that, the signals has spikes at the beginning and gradually it settles down in terms of voltage. From that we can infer that, the distortions which is extracted from the overshoot portion of analog signals (marked as a red box in Fig. 9) holds significant unique information which is vital in sender identification. And after that we have data points that are less informative. Based on that observation, an experiment was conducted where images were generated by varying the informative and uninformative data points. Then the images are fed into a MobileNetV2 model for evaluation. So, in order to verify that an simulation is designed where images generated by three approaches are tested against MobileNetV2 model and the validation accuracy are evaluated. They are,

- **Truncated sampling:** images generated using all the informative points.
- **Custom odd sampling:** images generated using all informative points and the odd sampling points of uninformative portions aka.

Table 4. Effect of environmental factors over the proposed framework

Training images			Testing images	Model performance
Noiseless	1% AWGN	2% AWGN	AWGN (%)	Accuracy (%)
Yes	No	No	0	98.34
Yes	No	No	1	68.11
Yes	No	No	2	61.09
Yes	No	No	3	55.81
Yes	No	No	4	51.05
Yes	No	No	5	47.00
Yes	No	No	10	33.47
Yes	Yes	No	1	94.77
Yes	Yes	No	2	95.56
Yes	Yes	No	3	92.31
Yes	Yes	No	4	85.45
Yes	Yes	No	5	80.94
Yes	Yes	No	10	52.84
Yes	Yes	Yes	2	96.82
Yes	Yes	Yes	3	95.99
Yes	Yes	Yes	4	94.65
Yes	Yes	Yes	5	89.13
Yes	Yes	Yes	10	61.52

– **5th sequence sampling:** images generated using all the informative points and the (0, 5, 10, 15 ...nth) sampling points in uninformative portions.

Table 5. Performance analysis on information aware down sampling

Sampling method	Accuracy(%)	Processing time (ms)
Truncated sampling	94.21	0.04
Custom odd sampling	95.05	0.06
5th sequence sampling	98.34	0.05

To conduct the above mentioned experiments, analog voltage samples for 8 ECUs are collected using the experimental setup described in subsection A and images are generated using the truncated sampling, the custom odd sampling and the 5th sequence sampling methods. The images are then trained, tested and validated with the MobileNetV2 deep learning architecture and evaluated based

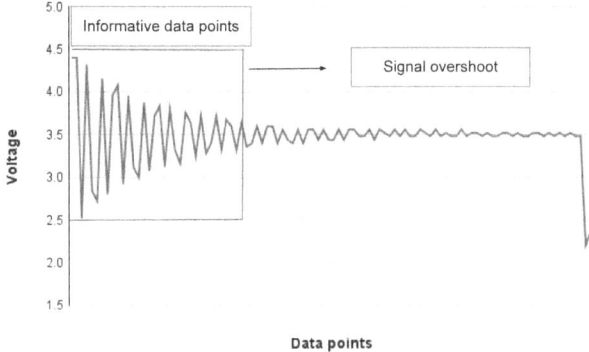

Fig. 9. Information aware down sampling technique

on validation accuracy that is shown in Table 5. According to the analysis, 5th sequence sampling achieved a validation accuracy of 98.34% which is better than the accuracy achieved while using truncated sampling and custom odd sampling. Where the truncated sampling is faster than the other two approaches in data processing by at least .01 ms.

5.6 Spoof Detection on Actual Vehicle Test Bench

To evaluate the performance on an actual vehicle test bench, an experiment is conducted where the goal is to detect spoofing attack using the proposed methodology. First an experiment is setup on a laboratory vehicle test bench (Model: GM Sierra, Year: 2020) shown as Fig. 10 to perform spoofing attack by changing the vehicle gear from park to drive mode using a raspberry pi 4 model B, where analog voltage data is captured using a picoscope 2205 A with a sampling rate of 25 MS/s. After that, analog voltage data is captured again with the same sampling rate of 25 MS/s when the vehicle is put to drive mode from park mode using the vehicle gearshift. Although the attacker ECU was sending the same data we can see form Fig. 11 the data send by the attacker has different analog voltage profile than the authorized ECU. The Fig. 11 shows that the two sets of benign CAN-H signal data from authorized ECU, marked in blue & orange and captured at different times, differ from the two sets of CAN-H data from the attacker, marked in green & purple and captured at different times. However, when sent from the same CAN ECU, the data from both the authorized ECU and the attacker appear almost identical. It is clear form the figure that, the ECUs has their own fingerprint in their CAN-H dominant bit analog data which is different from 3.5 v (marked as red in Fig. 11).

The analog voltages then prepossessed in a computer with 8 GM RAM and 416 images are generated from the distortions for both ECUs using Python programming language, where 240 were from authorized ECU and 176 were from attacker ECU. As the dataset is comparatively small we did data augmentation technique to flip each images from left to right and prepared a data size of 832

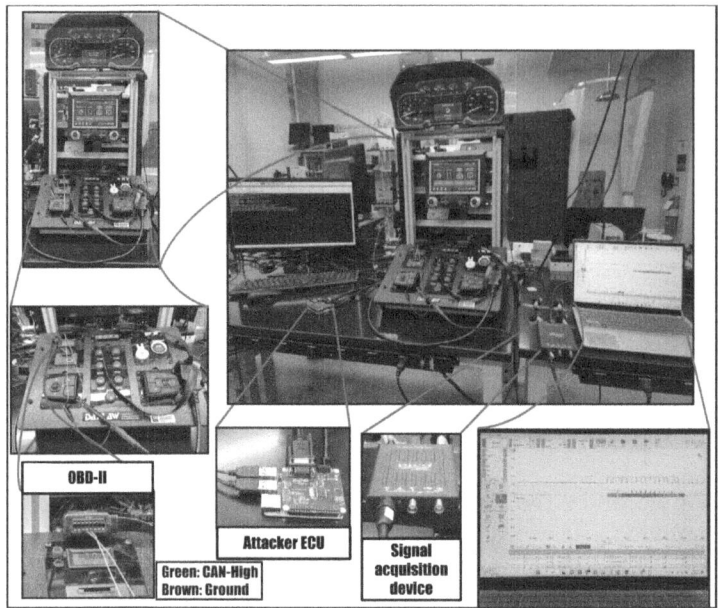

Fig. 10. Spoofing attack on vehicle test bench

Table 6. Performance under spoofing attack on laboratory vehicle bench

Data Sample	Accuracy (%)	Precision (%)	Recall (%)	F1 score (%)
No attack	97.78	100	97.78	98.87
Spoofing attack	100	96.72	100	98.33
Combined (spoof attack + no attack)	98.89	98.36	98.89	98.60

images. The images are then retrained using a pre-trained MobileNetV2 deep learning architecture where 70% data are used as training, 15% for tuning the model and 15% for testing the trained model. The performance of the proposed methodology is evaluated against metrics such as (1) attack detection accuracy, precision, recall, f1-score and (2) benign data detection accuracy, precision, recall and f1-score. The simulation result is summarized in Table 6 and it shows that the proposed methodology detects spoofing attack with 100% accuracy while it achieves a precision of 96.72%, recall of 100% and f1-score of 98.33%. On the other hand, it detects benign data with an accuracy of 97.78%, precision of 100%, recall of 100% and f1-score of 98.33%. So, finally, it can be concluded that, the proposed methodology is efficient and applicable to today's vehicles as it achieved an combined accuracy of 98.89% in detecting spoofed and benign data in the vehicle test bench.

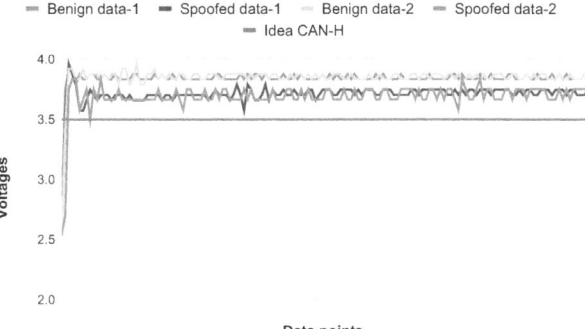

Fig. 11. Analog voltage data for an authorized ECU and an attacker ECU

Table 7. Comparison with the state-of-the-art

Approach	Accuracy (%)	Precision (%)	Recall (%)	F1 (%)	time (ms)
[5]	97.13	97.00	97.00	95.75	1.35
[14]	89.24	89.63	89.63	89.38	0.95
KNN	94.97	95.00	95.38	95.13	238
SVM	95.82	95.75	96.13	95.75	238
Proposed IDS	98.34	98.63	98.25	98.38	0.05

5.7 Comparison with the State-of-the-Art

Finally the proposed methodology is compared against the state-of-the-art sender identification methods [5] and [14] where [5] trains a Neural Network and [14] builds a Support Vector Machine (SVM) model using statistical features. In order to do so, the state-of-the-art [5] methodology extracts the distortion of the ECUs at first and handcrafts 11 statistical analysis based features including 6 time domain features i.e. maximum, minimum, mean, variance, skewness, kurtosis and 5 frequency domain features i.e. spectral standard deviation, spectral kurtosis, spectral skewness, spectral centroid, irregularity k to feed into an artificial neural network to evaluate the performance of their approach. While simulating their approach with data gathered from 8 ECU using the experimental setup described in subsection A, an validation accuracy of 97.13% as achieved while the proposed framework achieved 98.34%. But in terms of data processing (feature engineering), the state-of-the-art takes 27 times more than the proposed framework. The proposed methodology creates the a single recurrence plot for testing the model in 0.05 ms time using the 5th sequence information aware down sampling while, the state-of-the-art [5] generates 11 statistical features in 1.35 ms for identifying a single CAN sender which is computationally more expensive (see Table 7). Again the proposed approach is compared against [14] where the authors use signal characteristics by extracting 12 features such as maximum, mean, variance, skewness, kurtosis, centroid, flatness, power, irregularity, plateau, max plateau ratio and overshoot height to train machine learning model. While training & testing a SVM model, the state-of-the-art degrades by 9.1% in terms of sender identification accuracy than the proposed methodology and needs 0.95 ms to process features which is 19 times slower than the proposed approach. In addition to that, the proposed approach is compared with traditional machine learning based approaches i.e. K-Nearest Neighbors(KNN) and SVM where recurrence quantification parameters are used as features to fit machine learning models. Based on the data collected form 8 CAN ECUs, KNN achieved an accuracy of 94.97% and SVM achieved an accuracy of 95.82% while the feature generation took around 238 ms. It clearly shows that the proposed methodology is better both in terms of accuracy of identifying senders and feature processing time.

6 Conclusion and Future Work

The paper proposed a physical fingerprinting framework for solving the CAN protocol's inability to identify the sender by modeling the problem as an image classification problem. It introduces a novel approach for creating images utilizing uniqueness of the analog signal of CAN senders and it classifies the images using deep learning model to identify CAN ECUs. As, the proposed method only requires to generate an image for the identification of ECUs, it can put an end to the trend of handcrafted feature engineering process in CAN physical fingerprinting. As per contributing to the state-of-the-art, to the best of our knowledge the proposed methodology is the first ever work that utilizes the concept of computer vision in CAN sender identification problem. The experimental result shows that it is effective as it achieved an accuracy of 98.34% and efficient as it only requires an image to identify sender. In the future, we will investigate the effect of environmental factors like aging, temperature etc. on the proposed methodology as analog signals change by time and varying temperature.

Acknowledgement. This research work was supported by the National Science Foundation under the award # CNS-2035770 and award # 2414729.

References

1. Elkhail, A.A., Refat, R., Habre, R., Hafeez, A., Bacha, A., Malik, H.: Vehicle security: a survey of security issues and vulnerabilities, malware attacks and defenses. IEEE Access **9**, 162401–162437 (2021)
2. Greenberg, A.: Hackers remotely kill a jeep on the highway–with me in it. Wired **7**(2), 21–22 (2015)
3. Nie, S., Liu, L., Du, Y.: Free-fall: hacking tesla from wireless to can bus. Brief. Black Hat USA **25**, 1–16 (2017)
4. Research. Upstream Security (2022). https://www.upstream.auto/research/automotive-cybersecurity/?id=4710

5. Avatefipour, O., Hafeez, A., Tayyab, M., Malik, H.: Linking received packet to the transmitter through physical-fingerprinting of controller area network. In: 2017 IEEE Workshop on Information Forensics and Security (WIFS), pp. 1–6. IEEE (2017)
6. Halder, S., Conti, M., Das, S.K.: COIDS: A clock offset based intrusion detection system for controller area networks. In: Proceedings of the 21st International Conference on Distributed Computing and Networking, pp. 1–10 (2020)
7. Jichici, C., Groza, B., Ragobete, R., Murvay, P.S., Andreica, T.: Effective intrusion detection and prevention for the commercial vehicle SAE J1939 CAN bus. IEEE Trans. Intell. Transp. Syst. (2022)
8. Ishak, M.K., Khan, F.K.: Unique message authentication security approach based controller area network (CAN) for anti-lock braking system (ABS) in vehicle network. Procedia Comput. Sci. **160**, 93–100 (2019)
9. Nurnberger, S., Rossow, C.: vatiCAN-vetted authenticated CAN bus. In: Cryptographic Hardware and Embedded Systems: 18th International Conference Santa Barbara CA USA, 17–19 August 2016, Proceedings (2016)
10. Weisglass, Y., Oren, Y.: Authentication method for CAN messages. ESCAR Europe (2016)
11. Islam, R., Refat, R.U.D., Yerram, S.M., Malik, H.: Graph-based intrusion detection system for controller area networks. IEEE Trans. Intell. Transp. Syst. (2020)
12. Lee, H., Jeong, S.H., Kim, H.K.: OTIDS: a novel intrusion detection system for in-vehicle network by using remote frame. In: 2017 15th Annual Conference on Privacy, Security and Trust (PST), pp. 57-5709. IEEE (2017)
13. Hossain, M.D., Inoue, H., Ochiai, H., Fall, D., Kadobayashi, Y.: LSTM-based intrusion detection system for in-vehicle can bus communications. IEEE Access **8**, 185489–185502 (2020)
14. Kneib, M., Schell, O., Huth, C.: EASI: edge-based sender identification on resource-constrained platforms for automotive networks. In: NDSS (2020)
15. Gerdes, R.M.K.: Physical layer identification: methodology, security, and origin of variation. Iowa State University (2011)
16. Hafeez, A., Topolovec, K., Awad, S.: ECU fingerprinting through parametric signal modeling and artificial neural networks for in-vehicle security against spoofing attacks. In: 2019 15th International Computer Engineering Conference (ICENCO), pp. 29–38. IEEE (2019)
17. Xu, T., Lu, X., Xiao, L., Tang, Y., Dai, H.: Voltage based authentication for controller area networks with reinforcement learning. In: ICC 2019-2019 IEEE International Conference on Communications (ICC), pp. 1–5. IEEE (2019)
18. Bellaire, S., Bayer, M., Hafeez, A., Refat, R.U.D., Malik, H.: Fingerprinting ECUs to implement vehicular security for passenger safety using machine learning techniques. In: Proceedings of SAI Intelligent Systems Conference, pp. 16–32. Springer, Cham (2023)
19. Hafeez, A., Ponnapali, S.C., Malik, H.: Exploiting channel distortion for transmitter identification for in-vehicle network security. SAE Int. J. Transp. Cybersecur. Priv. **3**(11-02-02-0005), 5–17 (2020)
20. Refat, R.U.D., Elkhail, A.A., Malik, H.: Machine learning for automotive cybersecurity: challenges, opportunities and future directions. In: AI-Enabled Technologies for Autonomous and Connected Vehicles, pp. 547–567 (2023)
21. Sierota, A., Rungis, J.: Electrical insulating oils. I. Characterization and pretreatment of new transformer oils. IEEE Electr. Insul. Mag. **11**(1), 8–20 (1995)
22. Alani, M.M.: OSI model. In: Guide to OSI and TCP/IP Models, pp. 5–17. Springer, Cham (2014)

23. Sandler, M., Howard, A., Zhu, M., Zhmoginov, A., Chen, L.C.: MobileNetV2: the next generation of on-device computer vision networks. In: CVPR (2018)
24. Popa, L., Groza, B., Jichici, C., Murvay, P.S.: ECUPrint–physical fingerprinting electronic control units on CAN buses inside cars and SAE J1939 compliant vehicles. IEEE Trans. Inf. Forensics Secur. **17**, 1185–1200 (2022)
25. Lu, Z., Wang, Q., Chen, X., Qu, G., Lyu, Y., Liu, Z.: LEAP: a lightweight encryption and authentication protocol for in-vehicle communications. In: 2019 IEEE Intelligent Transportation Systems Conference (ITSC), pp. 1158–1164. IEEE (2019)
26. Hartkopp, C.R.O., Schilling, R.: MaCAN-message authenticated CAN. In: Proceedings of the 10th International Conference on Embedded Security in Cars (ESCAR) (n.d.)
27. Hazem, A., Fahmy, H.A.: Lcap-a lightweight can authentication protocol for securing in-vehicle networks. In: 10th escar Embedded Security in Cars Conference, Berlin, Germany, vol. 6, p. 172 (2012)
28. Schmandt, J., Sherman, A.T., Banerjee, N.: Mini-MAC: raising the bar for vehicular security with a lightweight message authentication protocol. Veh. Commun. **9**, 188–196 (2017)
29. Van Herrewege, A., Singelee, D., Verbauwhede, I.: CANAuth-a simple, backward compatible broadcast authentication protocol for CAN bus. In: ECRYPT workshop on Lightweight Cryptography, vol. 2011, p. 20. ECRYPT (2011)
30. Groza, B., Murvay, S., Herrewege, A.V., Verbauwhede, I.: LiBrA-CAN: a lightweight broadcast authentication protocol for controller area networks. In: International Conference on Cryptology and Network Security, pp. 185–200. Springer, Heidelberg (2012)
31. Shin, K.G., Cho, K.T.: U.S. Patent No. 11,044,260. U.S. Patent and Trademark Office, Washington, DC (2021)
32. Kneib, M., Huth, C.: Scission: signal characteristic-based sender identification and intrusion detection in automotive networks. In: Proceedings of the 2018 ACM SIGSAC Conference on Computer and Communications Security, pp. 787–800 (2018)
33. Choi, W., Joo, K., Jo, H.J., Park, M.C., Lee, D.H.: VoltageIDS: low-level communication characteristics for automotive intrusion detection system. IEEE Trans. Inf. Forensics Secur. **13**(8), 2114–2129 (2018)
34. Verma, K., Girdhar, M., Hafeez, A., Awad, S.S.: ECU identification using neural network classification and hyperparameter tuning. In: 2022 IEEE International Workshop on Information Forensics and Security (WIFS), pp. 1–6. IEEE (2022)
35. Ahmed, S., Juliato, M., Gutierrez, C., Sastry, M.: Two-point voltage fingerprinting: increasing detectability of ECU masquerading attacks. arXiv preprint arXiv:2102.10128 (2021)
36. Marwan, N., Kurths, J., Saparin, P.: Generalised recurrence plot analysis for spatial data. Phys. Lett. A **360**(4–5), 545–551 (2007)
37. Mathur, S., et al.: Exploiting the physical layer for enhanced security [security and privacy in emerging wireless networks]. IEEE Wirel. Commun. **17**(5), 63–70 (2010)
38. Dong, K., Zhou, C., Ruan, Y., Li, Y.: MobileNetV2 model for image classification. In: 2020 2nd International Conference on Information Technology and Computer Application (ITCA), pp. 476–480. IEEE (2020)
39. Koonce, B.: EfficientNet. In: Convolutional Neural Networks with Swift for TensorFlow, pp. 109–123. Apress, Berkeley (2021)

40. You, Y., Zhang, Z., Hsieh, C.J., Demmel, J., Keutzer, K.: ImageNet training in minutes. In: Proceedings of the 47th International Conference on Parallel Processing, pp. 1–10 (2018)
41. Tan, C., Sun, F., Kong, T., Zhang, W., Yang, C., Liu, C.: A survey on deep transfer learning. In: International Conference on Artificial Neural Networks, pp. 270–279. Springer, Cham (2018)
42. 8.0.0, P. Feb 21. (2021). https://pypi.org/project/PyRQA/
43. Aja-Fernández, S., Tristán-Vega, A.: A review on statistical noise models for magnetic resonance imaging. LPI, ETSI Telecomunicacion, Universidad De Valladolid, Spain, Technical report (2013)
44. Rawald, T., Sips, M., Marwan, N.: PyRQA–conducting recurrence quantification analysis on very long time series efficiently. Comput. Geosci. **104**, 101–108 (2017)
45. Hafeez, A., Ponnapali, S., Malik, H.: Exploiting channel distortion for transmitter identification for in-vehicle network security. SAE Int. J. Transp. Cybersecur. Priv. **3**, 5–17 (2020)

Short Paper: Software Bill of Materials Management for Embedded Vehicle Systems

Teddy Nyambe(✉), Rik Chatterjee, and Jeremy Daily

Colorado State University, Fort Collins, CO, USA
{teddy.nyambe,rik.chatterjee,jeremy.daily}@colostate.edu

Abstract. Modern vehicles are integral components of our daily lives, crucial for transportation and logistics. As such, ensuring cybersecurity in embedded vehicle systems is essential. Software Bills of Materials (SBOMs) are increasingly recognized as a valuable tool for enhancing software supply chain security. By providing transparency and traceability of software components, SBOMs facilitate the rapid identification and mitigation of vulnerabilities. Despite their benefits, implementing SBOMs for embedded vehicle systems presents unique challenges. This study investigates the challenges involved in the management of SBOMs for embedded vehicle systems, particularly focusing on a custom Yocto Linux distribution for a diagnostic platform in heavy-duty trucks. We highlight key issues and propose potential solutions. Our findings underscore the necessity for specialized tools and frameworks to effectively integrate SBOMs in embedded systems, thereby strengthening the cybersecurity posture of these critical systems.

Keywords: Bills of Materials · Cybersecurity · Embedded Systems · Supply Chain Transparency · Smart Vehicle Security

1 Introduction

Modern vehicles are an innate part of our daily lives, forming a crucial foundation for transportation and logistics. Embedded computers in these vehicles act as the central control units, managing most operations from engine performance to advanced driver assistance. However, in recent years, studies have highlighted significant cybersecurity vulnerabilities in these embedded vehicle systems [2,3,5], exposing them to potential attacks and exploitation. Further, the US Government through Executive Order (EO) 14028 of May 2021 [9] and the National Institute of Standards and Technology (NIST) Cybersecurity Supply Chain Risk Management special publication 800-161r1 [1] have recognized the complexity of the cybersecurity supply chain and the potential risks that can arise from malicious components.

In recent years, Software Bills of Materials (SBOMs) have become increasingly recognized as a vital tool for enhancing software supply chain security.

SBOMs provide a detailed inventory of the software components and their dependencies within a system, offering crucial insights into the provenance and integrity of software elements. By enabling organizations to track and manage software components effectively, SBOMs support the rapid identification of security issues and facilitate timely patching and updates.

However, there are obstacles to the widespread adoption of SBOMs for embedded systems. While academia and industry are actively involved in SBOM development, there remain practical challenges in implementing them [14].

The objective of this study is to identify challenges and recommend solutions associated with the development of SBOMs in embedded systems to enhance their security posture. The study aims to leverage insights gained from our experiences during the development of a Linux-embedded operating system based on a Yocto Linux Project for heavy-duty vehicle diagnostics and security.

2 Background SBOM

2.1 What is an SBOM

A Software Bill of Materials (SBOM) according to Executive Order 14028 Section 10 (j) is a formal record containing the details and supply chain relationships of various components used in building software. Software developers and vendors often create products by assembling existing open source and commercial software components [9] and by extension includes firmware [7]. In this regard, an SBOM is the only formal artifact that provides a use case for visibility into the software supply chain by detailing software dependencies between and among components, versions, licenses, and vendors.

As software development processes have evolved, they have increasingly incorporated third-party libraries and modules to expedite time-to-market and reduce costs. This reliance on external code increases security and compliance risks.

Producers have the responsibility of providing provenance and compliance of their software products to mitigate risks associated with vulnerabilities and compliance failures in software products. High-profile incidents, such as the Log4J [4] and SolarWinds vulnerabilities [13], exemplify how upstream vulnerabilities can have severe repercussions for downstream users.

It is important to clarify that while SBOMs do not eliminate vulnerabilities, they significantly enhance the triage process and facilitate the remediation of security incidents. By providing a detailed inventory of software components, SBOMs enable quicker responses to reported vulnerabilities, thereby improving overall security posture.

2.2 SBOM Production

Software Bill of Materials (SBOMs) are generated from metadata within the development environment, with source code serving as the most reliable source of metadata. However, many firmware and binaries are distributed without accompanying source code, rendering binaries—products of the toolchain process— opaque and lacking in valuable data for SBOM generation. While techniques exist for analyzing binaries through costly reverse engineering, these methods are neither practical nor repeatable.

In the context of the Software of Interest (SoI), the Yocto Project is employed to produce SBOMs utilizing Software Package Data Exchange (SPDX) tools. The SBOM generation process relies on metadata contained within recipes. A recipe (see Fig. 1) outlines a series of settings and tasks (instructions) necessary for building software packages, which subsequently contribute to the creation of a binary image for a custom-embedded operating system [10]. As illustrated in Fig. 2, the SBOM creation process begins when a producer initiates a sequence of tasks managed by the Yocto build system, specifically Bitbake. This build

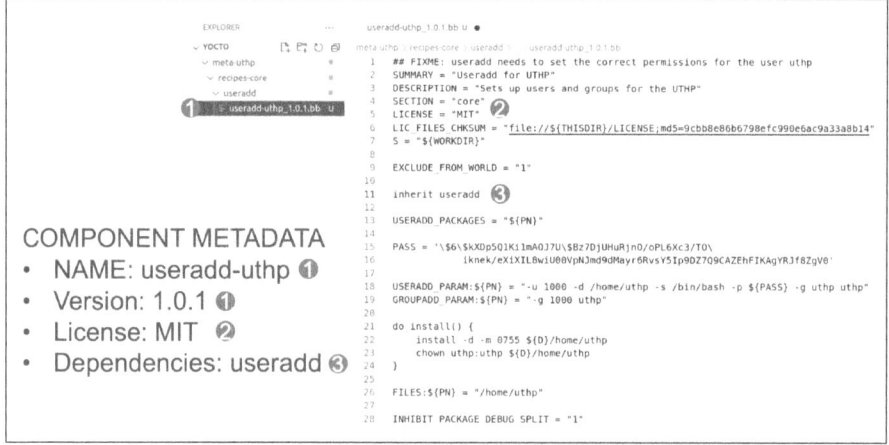

Fig. 1. Yocto recipe example with metadata

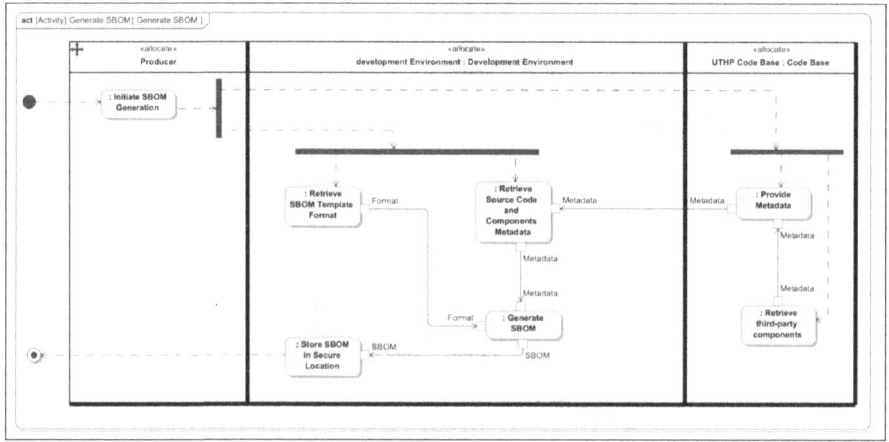

Fig. 2. Yocto Project SBOM Generation Activity Diagram

process activates the `do_create_spdx` task generator, which extracts metadata from the recipes (refer to Fig. 1) and enumerates them into an SPDX document.

2.3 SBOM Standards

The National Telecommunications and Information Administration (NTIA) as mandated by the EO 14028 recommended minimum elements requirements for Software Bills of Materials (SBOMs) to enhance transparency and security in the software supply chain [11]. The Software Package Data Exchange (SPDX) is the standard used in the SoI and is pioneered by the open-source community. Other noteworthy standards include CycloneDX, maintained by the Open Web Application Security Project (OWASP) community, and Software Identification (SWID) tags. While SWID primarily focuses on the allocation of unique identifiers for software components, its application is limited compared to the broader utility of SPDX and CycloneDX.

Yocto build environment uses the SPDX as its default format as illustrated in Fig. 3 depicting minimum recommended SBOM elements.

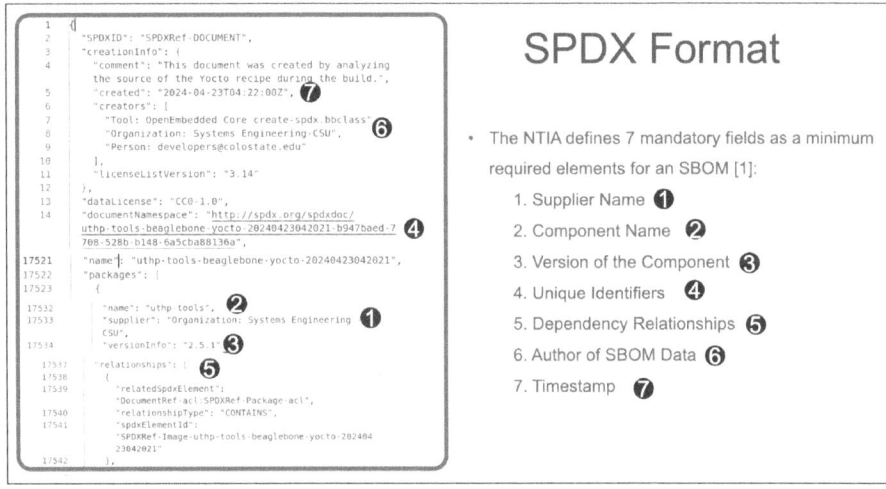

Fig. 3. SPDX Standard Format

2.4 Sharing SBOMs

Distributing software products alongside their SBOM is essential for effective risk management of vulnerabilities and compliance. The process should be automated to ensure downstream customers have continuous access to current and updated SBOMs. Sharing SBOMs must include tools for analyzing software components, their associated dependencies, licensing information, and vulnerability databases.

Currently, this SoI has not implemented tools for the analysis of SBOMs, however, it aims to demonstrate the anticipated challenges associated with analyzing complex, interdependent files produced by the Yocto build.

3 System of Interest

The Ultimate Truck Hacking Platform (UTHP) is a diagnostic tool for heavy-duty trucks based on the Beaglebone Black single board embedded computer. The operating system and built-in functions are built using the Yocto Project. Yocto is a widely recognized framework for constructing embedded Linux systems, extensively utilized in various applications within the automotive industry, such as telematics, infotainment, and navigation systems [6].

This platform's architecture involves integrating numerous open-source software components to support its diagnostic capabilities, making it an ideal case study for exploring the complexities of generating Software Bills of Materials (SBOMs) in embedded vehicle systems.

4 Challenges of SBOM Management

The following section discusses the experiences encountered during the management of SBOMs for the UTHP project. The following are highlighted:

4.1 Integration of Precompiled Binaries

The production of Software Bills of Materials (SBOMs) faces significant challenges, particularly concerning the limitations of binary files and the reliance on package managers. While tools such as bitbake included in the Yocto build environment and other package managers (apt, apk, dpkg, mvn, pip, npm, bitbake, and opkg) can list installed modules and applications, they often fail to account for software components added outside the package management system [7]. This oversight can lead to incomplete SBOMs, where critical components are omitted, leaving downstream users unaware of their existence and potential vulnerabilities.

When software is integrated directly into a project—whether through manual addition or by pasting code into the source—these components may not be captured by the package manager's inventory. Consequently, the generated SBOM may not reflect the full scope of the software's dependencies, which can hinder effective risk management and compliance efforts.

Moreover, the use of reverse engineering techniques to analyze opaque binary files presents additional challenges. These methods are often unreliable and not repeatable, complicating the reproducibility of SBOMs. In the context of the system of interest, there are instances where binaries are included in the build process through recipes that provide limited metadata (code Listing. 1.2 on Line 5). As a result, the SBOM in code Listing 1.2 on line 25 shows missing meta data, this diminishes the effectiveness of the SBOM in ensuring transparency and security.

Listing 1.1. An example of integrating a precompiled binary into Yocto

```
DESCRIPTION = "Integrating_a_precompiled_binary_into_Yocto"
LICENSE = "CLOSED"
LIC_FILES_CHKSUM = "file://LICENSE;md5=<checksum_here>"

SRC_URI = "file://path/to/precompiled-binary.tar.gz"
SRC_URI[md5sum] = "<md5_checksum>"
SRC_URI[sha256sum] = "<sha256_checksum>"

S = "${WORKDIR}/precompiled-binary"

do_unpack() {
    tar -xzf ${WORKDIR}/precompiled-binary.tar.gz -C ${S}
}

do_install() {
    install -d ${D}/usr/bin
    install -m 0755 ${S}/binary-file ${D}/usr/bin/
}

FILES:${PN} += "/usr/bin/binary-file"
```

Listing 1.2. UTHP SBOM with missing meta information

```
{
  "SPDXID": "SPDXRef-DOCUMENT",
  "name": "recipe-uthp-app-api",
  "packages": [
    {
      "SPDXID": "SPDXRef-Recipe-uthp-app-api",
      "copyrightText": "NOASSERTION",
      "description": "Flask_web_api",
      "downloadLocation": "https://github.com/SystemsCyber/UTHP",
      "externalRefs": [
        {
          "referenceCategory": "SECURITY",
          "referenceLocator": "cpe:2.3:a:*:uthp-app-api:1.0:*:*:*:*:*:*:*",
          "referenceType": "http://spdx.org/rdf/references/cpe23Type"
        }
      ],
      "licenseConcluded": "NOASSERTION",
      "licenseDeclared": "NONE",
      "licenseInfoFromFiles": [
        ""
      ],
      "name": "uthp-app-api",
      "summary": "uthp-app-api_application_to_set_the_device_tree_for_the_beaglebone_black",
      "supplier": "",
      "versionInfo": ""
    }
  ]
}
```

4.2 Large, Complex Builds and Dependency Management

Large Yocto builds present significant challenges in managing and visualizing software components, dependencies, licenses, and vulnerabilities. The scale and complexity of these builds often involve hundreds of packages, each with its own set of dependencies. For instance, as illustrated in Fig. 4, the "runtime-gpio-expansion-mapping" tool features a primary SPDX document linked to an SPDX index file that details its dependencies and associated Common Vulnerabilities and Exposures (CVEs).

This complexity is further exacerbated by the absence of dedicated, user-friendly tools designed to visualize these intricate dependencies. The lack of such tools hinders the efficient identification of potential vulnerabilities, making it difficult for developers and security teams to comprehensively and intuitively assess the security posture of their software.

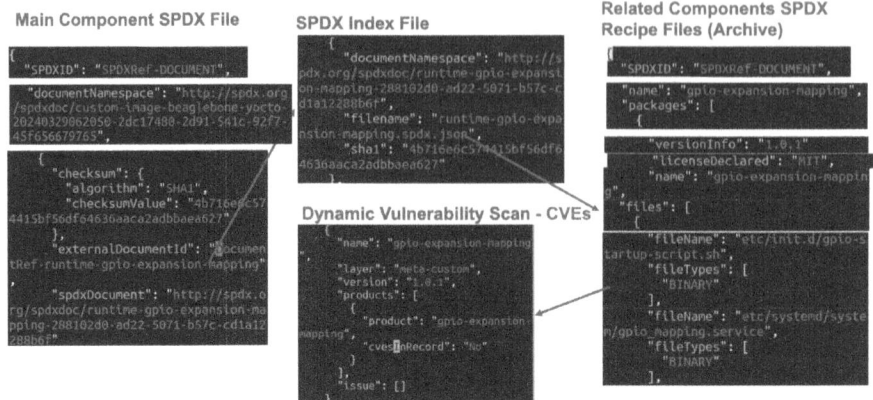

Fig. 4. Yocto Project SBOM File structure Complexity

5 Comparative Analysis of Current Works

Our study used experience-based observations using the Yocto Build. In addition, we undertook a desk study of current research which revealed challenges that exist in SBOM management among software producers and consumers. The research material reviewed include Stalnaker et al. [8] who discuss challenges based on their survey of 138 practitioners belonging to five stakeholder groups - practitioners familiar with SBOM, members of the critical open source projects, AI/ML, cyber-physical systems and legal practitioners. The other highly cited work on SBOM is Xia et al. [14] who conducted the first empirical study-survry that interviewed SBOM practitioners. Both findings, conducted over a space of 1 year apart, showed that SBOMs awareness has increased but absorption and utilization remain low. They attribute this to unresolved standardization of SBOM format and the lack of interoperable tools to integrate in the development environment. These research findings are consistent with NTIA's position as they also have not endorsed any SBOM standard even though ISO has standardized SPDX (ISO/IEC 5962:2021).

The material reviewed so far indicates limited study specifically on embedded vehicle systems SBOM management. While they address cyber-physical systems which include embedded systems there is a need to determine specific challenges faced by embedded vehicle systems and recommend feasible solutions. Our study specifically highlights inherent limitations with metadata extraction from binary files and limited tools to consume SBOM among the challenges of SBOM management. Despite similar outcomes to previous research, our study is focused specifically on embedded vehicle systems and will require empirical study and test bench analysis to further validate this study.

6 Recommendations for Enhancing SBOM Integration

To effectively leverage Software Bills of Materials (SBOMs) for improving cybersecurity in embedded systems, the community is encouraged to explore the following tailored strategies that address the unique challenges of this domain.

6.1 Embed Meta Data in Binaries

The Enhanced SBOM for Optimized Software Sustainment (E-BOSS) project [12] recommends the integration of metadata directly into compiled binaries to improve software patching and the generation of comprehensive binaries. To achieve this, it is essential to modify existing compiler infrastructures to support the inclusion of this metadata during the compilation and linking processes. Wide adoption of this model will not only achieve its objective but also provide necessary metadata for SBOM generation.

By pursuing these strategies, the community can significantly enhance the integration of SBOMs in software sustainment, leading to improved cybersecurity and more efficient software management.

6.2 Leveraging Community-Supported SBOM Tools

To effectively leverage Software Bills of Materials (SBOMs) in both SPDX and CycloneDX formats, it is crucial to develop tools that can consume these formats for visualization and analysis. By creating user-friendly tools that facilitate the visualization of SBOMs, consumers of SBOMs will gain deeper insights into software components, their dependencies, licenses, and associated vulnerabilities. This enhanced visibility will support informed decision-making regarding risk management and compliance within the software supply chain.

The development of such tools should focus on intuitive interfaces that allow users to navigate complex SBOM data easily. Integrating these tools into development pipelines, such as Yocto builds, will streamline SBOM generation and analysis, encouraging broader adoption in areas where SBOMs are not yet widely used.

6.3 Further Study on Embedded Systems

SBOMs apply to embedded systems as demonstrated in our study and review of recent works [8,14] which show that embedded vehicle system SBOM management remains under-explored. This therefore provides an opportunity to undertake further research in embedded vehicle systems SBOM management to validate the findings of our study through an empirical survey augmented with test bench results to contribute to ongoing research in embedded vehicle systems SBOMS management.

The anticipated outcome of the survey and test bench analysis is to recommend best practices for upstream firmware development by multilayered component manufacturers and streamline tools for SBOM management.

7 Conclusion

This study highlighted the challenges of managing SBOMs within the Yocto Linux environment for embedded vehicle systems. It underscored the complexities involved in accurately documenting software components and proposed improvements to streamline SBOM management. These enhancements aim to better integrate SBOM management into the development workflow in embedded vehicle systems, ultimately strengthening the security of embedded vehicle systems.

References

1. Boyens, J., Smith, A., Bartol, N., Winkler, K., Holbrook, A., Fallon, M.: Cybersecurity supply chain risk management practices for systems and organizations. Technical report, National Institute of Standards and Technology (2022)
2. Chatterjee, R., Green, C., Daily, J.: Exploiting diagnostic protocol vulnerabilities on embedded networks in commercial vehicles. In: Symposium on Vehicles Security and Privacy (VehicleSec). VehicleSec Symposium, San Diego (2024). https://doi.org/10.14722/vehiclesec.2024.23046
3. Chatterjee, R., Mukherjee, S., Daily, J.: Exploiting transport protocol vulnerabilities in sae j1939 networks. In: Proceedings of the Inaugural International Symposium on Vehicle Security & Privacy. Internet Society, San Diego (2023). https://doi.org/10.14722/vehiclesec.2023.23053
4. Feng, S., Lubis, M.: Defense-in-depth security strategy in log4j vulnerability analysis. In: 2022 International Conference Advancement in Data Science, E-learning and Information Systems (ICADEIS), pp. 01–04. IEEE (2022)
5. Jepson, J., Chatterjee, R., Daily, J.: Commercial vehicle electronic logging device security: unmasking the risk of truck-to-truck cyber worms. In: Symposium on Vehicles Security and Privacy (VehicleSec). VehicleSec Symposium, San Diego (2024). https://doi.org/10.14722/vehiclesec.2024.23047
6. Karacali, H., Donum, N., Cebel, E.: Enhancing workflow efficiency in yocto project: a build tool for fetch error detection and fixing. Eur. J. Res. Dev. **4**(2), 49–76 (2024)
7. Phillips, A., et al.: Software bills of materials for IoT and OT devices. IoT Security Foundation (2023)
8. Stalnaker, T., Wintersgill, N., Chaparro, O., Di Penta, M., German, D.M., Poshyvanyk, D.: Boms away! inside the minds of stakeholders: a comprehensive study of bills of materials for software systems. In: Proceedings of the 46th IEEE/ACM International Conference on Software Engineering, pp. 1–13 (2024)
9. The White House: Executive order 14028: Improving the nation's cybersecurity (2021). https://www.federalregister.gov/documents/2021/05/17/2021-10460/improving-the-nations-cybersecurity. Accessed 7 Sept 2024
10. The Yocto Project: What is a recipe? (2023). https://docs.yoctoproject.org/dunfell/overview-manual/overview-manual-yp-intro.html. Accessed 11 Sept 2024
11. United States Department of Commerce - National Telecommunications and Information Administration: The minimum elements for a software bill of materials (2021), https://www.ntia.gov/report/2021/minimum-elements-software-bill-materials-sbom. Accessed 12 Aug 2024

12. Wallach, D.: Enhanced sbom for optimized software sustainment (e-boss). DARPA. Online (2024). https://www.darpa.mil/program/enhanced-sbom-for-optimized-software-sustainment
13. Willett, M.: Lessons of the solarwinds hack. In: Survival April–May 2021: Facing Russia, pp. 7–25. Routledge (2023)
14. Xia, B., Bi, T., Xing, Z., Lu, Q., Zhu, L.: An empirical study on software bill of materials: where we stand and the road ahead. In: 2023 IEEE/ACM 45th International Conference on Software Engineering (ICSE), pp. 2630–2642. IEEE (2023)

ProvPredictor: Utilizing Provenance Information for Real-Time IoT Policy Enforcement

Michael Norris[1(✉)], Patrick McDaniel[2], Syed Rafiul Hussain[1], and Gang Tan[1]

[1] Penn State University, University Park, PA 16802, USA
man5336@psu.edu
[2] University of Wisconsin, Madison, WI 53706, USA
https://www.eecs.psu.edu/departments/EECS-Departments-Computer-Science-Engineering3.aspx,
https://www.wisc.edu/

Abstract. Internet of Things (IoT) platforms possess several properties that can potentially jeopardize the safety and security of users, such as distributed deployment, processing sensitive information, and exposed networks. This issue is further exacerbated by the physical nature of IoT allowing devices to compromise the confidentiality and integrity of both persons and property.Some IoT problems can be addressed by taking preventative measures before they happen, which requires making predictions about the future. We design ProvPredictor to gather provenance information in order to train a model to make predictions about potentially unsafe behaviors in the future. To demonstrate the effectiveness of ProvPredictor, we create a realistic deployment using IFTTT, a web-based IoT platform, using the most common IFTTT compatible services and applications in a home environment. We additionally use Agriculture datasets to show how ProvPredictor can operate in industrial systems. We train ProvPredictor on the generated provenance data and find that ProvPredictor can predict violations with over 90% accuracy. With ProvPredictor we demonstrate the advantage that provenance information provides to IoT and the feasibility of a provenance collector that focuses on predicting future behavior.

Keywords: Modeling and analysis of smart CPS · Threat modeling for CPS and smart vehicles

1 Introduction

As Internet of Things (IoT) develops and furter integrates with home, industrial, medical, and other environments, there is a growing requirement for ensuring reliability in IoT systems. With devices getting smarter, there is a significant increase in their access to sensitive information and control of physical environments. With this growing influence over environments, both the capability of devices to cause dangerous behavior and the difficulty of detecting this behavior expands beyond the capacity of current techniques.

Current techniques for IoT safety and security cannot prevent certain threats from manifesting. For example, forensic tools aim to identify why and how things occurred in the system rather than preventing them from happening. Conversely, static analysis based techniques [5,41,42] can identify possible violations in the system before runtime. While they can identify many potential problems, they cannot identify how likely those violations are. Also, they cannot take into account how the environment changes over time or when a device gets compromised during runtime, causing them to fail to prevent unseen attacks whose attack vectors (e.g., physical channels) are not considered during static analysis. On the other hand, dynamic analysis [2,3] can identify and block violations as they are about to happen at runtime. Despite this, dynamic analysis cannot predict and warn users about violations that can possibly occur in the future. Data provenance has also been shown to be effective for IoT systems either for forensic analysis [1,21] or to prevent undesirable behavior [19–21] such as through intrusion detection. However, most current techniques focus on identifying unwanted behavior either as it occurs (e.g., anomaly detection [36,39]) or immediately beforehand (e.g., dynamic analysis blocks a command that would violate a policy before it happens). Anomaly detection can recognize when unusual behavior occurs in the system, but it cannot always prevent or correct the behavior. Also, these techniques cannot determine what other effects that behavior will have on the system. These effects are most obvious in physical interactions, such as if an incorrectly open window lets water into the home, causing further problems even after the window is closed.

Furthermore, a common limitation among these techniques is that they observe only events and application level interactions. This misses out on the many ways that devices can be influenced either directly or indirectly by their environment or other devices. Some of these are more obvious physical interactions, such as those explored in IoTSafe [58]; e.g., the state of a heater having an effect on the temperature sensor. There are many more subtle interactions that go unrecognized by existing techniques. For example, the time of day and if the user has left work recently are a major indicator of how likely the user is to return home. How likely the user is to return home can then also affect many types of devices. Utilizing such information requires a design that can recognize interactions that can be highly context dependent and specific to the environment.

To introduce such a design, we propose ProvPredictor and aim to extend upon the recognition of how different device states in an environment interact outside of application code. First, we want to include interactions that do not have a direct physical channel to observe. For example, in a standard environment, a tracked change in a user's location does not physically change the state of a lock. However, if an application exists that unlocks the door when the user returns home, the fact that the user has recently left work does impact how likely the door is to unlock in the near future. All such complex interactions are difficult to capture through existing static or dynamic analysis, as many interactions simply alter the probability of an event rather than directly changing the environment in an observable way. We must also predict violations far ahead of when they occur to allow for policy enforcement when one or more device states of the environment are outside of the control of the system.

To achieve these goals, we follow a natural idea: use information of past events to train a machine learning model, in this case a Markov Model, capturing complex interactions among devices, applications, and environments; and use the model for predicting the likelihood of future unsafe behavior. In particular, we rely on past events' provenance information, which records both what occurred but also the context and the causal relation between events.

Leveraging provenance information for predicting the future in IoT environments using an ML model poses several challenges. Realistic IoT systems have a large number of heterogeneous devices, each of which can have a wide array of possible states. This results in an explosive growth in state information and possible system states. As supervised ML relies on observed patterns in training data, a large state space poses a scalability challenge, increasing time and model complexity as possible inputs and outputs grow [37,38]. The second challenge is to predict violations that have not occurred during training. This is an important challenge to address, as for various reasons the training data may not include instances of violations. Recognizing that this is not always possible for all policies, we aim to have a system that can achieve this as much as possible.

To address these challenges, we design and implement ProvPredictor to gather provenance information and provide predictions about future unwanted behavior. Our key idea in addressing aforementioned challenges is *model decomposition*. We first design a causal analysis of observed events in IoT systems to identify state information relevant to each type of event. Then, instead of combining all state information into a single large model that suffers from state explosion and scalability issues, we decompose that state space into a set of small models that each predict the probability of a single event. Each model contains only the state information relevant to its event, and the predictions of all models are combined to calculate the overall probability. In this way, the scalability challenge is addressed because each model discards information irrelevant to its predictions, reducing the state space. This also partially addresses a critical limitation of prior approaches, i.e., predicting unseen violations as it allows predicting behavior in contexts where it would violate a policy, even if there is no violation in the training data; more details on this later.

To ensure the challenges are properly met, we evaluate ProvPredictor on IFTTT, a popular platform that supports a wide range of services. We create a home IFTTT environment and use 22 services and over 40 applications as a testbed. We collect data representing over 10,000 events over the course of 5 months. On this data set, we find that ProvPredictor can predict all policy violations across the 12 policies we defined with strong accuracy. With the right parameters, it achieves 100% accuracy when predicting one minute ahead and remains around 90% accuracy up to 30 min ahead. It also has a small prediction latency under 4 s and negligible storage overhead of under 10 megabytes. While we choose to implement on IFTTT thanks to its higher availability of devices and applications, the ideas of ProvPredictor can be easily implemented in other smart home environments or even in industrial or health IoT environments. In this paper, we make the following contributions:

- We design a provenance tracking technique to classify provenance data into a scalable state space that identifies states which violate policies to train a Markov Model to predict future violations.
- We develop a causal analysis based Markov Model decomposition to allow our model predict unseen policy violations by making context aware predictions.
- We instantiate ProvPredictor for a popular IoT service platform and evaluate prediction accuracy and overheads of ProvPredictor in a realistic home deployment.

2 Background

2.1 Internet of Things Platforms

While the core purpose of IoT systems is to autonomously drive behavior in environments based on user rules and inputs from sensors and applications, each has its own important nuances. As an example, without loss of the generality, we discuss a popular web based platform IFTTT that we implemented ProvPredictor on. IFTTT allows users to install developer provided applications or create custom applications. These are termed *applets* and are comprised of one *trigger* and an number of *actions* that are performed when the trigger is detected. There are also optional *queries* that can provide information to actions and *filter*. A filter is TypeScript code (a JavaScript extension) and it can skip or modify actions based on trigger and query data. In this way, Triggers and Actions map to simple rules, such as those found in platforms like OpenHab [64]. On the other hand, Queries and Filters allow for more complicated code, as in platforms like SmartThings [65], which allows applications written in C code.

While IFTTT has similar capabilities to many platforms, there are a wide variety. Platforms such as Samsung SmartThings, Amazon Web Service (AWS) IoT, Microsoft Azure IoT, Openhab, Apple Homekit, Wink, Android Things, KaaIoT, IoTivity, and more all have their own unique architectures. *The fact there are a large number of platforms, most of which are closed-source, is the core motivation for developing techniques that do not require altering IoT platforms.*

2.2 Related Work

Fault Detection and Tolerance. Many works have aimed to increase fault detection [50,52,53] and fault tolerance [32,44–46,48] to protect the safety of users. There are also numerous tools for enforcing policies at runtime [27–29,49,51] or statically [5,41,42]. While these systems address many concerns about safety in IoT systems, none makes prediction to alert users of high-probability policy violations. Static analysis, such as Soteria [5] or iRuler [27], can enforce policies similar to our Forbidden Actions policies, but they cannot detect how probable violations are to occur or detect violations based on runtime information such as time. Dynamic tools like IoTGuard [51] do not utilize indirect correlation and cannot make predictions about violations significantly into the future. Another dynamic analysis tool IoTSafe [58] shares some similarities to ProvPredictor; its main goal is to detect non-explicit interactions between devices. While it can predict the likely change rate of analog values, such as temperature or humidity, it does not consider non-numeric states like location or a light to make predictions about the state of the entire system.

Provenance-Based Analysis. There also exists a large body of work towards retrieving, storing, and analyzing provenance information in IoT systems. The vast majority of this work focuses on analyzing individual events generated by deployed devices. This can be done by verifying common links between devices to check if a device is lying [23], tracking the event through a multi-hop network [26], using device metadata to mitigate faulty behavior [25], or other techniques. These techniques primarily leverage provenance data to verify the integrity of events.

While most research focuses on individual events, there are several existing works that aim to store and analyze provenance information collected across an entire system. DDIFT [1] and PDFC [24] aim to prevent improper information flow by using provenance information to enforce information flow policies. This is useful for ensuring confidentiality and integrity of data, but not protecting the actual physical state of IoT devices. ProvThings [18] allows code to be instrumented to identify sources and sinks to track provenance across applications and prevent certain sequences of actions. This is similar to ProvPredictor, however, they only consider a single sequence of why a certain event is occurring, not the overall system state. Moreover, they do not predict future policy violations, only identifying it the moment a violation happens.

3 Overview

3.1 Threat Model

The goal of ProvPredictor is to predict future behavior in the system to identify likely policy violations that require preventative measures. Therefore our model, targets the attack surfaces [15], where the correct behavior of the system can be manipulated by interference to physical devices through direct interaction or through application code. More specifically, the threat model is mostly comprised of both (i) non-adversarial behaviors caused by flaws in the system or unintentionally problematic actions, and (ii) adversarial behaviors such as sending messages through malicious applications or directly altering the physical environment. In the case of adversarial behaviors, we assume an adversary that can alter the states of devices, but cannot lie about the device state.

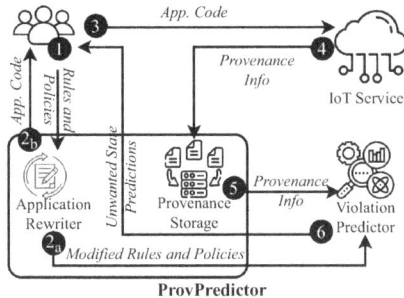

Fig. 1. ProvPredictor's architecture.

This is reasonable because the event information is observable and cannot be blocked or altered by the adversary as the event logs are captured and stored securely through secure logging mechanisms [16,18]. For instance, a malicious service could attempt to turn a heater on in order to raise temperature, causing a window to be opened and make the house insecure. Since these commands must be sent through the IoT platform, the changes are observable and securely logged. This example represents a basic example of the threats we target, with an adversary utilizing a less secure device (heater) to influence a device with more security implications (window). This is a realistic assumption and aligns with the prior work [16,18] detecting intrusions or performing forensic analysis of malware and Advanced Persistent Threats. ProvPredictor's aim is to predict all such violations by observing both direct and indirect interactions within devices and physical environment.

3.2 System Overview

ProvPredictor targets the core problem of predicting the probability that a future state will violate any of a given set of policies. To address this, our system makes predictions about the likelihood of paths that represent sequences of system behaviors. By predicting the total probability of all paths containing some violating behavior we get the probability of a policy violation. We define a set of policy types users can create that covers prior works [5,51] and more, as discussed later in Sect. 4.

To demonstrate how ProvPredictor solves this problem, we use Fig. 1 to show its overall architecture. There are two main components: **(1)** *Provenance Collector*, which is made up of the sub-components *Application Rewriter* and *Provenance Storage*, and **(2)** *Violation Predictor*. The Application Rewriter takes application code from the user and rewrites the applicationjs to insert code for tracking and sending provenance data to ProvPredictor. Once the user instals these applications on the IoT Service, the service will send provenance information to Provenance Storage. Storage receives, parses, formats, and stores provenance data. In addition to collecting event history, we can gather application-based behaviors, such as *a light will always turn on when there is motion*.

In contrast, the Violation Predictor focuses on device relationships not present in the application code. These are generally probabilistic, and so the predictor trains a model from the provenance data and makes predictions about behavior in the system, and if it will violate any policies. To use ProvPredictor, the user first inputs an application and the policies into the Application Rewriter. The Application Rewriter rewrites the application and forwards the policies to the Violation Predictor. The user installs the rewritten application, which during runtime sends provenance data to the Provenance Storage. Provenance Storage formats the provenance data into a useful format for the Violation Predictor, and sends it to the Violation Predictor. The predictor first trains on initially collected data, and afterwards reports any predicted policy violations and their causes to users. For this goal we make a prediction at each time step interval of of how likely a policy is to be violated within a time bound.

3.3 Motivating Examples

To demonstrate the potential utility of ProvPredictor, we will present a few practical IoT settings where static and dynamic analysis techniques fail to enforce policies but ProvPredictor can predict potential unsafe states far advance in time to enable user to prevent unwanted behavior.

A heater is used to heat a nursery. To prevent damage to the baby's health, a user has a policy: "Temperature should never rise above 85°". Depending on the type of the heater, temperature may continue to rise for some time after it is turned off. In this case, even if the potential violation was recognized right before the threshold was reached, turning off the heater would not resolve the issue and the health of the infant(s) could be put at risk. Additionally, it would be difficult to set a fixed level that the heater should be turned off to avoid reaching the threshold, as other environmental factors can affect how fast the heat rises and how long it continues to rise, such as outside temperature. In this IoT environment, problematic behavior cannot be stopped instantly, so the violation must be predicted ahead of time. This can be due to latency in the network, device needing time to change states, or residual effects of the device state. Although advanced devices with automatic failsafe, such as Proportional-Integral-Derivative (PID) controllers [4] to regulate the flow of temperature and pressure level, they are primarily used in industrial settings and not in smart homes as specialized knowledge and additional cost are required to operate them. Despite having such failsafe devices, there are always forces in the environment beyond the system's control, such as the actions of persons, device's system failure or compromise, changes in the environment or the weather. For example, there are many devices (e.g., motion detectors, temperature or humidity sensors) that can only detect the state but are not able to change it.

In these situations, predictions well before a violation occurs are the only favorable preventive measures or allows for more favorable prevention.

3.4 Challenges and Insights of ProvPredictor

Predicting behavior in an IoT system requires addressing several significant challenges. To achieve its goal, ProvPredictor must address **(1)** the scalability problem of state explosion in IoT caused by a large number of devices and device states, and **(2)** predict violations that are unseen in training data when possible.

3.4.1 Insights to Deal with Challenge 1. Since violations can occur among many different patterns of behavior, ProvPredictor should consider as much potential future activity as possible. Unfortunately, a common problem in IoT is that combining the numerous types of possible events causes an exponentially increasing state space. To address this challenge of state explosion, one observation is that whether an event happens or not may depend on only a small number of devices. For example, whether a light is on or not has no bearing on a rain sensor detecting that it is raining. Another observation is that for devices with a broad range of values, such as numeric states, we can reduce the number of states. For example, groups of similar temperature values are likely to have similar effects (e.g., a range of high temperatures such as 70–80 F all increase the odds a user opens a window or A/C is turned on) on a system and can be combined. Based on these observations, we design a causal-analysis method of *decomposing* our prediction model to a collection of component models each of which has a greatly reduced state space. To realize this, we develop a causal analysis technique for IoT systems that first utilizes the Granger causal analysis [7] to find relevant devices of given events. It then uses the Chi-Square analysis [14] to determine which of the device's states are relevant. Each model tracks and predicts only a small subset of possible system behavior. By combining the predictions of individual models, predictions have a much smaller state space.

3.4.2 Insights to Deal with Challenge 2. If ProvPredictor can only predict violations if they are observed first there can be negative ramifications for a user. Optimally, this merely causes user annoyance as the first few violations occur before they begin being predicted. It is also possible, though, that a violation may be catastrophic if not predicted before its first occurrence, such as if it destroys the involved device or causes significant harm to a user. While limited action policies can inherently be predicted in this case, unseen state and forbidden-chain policies are more difficult. Fortunately, our decomposition method also allows us to predict some types of violations even if they are so far unobserved. Specifically, this is possible for composite policies, whose violation require more than one type of behavior. In a prior example, we created a policy *"if it is raining, the window should not be open"*. It is quite possible to observe instances of the window being open without rain and rain occurring with the window closed during regular operations. Therefore, even if they have never occurred at the same time, if they are both individually likely it is possible to predict the violation.

4 Methodology

In this section, we detail the design of ProvPredictor's components and provide explanations for its design decision.

4.1 Provenance Collection

Our Provenance Collection is divided into two steps; (1) Application Rewriter, which alters user applications to track provenance information and (2) Provenance Collector, which parses information and stores it in a usable format.

4.1.1 Application Rewriter. To collect provenance information, we need to send messages to ProvPredictor every time a relevant event occurs in the system. These messages at minimum must contain (1) Event ID, which contains both the name of the event and ID of the relevant device, (2) Causal Event(s), if the event is an action there should be one or more Event IDs indicating what events caused the action, and (3) Timestamp, which indicates when the event occurred. In addition to this information, optional additional context information may be provided depending on the event and platform. For example, the user can determine static values setting up the application, such as what temperature threshold causes a trigger to fire. Conversely, there can be dynamic information determined once the event fires, which in this example would be what the exact temperature value is. As a core goal is to limit how much of our technique is specialized for use with the IFTTT platform, we design an application rewriter component to modify applications to send provenance data to ProvPredictor. This modifies user rules to send related data from them to ProvPredictor. In IFTTT, this is achieved by adding to all applets an action that sends all information about triggers, queries, filters, and actions to ProvPredictor. This design can be extended with minimal work to other IoT platforms, discussed in Sect. 7.

4.1.2 Provenance Collector. Due to mixing the diverse events of many heterogeneous devices, raw provenance data is too complex and intermixed for training a machine learning model. Therefore, in order for our provenance data to be useful, it must be converted into a form that is condensed and tractable. We utilize provenance data to generate a system state and update it as new messages are sent over time. Each time an event is fired the overall state is updated to match the new information. The new state includes the service name, trigger, and actions that were executed, as well as the context of the state, including the status of every service and the time when the state occurred. In this way, a state captures what the action is and the environment in which it occurred.

Because ProvPredictor predicts future events within a time bound, we split the state information into distinct time-steps. At each time-step, the is updated for any events that occurred during the step. These time-steps are a variable length based on how rapidly events are expected to occur in the system, but we choose 1 min for our implementation. This means the training data will consist of 1 min snapshots of the systems state at that time.

4.2 Violation Predictor

The goal of the violation predictor is to predict how likely a violation is to occur within some number of time steps. To do this, we utilize a machine learning model, which takes the current state and returns a prediction of what changes are likely in the next step. The predictor queries the model for every step and keeps track of the predicted overall state and violation probability from the model predictions. To address the previously discussed challenges of scalability and previously unseen violations, we analyze the training data to combine similar states and create decomposed models. By creating these Markov Models, and including additional information, such as history, we can accurately predict what will happen in the far future of a system.

4.2.1 Creating Individual Models.
The first step to creating a decomposed model is to determine the minimum set of information needed to make each of the predictions. For our purposes, the generated models must be able to predict all events the system can generate. Given this requirement, we choose to focus an individual model on predicting the probability of a single kind of event in the system (e.g., heater being turned on). So for each kind of event, we need a model that contains only the information necessary to predict that event.

To achieve this, we develop a causal analysis for IoT systems to determine what state information is closely linked with an event. For this, we leverage the Granger analysis [7] to find which devices and services are closely related. Granger is a better fit than other causal analysis algorithms such as PC [9] or Fast Causal Inference [10] for causal analysis in IoT for several reasons. Firstly, these algorithms require that the causal graph cannot have cycles; i.e., a node's causal relationships can never have an effect that can potentially affect the node through a causal relationship. This assumption does not hold in IoT, where such interactions are common and often intended through physical interactions and application code. Granger analysis does not make this assumption and can handle cyclic causal graphs. Granger also naturally considers that causal relationships may not always resolve immediately in time-series data and considers if examining extended history may reveal delayed causal relationships. These are common in IoT, where interactions could take some time to resolve, such as a physical interaction like a heater raising the temperature or implied interaction such as the user leaving work implying they will return home soon.

Granger first trains a model using a state trace to make predictions about the next most likely state based on previous device states. Granger requires a lag value, which is how many previous states it considers. We run experiments with increasing lag values up to 10 and use the value with the maximum Akaike information criteria (AIC) [11] and the Bayesian information criteria (BIC) [12]. Granger then measures how accurate this model is at predicting the device's next most likely state. Next Granger makes the same predictions, but this time using a new model that also trains on knowledge of another device's state history. We then calculate the *p-value* based on how much more accurate the second model is. This gives us the probability that the devices are not correlated. Following convention, we consider devices related if *p-value* is less than 5%.

Next we determine exactly which device states from those devices returned by the previous step have an effect on the event's probability. To do this, we perform a Chi-Square test for states of all devices against those found to be linked by Granger. Since we want causal relationships that can occur over time, we want to consider previous states similar to Granger. For this, we use the same max-lag Granger found for each device and observe how often each state was present when the given event occurred against how often it is expected to be. We analyze our Chi-Square results with the same p-value of 5% to determine what specific states drive the event behavior. Additionally, when an event's device has more than 2 states, we perform this analysis for the other states. We also perform Chi-Square against time values for each event to check if time is relevant.

During this step, we also address the issue of devices with excessive states, mainly numeric states. For example, the value in Fahrenheit of a temperature sensor can theoretically have many states on its own. For the model to scale, a reduced set of states must be selected. We first select any states present in policies or application codes, as those are important threshold values. Then we want to select additional events distributed across the range of observed values. We prioritize events that were seen frequently (up to a maximum of 10 values), as these are more useful in training. We enhance the dataset so that any event outside of this set is replaced with the closest selected event. Then, we perform our Granger and Chi-Square based causal analysis. If any of the selected events have no causal links, we check if there is another close event that has multiple occurrences and then use the Chi-Square test on it. Once there are no more valid targets or all events have causal links, we save those events as the device's events. This heavily reduces the number of events that need to be tracked.

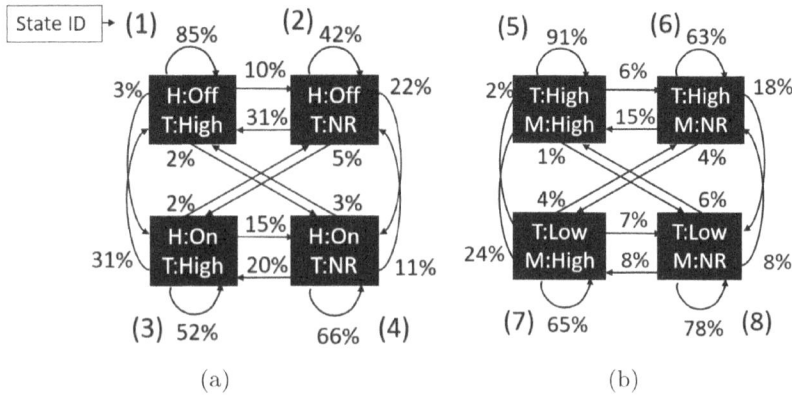

Fig. 2. Example models for predicting "Heater:On" (Left) and "Temperature: 64.1°–94.0" (Right) using only device states relevant to those events.

4.2.2 Incorporating History Into the Model. While a Markov model is useful for making predictions about state changes, it assumes that the probability of transitions is not influenced by previous states. In an IoT environment, however, previous states are important when predicting what will happen in the future. For example, if the user is home and in the last state the user was away, it implies that the user just arrived home and is less to leave home soon.

To address this, we use an N-gram Markov Model with the Kneser–Ney smoothing, similar to the tool Kramer [60]. Briefly described, the n-gram model considers the probabilities of state sequences of length n and less. However, it introduces a new scalability problem, since including history in effect increases the number of possible states at each step. For our implementation, we use history up to *three* previous states without over-inflating storage and latency overhead. We found that using histories longer than three in our implementation caused the prediction latency to be over a minute.

The use of this history is another major reason we adopted model decomposition. Without model decomposition, considering the history would dramatically increase the state space of making prediction into the future since each step has to consider a combination of the current state and the past history of all devices. With model decomposition, each individual model can track the history relevant to the model by recording the probabilities of all conditions it uses at each step. This allows for predictions to complete in a reasonable time, even when they are technically exploring billions or more paths.

4.2.3 Predicting Policy Violations. Once it has finished training, ProvPredictor makes predictions about potential policy violations. The user must define at what probability threshold (θ) they wish to be notified that a policy violation is likely. The trade-off is that the higher the threshold the more certain ProvPredictor has to be to notify the user. For this reason, a non-expert users may wish to begin with a low threshold, and adjust upward depending on their tolerance for false positives. A prediction is made at each time-step and compared to the threshold. When a violation is predicted, ProvPredictor also outputs the most likely paths for user diagnosis; in future work we plan to perform Root-Cause-Analysis to find common elements among predicted paths. Additionally, as discussed in Sect. 7, when devices are added or removed in the environment, the accuracy of prediction can drop until training can be re-performed.

Table 1. Example Set of Devices and States for a home environment to model.

Device	Possible States
Heater	On(H:On), Off(H:Off)
Temperature	34.0°–64.0°(T:Low), 64.1°–94.0°(T:High)
Humidity	25.0%–35.0%(M:Low), 35.1%–45.0%(M:Mid), 45.1%–55.0%(M:High)

To demonstrate this process, we will use a simple example system. In Table 1, we show a system with three devices, a Heater, Temperature Sensor, and Humidity Sensor. In the table are all of the states each can be in after Causal Analysis has been performed. So to calculate state probabilities, our training produces models from the traces such as those in Fig. 2. Using these models for prediction, let us assume we start at a timestep $t0$ in a state where Heater is On, Temperature is 64.1°–94.0°, and Humidity is 25.0%–35.0%. Assume that the policy requires that the Heater should not be Off while Temperature is below 64.1°. We know the policy violation probability is P({Heater:Off & Temperature:34.0°–64.0°}). So we then use the models to predict how the state will change as time moves from $t0$ to $t1$. Using the first model in Fig. 2, we can calculate $P(Heater:Off_{t1})$ using the related equation. Since we know that at $t0$ Heater is On and Temperature is 64.1°–94.0°, the system is initially in state 3. Since we are interested in whether Heater is on at $t1$, and state 1 and 2 are Heater off states, the equation is $(P(3 \rightarrow 1) + P(3 \rightarrow 2)) * (P(3_{t0}) = (31\% + 2\%) * 100\% = 33\%$. Then we calculate P(Temperature: 64.1°–94.0°$_{t1}$) from a model for predicting the temperature being in that range, such as the second model in Fig. 2. Using this model, P(Temperature:64.1°–94.0°$_{t1}$) = $(P(6 \rightarrow 5) + P(6 \rightarrow 6)) * (P(6_{t0}) = (15\% + 63\%) * 100\% = 78\%$. The probability of policy violation is then P({Heater:Off & Temp:34.0°–64.0°}) = 33% * (1–78%) = 25.74%.

Once all device state probabilities are calculated, we can calculate the probabilities of the states in timesteps beyond $t1$ in a similar way. So when we calculate $P(Window:Open_{t2})$, we use the probabilities of each state in the previous step.

Finally, performing predictions in this way helps address another key challenge, which is predicting unseen policy violations. This is because, when a novel or uncommon state is entered, a normal Markov Model cannot make any predictions about likely transitions. Our decomposed model seperates the state into only relevant chunks of information, and can therefore make predictions based on previously observed individual device states even in novel overall system states. For example, the violation {Heater:Off & Humidity:45.1%–55.0%} can feasibly be predicted even if not previously observed, since Heater and Humidity are not related and their behaviors are predicted from separate devices. Conversely, violating events where conditional events have strong causal relationships to each other are far less likely to be predicted. For example in our Fig. 2, if the condition {Heater:Off & Temperature: 64.1°–94.0°} violates a policy, it will be difficult to predict before it is observed as the device states share a strong causal relationship. ProvPredictor also still cannot predict behavior that was never or very infrequently observed previously or behavior that varies wildly from previous patterns. This is a weakness core to ML, which relies on observed training data, and so lies outside of the scope of our work. Overall, this method not only allows for predictions on data far too expansive for a normal Markov Model, but also potentially allows predicting states before they have occurred in the system.

4.3 Policy Definitions

In order to protect users from unwanted behaviors, we must define a set of policies that are both intuitive and cover many possible activities a user would want to prevent. We support three kinds: state policies, limited-repetition policies, and forbidden-chain policies. Their grammar is shown in Fig. 3. In this grammar, we have several base nonterminals: A <Node ID> represents the name of a device or virtual service; a <Node Property> is the value of a property the service has, such as ON/OFF; a <Time Bound> is either a single Hour:Minute value or a pair of Hour:Minute values separated with a dash. We format these policies in natural language to allow non-expert users to interact with the system. While complex policies are possible, any user should be able to create intuitive policies, such as the door not being unlocked while they are away.

⟨Policy⟩	:=	⟨SP⟩ | ⟨LP⟩ | ⟨FP⟩
⟨SP⟩	:=	[NOT] ⟨State Element⟩ [⟨Time Limit⟩]
	|	⟨SP⟩ ⟨Operator⟩ [NOT] ⟨State Elem⟩ [⟨Time Limit⟩]
	|	⟨SP⟩ ⟨Operator⟩ [NOT]Time-⟨Time Bounds⟩
	|	(⟨SP⟩)
⟨State Elem⟩	:=	⟨Node ID⟩-⟨Node Property⟩
⟨Time Limit⟩	:=	⟨Days⟩:⟨Hours⟩:⟨Minutes⟩:⟨Seconds⟩
⟨Operator⟩	:=	AND | OR | NAND | NOR | XOR | XNOR
⟨LP⟩	:=	⟨State Element⟩,⟨Time Limit⟩,⟨Int⟩
⟨FP⟩	:=	⟨Node ID⟩,[⟨Node Property⟩],⟨Elem List⟩
⟨Elem List⟩	:=	⟨Node ID⟩ | ⟨Node ID⟩-⟨Action Name⟩
	|	⟨Element List⟩,⟨Element List⟩

Fig. 3. BNF grammar for ProvPredictor's policies.

State policies define states that the user does not want to occur in the system. For example, that the heater must not be on while temperature is over 75°. The user can also specify whether a state violates the policy only at certain times, such as turning on the lights at night. Additionally, to provide more flexibility to users we support the full set of logical operations; AND, OR, NOR, XOR, XNOR, NAND, and NOT. We also allow users to define policies based on the duration that nodes have been in certain states by giving a length of time.

A *Limited-repetition policy* gives a defined limit to the number of times an action should be performed in a given time period. For example, a user may want to limit the number of automatic donations they make in a month to 5: {"DonateService"–"Make-a-Donation", 30:0:0:0, 5}.

A *forbidden-chain policy* defines a specific event chain that should not happen. The user defines a driving Node ID that causes the event chain and optionally what property of the node needs to be involved to violate the policy. The policy also specifies the event chain, in terms of what nodes and actions are involved. This kind of policy is most useful when there are very specific event chains that the user wants to prevent.

While our grammar is highly versatile, our policy language does not cover all possible policies in an IoT environment. For example, there is information we do not gather, such as packet metadata or changes in application code, which cannot be used in policies. Additionally, some complex policies cannot be expressed in our language. For example, complex policies could theoretically be made by combining limited action and state policies. This could make a policy such as "If it has rained twice in the last 24 h do not activate sprinklers." Creating a provably sound and complete coverage of potential policies is a difficult problem beyond the scope of this paper. Regardless, we argue our set of policies is sufficient for two reasons: (1) It covers and expands upon the sets of policies contained in previous papers [5,51], and (2) Our simple grammar definition is flexible enough to be expanded to cover additional policies as the need arises.

Table 2. List of polices used in evaluation.

Policy ID	Description
1	Lights should not be turned on after 10 PM as user is asleep
2	Heater should not be on while temperature is above 73°
3	Heater should not be off while temperature is below 71°
4	Should not send > 4 texts to user over 1 h to avoid annoyance
5	Should never leak location data except through SMSs to user
6	Should never leak finance data except SMSs to user
7	Make It Donate should not donate more than $5 at a time
8	Make It Donate should not donate more than 4 time a month

5 Implementation

ProvPredictor is implemented using ∼4000 lines of code in addition to the IFTTT service. For our application rewriter, IFTTT rules input by user are parsed into triggers, queries, actions, and filter code (as discussed in Sect. 2). A new IFTTT applet is then generated as an artifact. The generated IFTTT applet contains provenance tracking code that sends the provenance information (including the trigger, the actions, the query, and the filter) to be stored in an external file each time the applet is triggered; additionally, the IFTTT ingredient "CreatedAt" is inserted into the applet to record the timestamp when the applet is fired. The recorded provenance data is then read by ProvPredictor and used to train a Markov model. As IFTTT is a web-based platform with no Hub device, we run ProvPredictor's model training and prediction on a local Linux machine where it stores and processes all data. This machine runs a 64 bit version of Ubuntu 20.4 with 8 1.5 GHz CPUs and 32 GB of memory.

6 Evaluation

In our evaluation, we aimed to answer (1) *Effectiveness:* can ProvPredictor make accurate predictions about whether a policy will be violated, and (2) *Efficiency:* does ProvPredictor have low enough prediction latency.

Testbed. To produce provenance data for evaluation, we created a smart home deployment with 24 services, 7 physical devices, and 17 virtual services. We installed 45 applications to drive smart home behavior. Most applications were from the recommendations IFTTT provides for each service to ensure they were representative. These applications contained a mix of simple Trigger-Only applications as well as applications with more complex conditions. We also added 4 custom applications, including 2 malicious ones.

Dataset. To evaluate our system, we aimed to gather a sufficient amount of data from a realistic environment. We performed data collection for five months and then used the collected data to train the Violation Predictor and make predictions. Over this time we recorded 10,963 events. We used up to 75% of the data to train and the rest of 25% to make predictions. We select 1 min for our time-step size as it is rare for multiple events to occur within one minute unless they are part of the same event chain. Much shorter and paths would be too fine-grained to predict far into the future. Much longer and transitions would have to account for many events occurring between time-steps. Since it is generated primarily from applications suggested by IFTTT, we can assume it is unbiased and represents an environment similar to that most users would have. The exceptions are the 4 custom malicious and vulnerable applications we inserted to ensure policies were violated in our training; however, that is only 4 out of 45 applications and they were still modeled after suggested applications (but modified to produce the desired behavior). This provides us with a reasonable dataset for testing ProvPredictor in a home environment.

6.1 Effectiveness

Our first evaluation is to test if ProvPredictor can accurately predict when violations will occur. We define a variety of policies to be violated during data collection. Some violations occur naturally from normal application interactions and some due to malicious and vulnerable applications.

Policy Instances. Table 2 shows policies we selected. We selected these to be realistic for a typical user's environment. Some policies were selected to prevent accidental improper behavior. For example, policies 1–4 are selected for user convenience such as limiting SMS per hour in policy 4 or preventing light from turning on while they are asleep in policy 1. We did not explicitly design any applications in our testbed to violate these policies. Due partly to user behavior

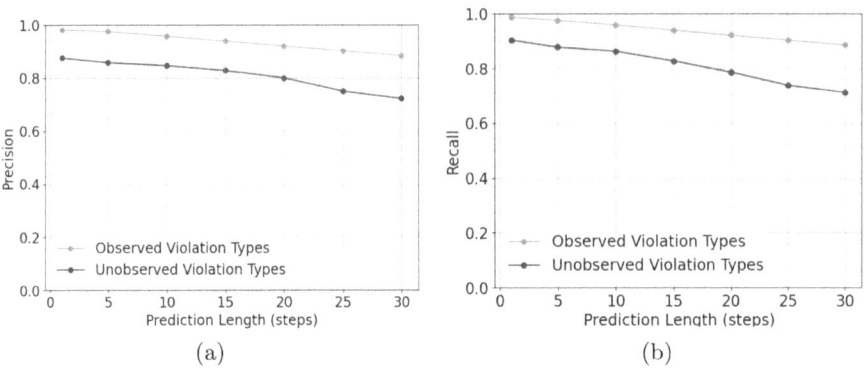

Fig. 4. The Precision and Recall of Provpredictor a Prediction Length is changed. We also compare accuracy difference between when violations occurred in training data and when they did not.

and primarily from installed applications, all policies are violated at least a few times over the course of training and evaluation except for policy 3. This is because we did not construe specific applications to force that violation and it did not happen to occur naturally. As stated in the implementation, all but 4 applications were chosen from the recommended applications on IFTTT to ensure it was a realistic environment.

Malicious applications considered in our experiment setup aimed to violate two of our selected policies. For this, the malicious applications considered 2 target services for possible adversarial attacks. The Finance service has a data leak in policy 6 and the Make It Donate maximum donation amount in policy 7, which are both violated by malicious applications we installed.

Parameters. ProvPredictor requires several parameters that can change the accuracy and latency of predictions. The *prediction length* dictates how far into the future ProvPredictor makes predictions; intuitively, a larger prediction length usually decreases accuracy as noise and potential futures are increased. The tradeoff is it provides the user more advance warning of potential violations. The *number of events* in the training data alters how accurate ProvPredictor's model is. A larger value of the number of events increases accuracy, but requires more time for ProvPredictor to observe the environment, during which violations may occur. We alter one of these parameters over evaluations to show the tradeoff in accuracy, the default values are 30 min prediction length and 1000 events.

Prediction Accuracy. For accuracy evaluation, we execute ProvPredictor so that at each time step it makes a prediction of how likely each policy is to be violated within a given prediction length. We note that from test data we cannot determine whether a violation definitely occurs or definitely not occur at a specific state and time step. The reason is that the test data contains multiple traces (i.e., multiple possible futures) from the state and time step; in some traces violations of the policy occur, while in others violations do not occur. The only ground truth we have from test data is the probability that violations occur from a state and a time step (within the prediction length). Given that, we need to choose a threshold for both ProvPredictor's prediction and the ground truth to be able to compute standard metrics including precision and recall. Particularly, if ProvPredictor's calculated probability is greater than threshold, we deem it a positive case; otherwise, it is negative. The same goes for the ground truth.

For our experiments, we define this threshold empirically, such as in anomaly detection [36]. In our case, we determine a threshold for each evaluation by choosing the threshold that would result in the highest F-Score in training data. This is then used during evaluation to differentiate between positives and negatives in predictions. In this way, we can evaluate how accurately ProvPredictor predicts positives only when the ground truth is above this threshold.

Results of Violations Observed in Training (Seen Violations). We first evaluate our accuracy when predicting violations that occurred in training data and should therefore be easier to predict. The green line in Fig. 4(a) and (b) shows how precision and recall change as prediction length is increased, which determines how many minutes into the future ProvPredictor predicts. As shown in the graph, when the prediction length is small, the precision is nearly 100%. The further ProvPredictor is required to predict ahead, the more likely an unexpected event or lack of predicted event is to interfere with the prediction. Despite this, ProvPredictor maintains a greater than 90% average precision and recall while predicting previously observed violations.

We also evaluated how accuracy improved with observed events in training data. We found Initial accuracy on less than 1000 events was low, around 60–80% as it increased. Once 1000 events were observed, though, accuracy reached over 90%, and climbed slowly from there. It is difficult to say how long data gathering takes from these results. The time it takes to gather data varies between environments. In our environment, we gathered around 3000 events per month due to our many services and apps. A different environment can take more or less time to generate events based on complexity and activity, and environments with less complexity would also take less training data to increase precision.

Violations Not Observed in Training (Unseen Violations). In order to test how accurately ProvPredictor could predict violations that did not occur during training, we reran the 6 previous experiments. Each experiment would be run once for each of our policies. Each time, all violations of that policy in training data would be removed. When a violation occurred in training, the trace would end right before the violation, and a new trace would begin immediately after it ended. In this way, violations were removed without adding incorrect training to the model. The predicted probabilities of these removed violations are averaged to determine how close they are to the baseline compared to seen violation predictions. These results are represented by the blue lines in Fig. 4.

Observing prediction length, the Precision and Recall are initially just above 90%, so about 10% less accurate than observed violation types. Unseen violations also lose accuracy faster than observed violations. A similar drop in accuracy occurs when evaluating over training data, with accuracy never managing to reach 80%. Since unseen violations will naturally have a lower probability of occurring, predicting them further out becomes more difficult. For training data, unobserved violations gain Precision slower as the number of events in training data increases. This is because removing violations reduces instances of the violating behavior occurring, resulting in our model having less confidence.

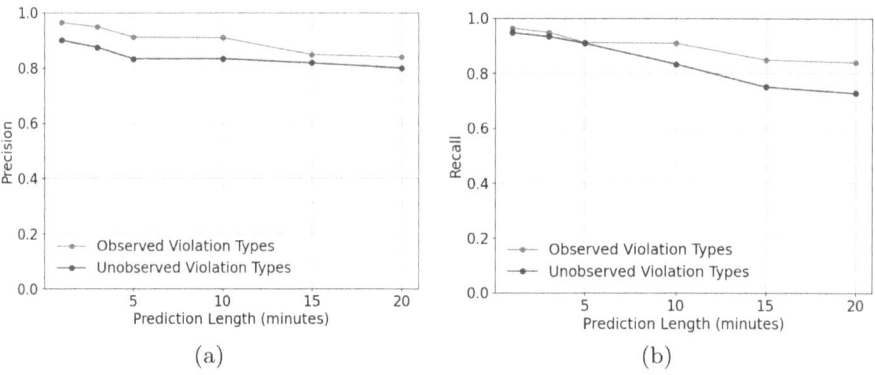

Fig. 5. The Precision and Recall of ProvPredictor on an agricultural dataset.

ProvPredictor in Industrial Environments. While home environments are the most numerous types of IoT environments, they differ greatly from other environments, such as industrial environments. While often containing complex devices, industrial environments can also exhibit less randomness than home environments. We used data gathered from a Purdue study using agricultural IoT equipment to measure values related to soybean and corn crops, such as soil temperature and water content [13]. We split the dataset into 75% training and 25% evaluation. We defined policies such that violations occurred if either soil temperature or water content left healthy ranges for either type of crop. We required multiple sensors reach unsafe values to corroborate the result. These devices were also set to query states every 15 min; so we chose that as the timestep length. Performing evaluations using the methodology discussed above, we produced results shown in Fig. 5. The accuracy is similar to our home environment evaluations, beginning at near 100% and dropping off as prediction length is increased. The most notable difference is that unobserved violations are closer in accuracy to observed violations. This is likely because the temperature and water level values are highly correlated; so when one of them reaches a violating value it becomes easier to predict if the other will as well. These results show that

ProvPredictor can be useful even in environments without application code to provide strong provenance connections and in environments with large numbers of devices with continuous states.

6.2 Efficiency

Time Overhead. The query latency of ProvPredictor must be low enough so that a prediction can be made before another state transition occurs in the system to change its state. To ensure ProvPredictor meets this expectation, we measured how adjusting ProvPredictor's parameters changes the average query latency. When analyzing the data, we find that query latency is only sensitive to prediction length, and so we ensure this latency is acceptable.

We evaluated how average query latency scales with prediction length. We found latency initially starts under 1 s when ProvPredictor only considers a single step ahead, but increases to almost 5 s in latency when 5 steps ahead are considered. This latency continues to grow, but remains primarily linear due to our efficiency improvements. In normal Markov Model exploration, latency would scale exponentially as it explores every possible path. However, since we limit ourselves to only tracking information usable by each model, latency scales closer to linear, depending on how many dependencies are present between events. As this growth remains linear, predictions can be made far ahead without taking too long to be useful.

7 Limitations and Discussion

Amount of Provenance Data vs. Modification of IoT Platforms. A limitation of ProvPredictor is how much information we are able to collect. Currently, we gather only information when an applet fires, but some data would be more useful if it was collected routinely. If ProvPredictor was implemented such that it could arbitrarily request current states, it could be more accurate. We avoided this implementation to prefer an implementation that does not modify an IoT platform's infrastructure, but the benefits may be worth a separate study. We also assume relevant behaviors occur sufficiently for training, though this is an assumption common among ML techniques.

Online Model Training. Another limitation is we do not perform online model training, so if the environment is modified, training must be repeated. In the event a device is removed, it is as simple as removing it from the list of devices in ProvPredictor and repeating training. In the case a device is added, ProvPredictor will have to wait until a reasonable number of events related to the device have occurred before, but could be corrected by adding logic to the model to update the transition matrix and state space without entirely retraining.

Application Rewriter. Our application rewriter automatically rewrites an input rule by adding an action that sends the rule to ProvPredictor when the corresponding rule is triggered. Such a design does not require the user to modify the application code and can be easily deployed to other platforms, such as SmartThings, in which applications can be instrumented by inserting code around the event handlers and command lines. This means future work can extend ProvPredictor to additional platforms with minimal engineering work. The most notable exception is distributed IoT environments [32,43] where there is no centralized hub; it would require more significant alterations.

8 Conclusion

As the influence of IoT grows in people's day to day life, more and more ways for it to cause harm through adversarial attacks or mismanagement appear. Safeguards to prevent misbehavior need to be implemented across the various platforms that manage IoT systems. Previous systems have created a variety of ways to enforce policies, but they are insufficient for when a problem cannot be prevented as it happens due to forces outside of the platform's control. For this reason, we design and implement ProvPredictor, a tool that gathers provenance information and uses it to make predictions about what will occur. Being able to inform the user of what will violate their policies ahead of time allows the user to take preventative actions. We find that ProvPredictor can accurately predict what activity is likely to occur in the system with moderate overhead. ProvPredictor is a good step towards protecting users from the harm that IoT devices can cause, and knowing what behaviors will cause policy violations will allow for a more automated approach to IoT protection moving forward.

References

1. Sapountzis, N., Sun, R., Oliveira, D.: DDIFT: decentralized dynamic information flow tracking for IoT privacy and security. In: Workshop on Decentralized IoT Systems and Security (DISS). NDSS, San Diego (2019)
2. Jeon, J., Park, J.H., Jeong, Y.S.: Dynamic analysis for IoT malware detection with convolution neural network model. IEEE Access **8**, 96899–96911 (2020)
3. Ficco, M.: Detecting IoT malware by Markov chain behavioral models. In: 2019 IEEE International Conference on Cloud Engineering (IC2E), pp. 229–234. IEEE, Prague (2019)
4. Eurotherm, Principles of PID Control and Tuning (2023). https://www.eurotherm.com/temperature-control/principles-of-pid-control-and-tuning/
5. Celik, Z.B., McDaniel, P.,Soteria, G.T.: Automated IoT safety and security analysis. In: 2018 USENIX Annual Technical Conference (USENIX ATC 18), Boston, MA, pp. 147–158. USENIX Association (2018). ISBN 978-1-939133-01-4. https://www.usenix.org/conference/atc18/presentation/celik
6. Palacio, D.N., McCrystal, D., Moran, K., Bernal-Cárdenas, C., Poshyvanyk, D., Shenefiel, C.: Learning to identify security-related issues using convolutional neural networks. In: 2019 IEEE International conference on software maintenance and evolution (ICSME), pp. 140–144. IEEE (2019)

7. Granger, C.W.: Some recent development in a concept of causality. J. Econometr. **39**(1–2), 199–211 (1988)
8. Pawlicki, M., Kozik, R., Choras, M.: A survey on neural networks for (cyber-) security and (cyber-) security of neural networks. Neurocomputing **500**, 1075–1087 (2022)
9. Le, T.D., Hoang, T., Li, J., Liu, L., Liu, H., Hu, S.: A fast PC algorithm for high dimensional causal discovery with multi-core PCs. IEEE/ACM Trans. Comput. Biol. Bioinf. **16**(5), 1483–1495 (2016)
10. Spirtes, P.: An anytime algorithm for causal inference. In: International Workshop on Artificial Intelligence and Statistics, pp. 278–285. PMLR (2001)
11. Cavanaugh, J.E., Neath, A.A.: The Akaike information criterion: background, derivation, properties, application, interpretation, and refinements. Wiley Interdisc. Rev. Comput. Stat. **11**(3), e1460 (2019)
12. Neath, A.A., Cavanaugh, J.E.: The Bayesian information criterion: background, derivation, and applications. Wiley Interdisc. Rev. Comput. Stat. **4**(2), 199–203 (2012)
13. Dataset: Data Collection and Analytics from Soil Sensors and Weather Stations at Production Farms. https://purduewhin.ecn.purdue.edu/dataset2021
14. Pearson, K.: X. On the criterion that a given system of deviations from the probable in the case of a correlated system of variables is such that it can be reasonably supposed to have arisen from random sampling. Lond. Edinburgh Dublin Phil. Maga. J. Sci. **50**(302), 157–175 (1900)
15. Deogirikar, J., Vidhate, A.: Security attacks in IoT: A survey. In: 2017 International Conference on I-SMAC (IoT in Social, Mobile, Analytics and Cloud)(I-SMAC), pp. 32–37. IEEE (2017)
16. Bates, A., Pohly, D.J., Butler, K.: Secure and trustworthy provenance collection for digital forensics. In: Wang, C., Gerdes, R.M., Guan, Y., Kasera, S.K. (eds.) Digital Fingerprinting, pp. 141–176. Springer, New York (2016). https://doi.org/10.1007/978-1-4939-6601-1_8
17. Yang, C., Yang, G., Gehani, A., Yegneswaran, V., Tariq, D., Gu, G.: Using provenance patterns to vet sensitive behaviors in android apps. In: Thuraisingham, B., Wang, X.F., Yegneswaran, V. (eds.) SecureComm 2015. LNICST, vol. 164, pp. 58–77. Springer, Cham (2015). https://doi.org/10.1007/978-3-319-28865-9_4
18. Wang, Q., Hassan, W.U., Bates, A., Gunter, C.: Fear and logging in the internet of things. In: Network and Distributed Systems Symposium. NDSS, San Diego (2018)
19. Bhuyan, F.A., Lu, S., Reynolds, R., Zhang, J., Ahmed, I.: A security framework for scientific workflow provenance access control policies. IEEE Trans. Serv. Comput. **15**(1), 97–109 (2019)
20. , Pasquier, T., et al.: Runtime analysis of whole-system provenance. In: Proceedings of the 2018 ACM SIGSAC Conference on Computer and Communications Security, pp. 1601–1616. ACM, Toronto (2018)
21. Xie, Y., Feng, D., Tan, Z., Zhou, J.: Unifying intrusion detection and forensic analysis via provenance awareness. Future Gener. Comput. Syst. **61**, 26–36 (2016)
22. Belhajjame, K., et al.: Prov-dm: the prov data model. In: W3C Recommendation, 2013, World Wide Web Consortium (W3C) (2013)
23. Aman, M.N., Basheer, M.H., Sikdar, B.: A lightweight protocol for secure data provenance in the Internet of Things using wireless fingerprints. IEEE Syst. J. **15**(2), 2948–2958 (2020)
24. Rong-na, X., Hui, L., Guo-zhen, S., Yun-chuan, G., Ben, N., Mang, S.: Provenance-based data flow control mechanism for Internet of things. Trans. Emerg. Telecommun. Technol. **32**(5), e3934 (2021)

25. Aktas, M.S., Astekin, M.: Provenance aware run-time verification of things for self-healing Internet of Things applications. Concurr. Comput. Pract. Exp. **31**(3), e4263 (2019)
26. Kamal, M.: Light-weight security and data provenance for multi-hop Internet of Things. IEEE Access **6**, 34439–34448 (2018)
27. Wang, Q., Datta, P., Yang, W., Liu, S., Bates, A., Gunter, C.A.: Charting the attack surface of trigger-action IoT platforms. In: Proceedings of the 2019 ACM SIGSAC Conference on Computer and Communications Security, pp. 1439–1453. ACM, London (2019)
28. Siddiqui, F., Hagan, M., Sezer, S.: Embedded policing and policy enforcement approach for future secure IoT technologies. IET, London (2018)
29. Sicari, S., Rizzardi, A., Grieco, L.A., Piro, G., Coen-Porisini, A.: A policy enforcement framework for Internet of Things applications in the smart health. Smart Health **3**, 39–74 (2017)
30. Gnad, D.R., Krautter, J., Tahoori, M.B.: Leaky noise: new side-channel attack vectors in mixed-signal iot devices. IACR Trans. Cryptogr. Hardw. Embed. Syst. **2019**, 305–339 (2019)
31. Sayakkara, A., Le-Khac, N.A., Scanlon, M.: A survey of electromagnetic side-channel attacks and discussion on their case-progressing potential for digital forensics. Dig. Invest. **29**, 43–54 (2019)
32. Ardekani, M.S., Singh, R.P., Agrawal, N., Terry, D.B., Suminto, R.O.: Rivulet: a fault-tolerant platform for smart-home applications. In: ACM Middleware Conference, Las Vegas, Nevada, pp. 41–54. ACM, Las Vegas (2017). https://doi.org/10.1145/3135974.3135988
33. Ghafir, I., et al.: Hidden Markov models and alert correlations for the prediction of advanced persistent threats. IEEE Access **7**, 99508–99520 (2019)
34. Sarker, I.H., Colman, A., Han, J., Khan, A.I., Abushark, Y.B., Salah, K.: BehavDT: a behavioral decision tree learning to build user-centric context-aware predictive model. Mobile Networks and Applications **25**(3), 1151–1161 (2019). https://doi.org/10.1007/s11036-019-01443-z
35. Ammar, M., Russello, G., Crispo, B.: Internet of Things: a survey on the security of IoT frameworks. J. Inf. Secur. Appl. **38**, 8–27 (2018)
36. Rieger, P., et al.: ARGUS: Context-Based Detection of Stealthy IoT Infiltration Attacks. arXiv preprint arXiv:2302.07589 (2023)
37. Hussain, F., Hussain, R., Hassan, S.A., Hossain, E.: Machine learning in IoT security: current solutions and future challenges. IEEE Commun. Surv. Tutor. **22**(3), 1686–1721 (2020)
38. Chen, X.W., Lin, X.: Big data deep learning: challenges and perspectives. IEEE Access **2**, 514–525 (2014)
39. Hasan, M., Islam, M.M., Zarif, M., Hashem, M.: Attack and anomaly detection in IoT sensors in IoT sites using machine learning approaches. Internet Things **7**, 100059 (2019)
40. Shih, C.S., Chou, J.J., Reijers, N., Kuo, T.W.: Designing CPS/IoT applications for smart buildings and cities. IET Cyber-Phys. Syst. Theory Appl. **1**(1), 3–12 (2016). https://doi.org/10.1049/iet-cps.2016.0025
41. Ferrara, P., Mandal, A.K., Cortesi, A., Spoto, F.: Static analysis for discovering IoT vulnerabilities. In. J. Softw. Tools Technol. Transf. **23**(1), 71–88 (2020). https://doi.org/10.1007/s10009-020-00592-x
42. Ngo, Q.D., Nguyen, H.T., Le, V.H., Nguyen, D.H.: A survey of IoT malware and detection methods based on static features. ICT Express **6**(4), 280–286 (2020)

43. Javed, A., Heljanko, K., Buda, A., Framling, K.: Cefiot: a fault-tolerant iot architecture for edge and cloud. In: 2018 IEEE 4th World Forum on Internet of Things (WF-IoT), Singapore, pp. 813–818. IEEE, Manhattan (2018). https://arxiv.org/pdf/2001.08433.pdf. https://doi.org/10.1109/WF-IoT.2018.8355149
44. Norris, M., et al.: IoT repair: flexible fault handling in diverse IoT deployments. ACM Trans. Internet Things (2022)
45. Norris, M., et al.: IoT Repair: Systematically addressing device faults in commodity IoT. In: 2020 IEEE/ACM Fifth International Conference on Internet-of-Things Design and Implementation (IoTDI), pp. 142–148. IEEE, Sydney (2020)
46. Misra, S., Gupta, A., Krishna, P.V., Agarwal, H., Obaidat, M.S.: An adaptive learning approach for fault-tolerant routing in IoT. In: Wireless Communications and Networking Conference, pp. 815–819. IEEE, Paris (2012)
47. Celik, Z.B., McDaniel, P.D., Tan, G.: Soteria: automated IoT safety and security analysis (2018). CoRR http://arxiv.org/abs/1805.08876. https://dblp.org/rec/journals/corr/abs-1805-08876.bib
48. Su, P.H., Shih, C.-S., Hsu, J.Y.-J., Lin, K.-J., Wang, Y.-C.: Decentralized fault tolerance mechanism for intelligent iot/m2m middleware. In: World Forum on Internet of Things (WF-IoT), WF-IoT '14, Seoul, South Korea, pp. 45–50. IEEE, New York City (2014). https://doi.org/10.1109/WF-IoT.2014.6803115,10.1109/WF-IoT.2014.6803115
49. Alkhresheh, A., Elgazzar, K., Hassanein, H.S.: CAPE: continuous access policy enforcement for IoT deployments. In: 2019 15th International Wireless Communications & Mobile Computing Conference (IWCMC), pp. 1576–1581. IEEE, Tangier (2019)
50. Atakli, I.M., Hu, H., Chen, Y., Ku, W.S., Su, Z.: Malicious node detection in wireless sensor networks using weighted trust evaluation. In: Spring Simulation Multiconference, SpringSim '08, San Diego, California, pp. 836–843. SpringSim, San Diego (2008). https://doi.org/10.1145/1400549.1400686,10.1145/1400549.1400686
51. Celik, Z.B., Tan, G., McDaniel, P.D.: IoT guard: dynamic enforcement of security and safety policy in commodity IoT. In: 26th Annual Network and Distributed System Security Symposium, NDSS 2019, San Diego, California, USA, 24–27 February 2019. The Internet Society, San Diego (2019). https://www.ndss-symposium.org/ndss-paper/iotguard-dynamic-enforcement-of-security-and-safety-policy-in-commodity-iot/. https://dblp.org/rec/conf/ndss/CelikTM19.bib
52. Choi, J., et al.: Detecting and identifying faulty IoT devices in smart home with context extraction. In: 48th IEEE/IFIP Dependable Systems and Networks (DSN), Luxembourg City, Luxembourg, pp. 610–621. IEEE, New York City (2018). https://doi.org/10.1109/DSN.2018.00068
53. Alfandi, O., Otoum, S., Jararweh, Y.: Blockchain solution for iot-based critical infrastructures: byzantine fault tolerance. In: NOMS 2020-2020 IEEE/IFIP Network Operations and Management Symposium, Budapest, Hungary, pp. 1–4. IEEE, Budapest (2020). https://arxiv.org/pdf/2001.08433.pdf. https://doi.org/10.1109/WF-IoT.2018.8355149
54. Jabbar, W.A., et al.: Design and fabrication of smart home with Internet of Things enabled automation system. IEEE Access **7**, 144059–144074 (2019)
55. Boyes, H., Hallaq, B., Cunningham, J., Watson, T.: The industrial internet of things (IIoT): an analysis framework. Comput. Ind. **101**, 1–12 (2018)
56. Mieronkoski, R., et al.: The Internet of Things for basic nursing care–a scoping review. Int. J. Nurs. Stud. **69**, 78–90 (2017)

57. Han, J., et al.: Do you feel what I hear? Enabling autonomous IoT device pairing using different sensor types. In: IEEE Symposium on Security and Privacy (SP), SP '18, San Francisco, CA, pp. 836–852. IEEE, New York City (2018). https://doi.org/10.1145/567752.567774
58. Ding, W., Hu, H., Cheng, L.: IoT safe: enforcing safety and security policy with real IoT physical interaction discovery. In: 28th Annual Network and Distributed System Security Symposium, NDSS 2021, Virtually, 21–25 February 2021. The Internet Society (2021). https://www.ndss-symposium.org/ndss-paper/iotsafe-enforcing-safety-and-security-policy-with-real-iot-physical-interaction-discovery/. https://dblp.org/rec/conf/ndss/DingH021.bib
59. Yao, Y., Zhou, W., Jia, Y., Zhu, L., Liu, P., Zhang, Y.: Identifying privilege separation vulnerabilities in IoT firmware with symbolic execution. In: Sako, K., Schneider, S., Ryan, P. (eds.) ESORICS 2019. LNCS, vol. 11735, pp. 638–657. Springer, Cham (2019). https://doi.org/10.1007/978-3-030-29959-0_31
60. Kramer, M.W.: https://github.com/martingwhite/kramer. Accessed 15 Oct 2021
61. Kang, H.J., Sim, S.Q., Lo, D.: IoT box: sandbox mining to prevent interaction threats in IoT systems. In: 2021 14th IEEE Conference on Software Testing, Verification and Validation (ICST), pp. 182–193. IEEE (2021)
62. Li, W., Song, H., Zeng, F.: Policy-based secure and trustworthy sensing for internet of things in smart cities. IEEE Internet Things J. **5**(2), 716–723 (2017)
63. Helion, S.S.: https://github.com/helion-security/helion. Accessed 11 Aug 2022
64. OpenHAB, OpenHAB: Open Source Automation Software (2018). https://www.openhab.org/
65. SmartThings, Samsung SmartThings Developer Documentation (2018). https://docs.smartthings.com/

A Case Study of API Design for Interoperability and Security of the Internet of Things

Dongha Kim●, Chanhee Lee●, and Hokeun Kim(✉)●

Arizona State University, Tempe, AZ 85281, USA
{dongha,chanheel,hokeun}@asu.edu

Abstract. Heterogeneous distributed systems, including the Internet of Things (IoT) or distributed cyber-physical systems (CPS), often suffer a lack of interoperability and security, which hinders the wider deployment of such systems. Specifically, the different levels of security requirements and the heterogeneity in terms of communication models, for instance, point-to-point vs. publish-subscribe, are the example challenges of IoT and distributed CPS consisting of heterogeneous devices and applications. In this paper, we propose a working application programming interface (API) and runtime to enhance interoperability and security while addressing the challenges that stem from the heterogeneity in the IoT and distributed CPS. In our case study, we design and implement our application programming interface (API) design approach using open-source software, and with our working implementation, we evaluate the effectiveness of our proposed approach. Our experimental results suggest that our approach can achieve both interoperability and security in the IoT and distributed CPS with a reasonably small overhead and better-managed software.

Keywords: Internet of Things · Interoperability · Security · API Design

1 Introduction

Heterogeneous distributed systems, including the Internet of Things (IoT) or distributed cyber-physical systems (CPS), have been rising, especially with the benefits of edge computing, such as low latency, privacy protection, and scalability [35,52]. As more IoT and distributed systems running on the edge are used in many different domains, it is increasingly challenging to support a diversity of communication models because of the heterogeneity in edge-computing environments where the heterogeneous devices and applications interact with one another [8]. One representative such application is the smart city applications [9,23] that work across the boundaries of different domains, including smart homes and buildings with various purposes under heterogeneous environments.

Thus, the problem of *interoperability* [3] between various components in the IoT and distributed CPS has become one of the major obstacles blocking further deployment of such systems. This interoperability problem includes dealing with the different security requirements depending on the specific applications and the characteristics of their data [51,53]. For example, safety-critical applications such as transportation or medical applications will require stronger security guarantees [49], while environmental sensing systems may require minimal security, such as data integrity while prioritizing low power consumption over security [46].

There have been research efforts to address the interoperability and security problems in the IoT and distributed CPS, such as a semantic-based collaboration platform [47], networking strategies for interoperability in real-time systems [12], or a secure network for mobile edge computing [24]. As programming models and APIs play a critical role in the development and deployment of edge-based IoT and CPS [11,26], programming models dedicated to edge-based systems [28] have been proposed. However, to the best of our knowledge, there has not been enough research work on programming models or APIs tailored to the IoT and distributed CPS with a working implementation trying to address both interoperability and security problems as a single integrated platform.

In this paper, we design an API and conduct a case study with the working implementation of the runtime based on the designed API to address the interoperability and security issues of the IoT and distributed CPS. Our contributions are threefold:

1. We design an API that supports multiple communication models, including point-to-point (e.g., client-server) and publish-subscribe paradigms, to facilitate seamless interaction between heterogeneous devices.
2. We incorporate a flexible security framework that can be adaptively applied based on varying security requirements.
3. We develop a working runtime system using open-source platforms to demonstrate the practicality of the API, and evaluate its performance, showing that it achieves interoperability and security with reasonably small overhead while simplifying software development and enhancing maintainability.

2 Related Work

Extensive surveys and reviews have been given by the literature [2,4,13,25,36,43] on the interoperability and security issues in the IoT and distributed CPS. Abounassar *et al.* [2] point out that the IoT, especially in the healthcare sector, still suffers from security and interoperability challenges even now in the 2020s. Lee *et al.* [25] provide a survey on current standards and efforts to bring interoperability and security to the IoT; however, they also present a number of remaining challenges, including the lack of developer support for enabling interoperability and limited security considerations in standards. A survey by Amjad *et al.* [4] reviews the interoperability and security challenges in industrial IoT (IIoT) with a focus on data transfer and application protocols, including the security vulnerabilities in the widely used publish-subscribe protocols [43]. Noura *et al.* [36] discuss the interoperability issues in the IoT from various perspectives, including, network interoperability, syntactical/semantic interoperability, and platform interoperability. Gürdür and Asplund [13] emphasize the importance of interoperability in CPS under distributed environments.

There have been several efforts and approaches to dealing with interoperability and security in the IoT and distributed CPS. Oh *et al.* [38] identify the security threats in the widely used interoperability protocol for heterogeneous IoT platforms, the OAuth 2.0 framework. Margarita *et al.* [33] consider interoperation among various entities, including IoT devices and production lines focused on Industry 4.0; however, this approach primarily targets the REST communication model and does not support various network protocols or security concerns. Pereira *et al.* [40] propose an ontology-based middleware to mitigate interoperability issues, also in industrial IoT (IIoT). The OPC unified architecture (OPC UA) [39] has been designed to support the interoperability between the smart sensors and the cloud by embracing heterogeneous communication protocols.

Some research work focuses on encryption or authentication to ensure security in the IoT and distributed CPS. Kannan *et al.* [16] apply encryption and digital signatures only targeting military applications to secure communication in an IoT environment. Sensor data, commonly used in military equipment, are collected and tested using encryption to show the effectiveness against data modification. This literature [16], however, does not consider the authentication and authorization process. Quadir *et al.* [41] propose an authentication framework targeting consumer electronics in medical, home, and personal applications. The framework proposed by Quadir *et al.* [41] mainly depends on a centralized server to communicate the distributed device via TLS. Secure Swarm Programming Platform (SSPP) [19] is a conceptual platform for providing security at the programming platform level, however, yet without a concrete implementation.

Also, some prior arts address the security of edge-based IoT and distributed CPS running on Robot Operating System (ROS). Sanchez *et al.* [48] integrate a smart personal protective system into ROS with edge computing. For communication between devices, only the MQTT protocol is provided with a Mosquitto message broker, and encryption based on SSL is considered only. Zhang *et al.* [54] utilize ROS2, where data distribution service (DDS) is used only to provide authentication and access control. This approach, however, depends on both cloud computing and edges. It causes user privacy problems immediately across the cloud and edge and can experience performance degradation due to communication overhead.

3 Proposed Approach

This section introduces our proposed API and runtime, designed based on our observation of different modes of communication and security requirements that are common to the IoT and distributed CPS.

For the discussion of our API design, we define two types of nodes, i.e., *Listener* and *Connector* (shown in Fig. 1), to represent typical nodes in the IoT and distributed CPS. We note that these terms (*Listener* and *Connector*) or

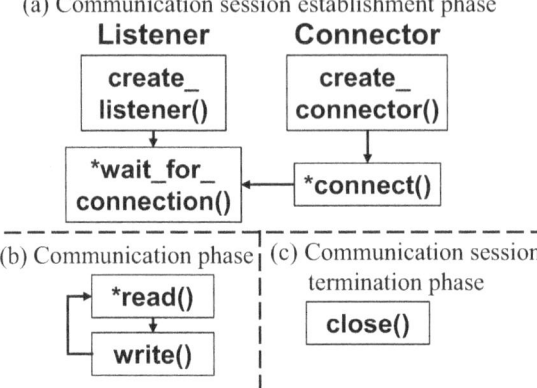

Fig. 1. Proposed API functions. Functions with * are blocking functions.

their concept are not novel, but we use these terms to capture the important and common characteristics of the nodes used in networked and distributed systems, mainly for our discussion in this paper. A *Connector* requests a connection to be waited and accepted by a *Listener*. These nodes are defined so that we can flexibly and seamlessly support different modes of communication and optional security guarantees.

3.1 Design of API Functions

Our proposed API consists of seven functions for *Listeners* and *Connectors*. These functions are used to establish connections and transfer data between different communication models. The API is roughly divided into three phases: communication session establishment phase, communication and data transfer phase, and communication session termination phase.

The first phase involves establishing a single session between two nodes as shown in Fig. 1a. A *Connector* and a *Listener* initiate the process by creating their respective objects. The *Listener* calls *wait_for_connection()*, while the *Connector* calls *connect()*, which are both blocking functions, to establish a communication session. The second phase is for data transfer between the *Listener* and the *Connector* as shown in Fig. 1b. For this purpose, we use *read()* and *write()*, which encapsulate the application layer protocol's data transfer. The final phase is disconnecting the session, using *close()*, as shown in Fig. 1c. This *close()* function sends a disconnection request to the other node and releases the resources associated with the session.

3.2 Interoperability Support

Our proposed API is designed to be applied to various communication models. We apply our API to two representative communication models that are most commonly used in the IoT and distributed CPS.

A Case Study of API Design for Interoperability and Security of the IoT   139

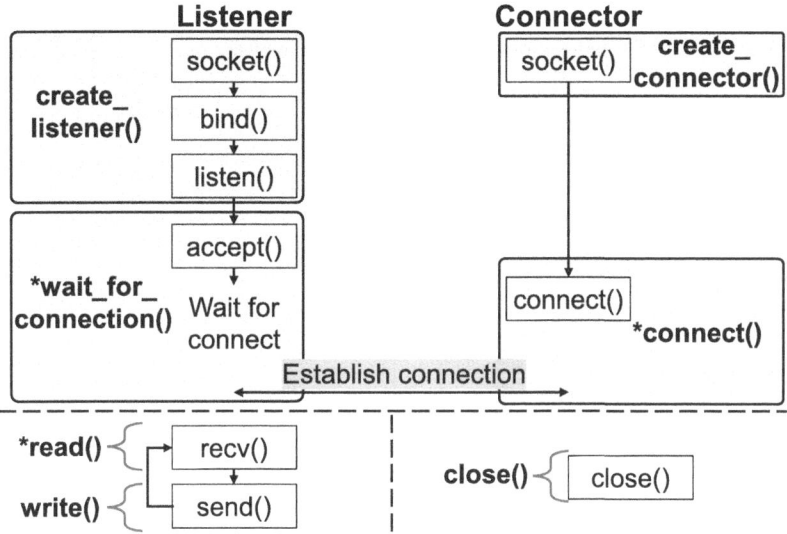

Fig. 2. Proposed API functions. Functions with * are blocking functions. These seven functions are used for the following three communication phases of networked nodes: (a) Communication session establishment (initialization) phase, (b) communication (data transfer) phase, and (c) communication session termination (closing) phase.

Point-to-Point Communication. Figure 2 shows how our API functions work for point-to-point communication, for example, a client-server model such as TCP or FTP.

Here, we take TCP as an example. A server and a client in TCP correspond to a *Listener* and a *Connector* in our proposed API, respectively. On the server side, our API function *create_listener()* handles the server socket creation, binding, and listening of the socket. The API function *wait_for_connection()* waits and accepts a client's request with blocking. On the client side, our API function *create_connector()* initializes the client socket, and another API function *connect()* calls the TCP socket function, connect(), which establishes a TCP connection with the server. After the session is set up, API functions *read()* and *write()* perform TCP socket functions recv() and send() to transfer data. Finally, *close()* is used to shut down and close the TCP sockets, disconnecting the server and the client.

Publish-Subscribe Communication. Our proposed API design also supports pub-sub communication as illustrated in Fig. 3. The main challenge in adopting pub-sub into our API is the session establishment. For the discussion in this paper, we call an application running on nodes of an distributed system running on the edge a *federation*. We call a unique identifier of a federation *federationID*. In our API, we assume that all distributed nodes know *federationID* as well as the IP address/port number of the centralized coordinator before the federation

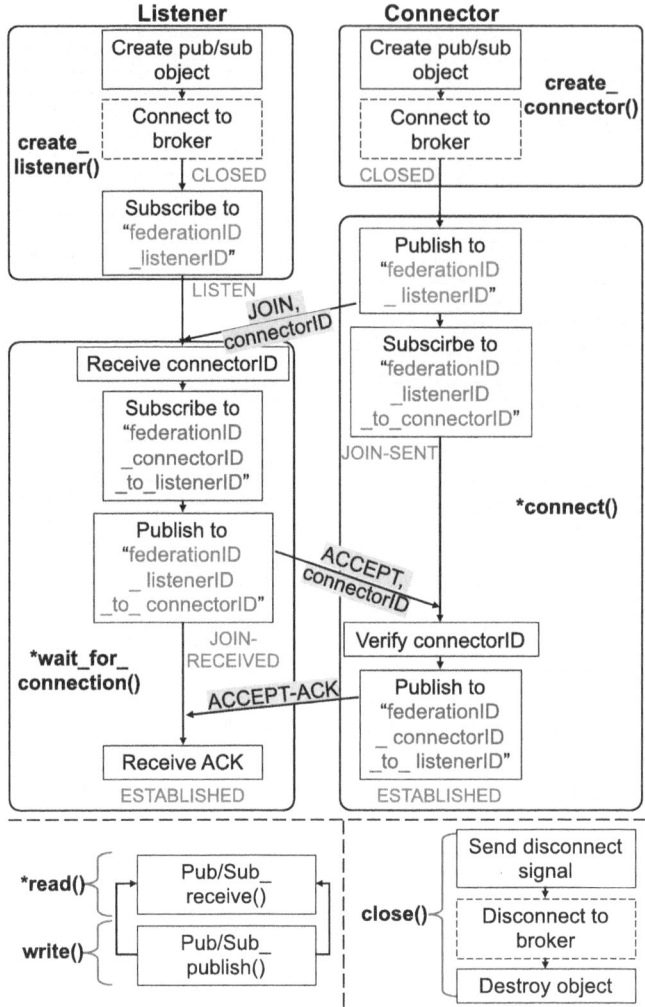

Fig. 3. Execution model of the proposed API for publish-subscribe communication. Blocking functions are marked *.

starts. Prior arts such as Robot Operating System 1 (ROS1) [42] are based on a similar assumption; for example, the ROS Master's IP address and port number are known to all nodes before the runtime starts.

As shown in Fig. 3, *create_listener()* and *create_connector()* create *Listener* and *Connector* nodes with an initialized pub/sub object. If the underlying pub-sub communication uses a centralized third party message broker such as the MQTT [1] broker, it will also connect to the broker. It is important to note that this broker is not considered a *Listener* or *Connector* node, as it is a feature of the pub-sub protocol itself rather than part of our design. Therefore, this connection step is bypassed in decentralized pub-sub protocols like Data Distribution Service (DDS) [37]. Both *Listener* and *Connector* start from a CLOSED state. The *Listener* node additionally subscribes to a topic named "*federationID_listenerID*" and enters a LISTEN state.

After that, the *Listener* and the *Connector* establish a session through a three-way handshake. First, the *Connector* node publishes to the topic "*federationID_listenerID*", sending a JOIN message with its *connectorID*. The *listenerID* should be known to the *Connector*, such as the topic name should be known for broadcasting communication. The *Connector* also subscribes to the topic "*federationID_listenerID_to_connectorID*". The *Connector* enters a JOIN-SENT state.

The *Listener* node receives the *connectorID*, and subscribes to the topic "*federationID_connectorID_to_listenerID*". Then it publishes an ACCEPT message and its *listenerID* to the topic "*federationID_listenerID_to_connectorID*". The *Listener* enters a JOIN-RECEIVED state.

The *Connector* verifies the received *connectorID*, publishes an ACCEPT-ACK message to "*federationID_connectorID_to_listenerID*" which is the specific topic for the *Listener*, and enters an ESTABLISHED state. The *Listener* finally receives the ACCEPT-ACK message and also enters an ESTABLISHED state. Finally, there are two topics for one *Connector-Listener* session, which are one-way for each.

The *read()* operation is a synchronous call, causing the node's execution to block until data is received from the corresponding *write()* operation on another node. The *close()* sends disconnect signals to the other node, disconnects to the broker if exists, and destroys the pub/sub object.

3.3 Security Support

For the authentication of each node, we assume that there is a key distribution center (KDC), which is a trusted third party nor Listener or Connector. KDC is only responsible for generating and distributing encryption keys to multiple *Listeners* and *Connectors* over a network, and do not directly participate in the system's communication. For example, a well-known network authentication protocol, Kerberos [45], uses a centralized authentication server (AS) as its KDC. The KDC issues tickets containing encrypted session keys, which are symmetric cryptographic keys as a common secret between two nodes, allowing nodes to authenticate and establish secure communication.

Figure 4 illustrates an overview of how our API design seamlessly applies security. The *create_listener()* and *create_connector()* first initialize the *Listener* and *Connector* object with configurations, including any certificates necessary for authentication with the KDC. When *connect()* is invoked, the *Connector* requests a session key from the KDC and sends a connection request to the *Listener*. Next, the *Listener* will call *wait_for_connection()* and wait for a connection request from the *Connector*. When the request arrives, the *Listener* must obtain the same session key from the KDC to ensure secure communication with the

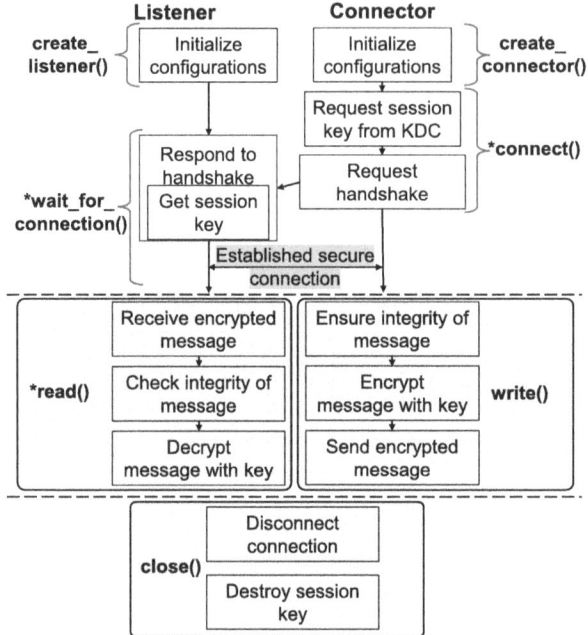

Fig. 4. Execution model of the proposed API with security enabled. Functions with * are blocking functions.

Connector. The connection request includes a handshake to ensure that both ends have the same session key and session nonce to encrypt the traffic.

When security is enabled, the *write()* function internally encrypts the message to ensure confidentiality and adds authentication methods, such as Message Authentication Codes (MAC), to maintain message integrity. The *read()* function checks the integrity of received messages through the MAC and decrypts the message content using the session key. Finally, *close()* disconnects the connection and cleans up the resource related to the secure session, such as the session key.

4 A Case Study: Design and Implementation

This section provides a detailed overview of the design and implementation for a case study of the proposed API design, including example code. For efficient implementation, we leverage open-source software libraries and runtime.

4.1 Open-Source Software Used for Implementation

We leverage existing open-source software projects for our case study. For the communication and coordination among distributed nodes in the IoT or distributed CPS, we use an open-source coordination language and runtime, Lingua Franca (LF) [30], and for security support, Secure Swarm Toolkit (SST) [21].

Lingua Franca. Lingua Franca (LF) is a coordination language designed to guarantee deterministic concurrency with *reactors* [32]. Reactors are lightweight concurrent entities that communicate with each other via timestamped messages. LF's C-runtime supports *federated execution* [6], allowing reactors to run across distributed systems and communicate over networks. The resulting networked system is called a *federation*, with each individual component known as a *federate*. The LF compiler generates separate code for each federate.

LF also provides a separate run-time infrastructure (RTI), which manages time synchronization among federates, ensuring a deterministic message flow and coordinating startup and shutdown. RTI also facilitates message exchange among federates, working as a message broker in *centralized coordination*.

Secure Swarm Toolkit. The Secure Swarm Toolkit (SST) is an open-source toolkit framework designed to offer robust authorization and authentication mechanisms for distributed environments. The local entity *Auth* [22], provides authentication and authorization for its locally registered entities, and also supports resilience to migrate trust to another Auth [20]. SST supports a C API [17] designed to support resource-constrained devices. It has also been used for access control of decentralized and distributed file systems [15].

4.2 Software Design Considerations

We chose LF as the runtime environment of our target systems primarily due to LF's compatibility with a wide range of embedded platforms [14], including Arduino [5], RP2040 [44], Zephyr [29], and also bare metal devices. However, LF has some limitations on network communication, relying on TCP sockets for message exchanges between RTI and federates. This reliance on TCP restricts interoperability, especially in heterogeneous environments.

For authentication of the federates, LF has two options, basic *federationID* check, and key-hashed message authentication code (key-hashed MAC or HMAC) authentication. As default, LF currently supports the basic *federationID* check when a federate joins a federation although there is no encryption or message integrity involved. A federate sends its *federationID* in plaintext to the RTI, which verifies if the *federationID* is correct. However, this allows malicious entities to join the federation if they can eavesdrop the *federationID* sent over the network. To address this vulnerability, HMAC is used to secure the connection between devices [18], but LF still does not provide message encryption over the network.

To address the aforementioned security gaps, we integrated SST into LF. SST is well-suited for large-scale distributed environments, with the Auth providing decentralized authentication and authorization. SST's flexibility in supporting lightweight cryptography and hash algorithms allows for the customization of security levels to meet different requirements for heterogeneous devices. The C APIs in SST made integration with LF's C runtime straightforward, thus enhancing both interoperability and security within our distributed system.

4.3 Code-Generation and Compilation

To illustrate how we enhance interoperability and security in LF, a simple LF program is organized as shown in Fig. 5. Figure 5a represents a diagram of the example LF program where an integer value is sent from a `Source` node to a `Destination` node. Note that the diagram is automatically generated in either Visual Studio Code or Eclipse IDE after LF plug-in is installed. In Fig. 5b, the LF program named `HelloDistributed` has two reactors `Source` and `Destination`. With this LF code, LF compiler generates the executable binaries of two federates for two reactors automatically.

Line 2 indicates that the whole execution of the generated executable binaries is performed with *centralized coordination*, which means that the RTI mediates all messages between the generated federates. Line 5 sets the federate joining process to use the HMAC authentication. With *centralized coordination*, in line 12, the message sent from `Source` reactor is transferred to RTI first and then forwarded to the `out` port of the `Destination` reactor. During communication between federates, the RTI acts like a message broker, providing deterministic timing controls and traceability of all messages among all federates.

In addition to the existing target properties in LF, we newly add a property to select the underlying communication stack as part of our proposed approach. In line 3, the `comm-type` keyword enables users to specify a network protocol selectively to be used. Currently, this new feature supports three communication stacks: TCP for client-server, MQTT for pub-sub, and SST for a security model. Note that other communication protocols can be easily added to this feature based on our proposed API design.

The LF program is conditionally compiled to ensure federates install only the libraries they need, avoiding unnecessary dependencies. To enable this flexibility, LF includes a network abstraction layer called the *network driver* (netdriver), which connects low-level communication protocols to the high-level API. The netdriver's design allows for polymorphism, enabling it to switch between different communication methods without additional code changes. As shown in Fig. 6, the netdriver contains a void pointer which is cast to another struct pointer depending on the communication type. This enables providing a unified interface for multiple communication protocols. The details of *remaining_bytes* will be explained in Sect. 4.5. To avoid confusion from the POSIX system calls, the defined APIs include a *netdrv_* prefix, such as *netdrv_read()*.

4.4 Runtime Implementation with Proposed API

We illustrate how our API design supports network interoperability and security.

Client-Server Runtime. We implement the client-server model using TCP sockets, which establish connections and enable data transfer between the RTI and fedcrates. Since functions of TCP protocol can be easily mapped to the proposed API, the implementation is straightforward.

A Case Study of API Design for Interoperability and Security of the IoT 145

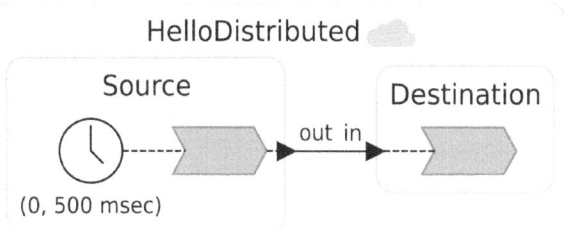

(a) Diagram of a simple federated Lingua Franca program with two federates, Source and Destination.

```
target C {
  coordination: centralized,
  comm-type: MQTT,
  timeout: 500 sec,
  auth: true
}

reactor Source {
  output out: int
  timer t(0, 500 msec)
  reaction(t) -> out {=
    lf_set(out, 0);
  =}
}

reactor Destination {
  input in: int
  reaction(in) {=
    lf_print("Dest received: %s", in->value);
  =}
}

federated reactor HelloDistributed{
  s = new Source()
  d = new Destination()
  s.out -> d.in
}
```

(b) Lingua Franca code for HelloDistributed.lf.

Fig. 5. Example of a simple Lingua Franca program.

Publish-Subscribe Runtime. We build the pub-sub model based on the MQTT protocol [1], utilizing the Eclipse Paho C library [10], a widely used MQTT client library for C.

The proposed three-way handshake in Sect. 3.2, establishes a unique session for each connection. The RTI acts as the *Listener* with a designated *Listener* ID of 'RTI', while each federate has a unique *Connector* ID. Setting the MQTT

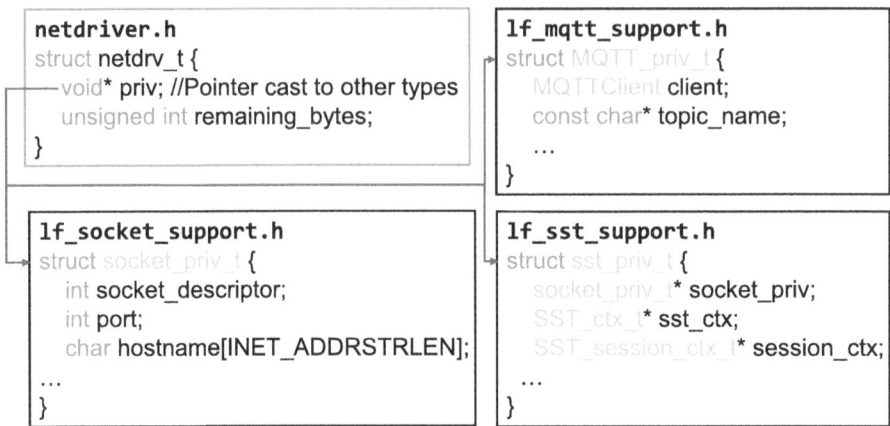

Fig. 6. The structure of the *netdriver*. The void pointer in `netdrv_t` is cast to another struct depending on the network type for polymorphism.

Quality of Service (QoS) level to 2 ensures that each message is delivered exactly once. Moreover, synchronous send and receive operations are critical in LF, providing reliability to message exchanges.

Security. To improve security, we integrated SST into LF. When a federate wants to establish a secure session with the RTI, it initiates the connection through *netdrv_connect()*. The federate requests for a session key from the *Auth*, which acts as a KDC. The Auth authenticates the federate through a three way handshake, and then sends out a session key. The federate now sends a connection request to the RTI sending the received session key's ID. The RTI receives the request from the federate, and requests the Auth the same session key with the matching key ID. Once the RTI and federate have the same key, they complete the handshake, establishing a secure session. Subsequent messages between the federate and the RTI is encrypted and decrypted using the session key.

4.5 Addressing Further Implementation Challenges

One of the significant challenges in implementation of our proposed API decouples network-related components that are tightly integrated with the LF code base. Many existing software platforms use a single communication model, thus, their implementation is deeply coupled with the underlying network protocol for various optimizations, for example, using the socket-level buffers for parsing network messages. Similarly, the current LF implementation is also integrated with TCP. We discuss how we tackle this problem in our implementation of the proposed API design.

To receive messages, LF uses the POSIX *read()* functions. However, in some cases, it calls *read()* multiple times to process a single message, first to retrieve

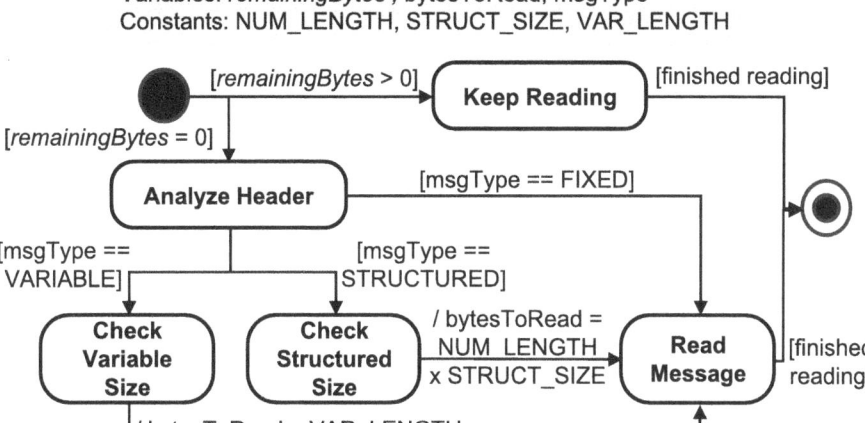

Fig. 7. The simplified state machine for the *netdrv_read()*. The boldface are the states and italic is the input.

the message type and then again to collect the rest of the payload. Similarly, LF sometimes sends messages in multiple chunks with separate *write()* calls. These inconsistent number of system calls for a single message increases code complexity, requiring additional logic to manage these operations. Also, it increases the chance to cause I/O errors multiple times and makes it difficult to trace errors.

The proposed *netdrv_read()* and *netdrv_write()* are designed to receive and send complete messages. It is critical to establish a clear mechanism to manage message boundaries and length. Protocols like MQTT and SST operate at the application layer and include information on the message length to ensure message boundaries are maintained. However, TCP is a stream-based transport layer protocol, meaning data is transmitted as a continuous stream of bytes without explicit message boundaries. This characteristic introduces challenges when trying to receive complete messages.

Managing message boundaries and lengths can be straightforward when each message explicitly indicates its total length. However, like other existing software platforms, not all LF messages have the payload length in their header. We classify LF messages into three types: *fixed-length*, *structured variable length*, and *variable length*, which can also be generalized for other distributed-system platforms.

Fixed-length messages have constant-sized payloads as defined by the message type. For example, MSG_TYPE_TIMESTAMP, which is one of the message types in LF for sending the start time of the RTI and federates, has a fixed payload of 8 bytes. Next, structured variable-length messages have a variable number of fixed-length structures. So, the message consists of an integer indicating the number of structures of 'n,' followed by 'n' fixed-length structures. Finally, the variable length has an integer 'n' indicating the length of the payload, followed by the variable payload.

To address this challenge, the TCP model's *netdrv_read()* uses a state machine to handle each type of message. Figure 7 shows a simplified version of the state machine. The *netdrv_read()* starts with an INITIAL state denoted by a filled black circle. Depending on the size indicated by *remainingBytes*, the state goes to either `Analyze Header` or `Keep Reading` state. When the *netdrv_read()* function is called and the *remainingBytes* is more than zero, the state machine enters a `Keep Reading` state. In this state, *netdrv_read()* reads as much data as the buffer allows, updating *remainingBytes*. This design provides flexibility for processing variable-length messages and reduces the risk of buffer overflows.

If the state machine enters an `Analyze Header` state, it calls the POSIX *read()* to receive only one byte to check a message type. Regarding the message type, *bytesToRead* is set to indicate the number of bytes of the payload. If the message type is a fixed-length message, it reads exactly the amount according to the message type header after entering into a `Read Message` state. For structured length messages, it calculates *bytesToRead* with NUM_LENGTH and STRUCT_SIZE, i.e., the number of structures and a struct size, respectively. For variable length messages, it first reads the integer 'n' indicating the variable length, and then reads 'n' bytes. The netdriver structure keeps track of the additional bytes to be read through an integer, *bytesToRead*, until it reaches to the end state from `Read Message` state.

5 Evaluation

In this section, we evaluate our API's overhead in terms of communication time, binary size, and message lengths. We also clarify the strong points of the proposed API design qualitatively through a software design analysis.

5.1 Experimental Setup

An IoT or distributed CPS that runs the LF program in Fig. 5, where the federate `Source` sends a message to the federate `Destination` is used for the evaluation of our API. The RTI, Mosquitto message broker [27] for MQTT, and Auth for SST run on a workstation as the edge, equipped with an i9-13900 CPU and 128 GB memory. The federates are deployed on two Raspberry Pi 4 (Model B 4GB RAM) devices. The workstation and the devices communicate over Wi-Fi, and the average end-to-end round-trip latency from the Raspberry Pi to the workstation, measured by a simple ping, is 13.60 milliseconds. The federates joining the federation use HMAC authentication mode as a baseline.

5.2 Communication Time Overhead

To estimate the communication time overhead, we measure a *lag*, the time difference between the physical time and logical time. Specifically, we measure the

A Case Study of API Design for Interoperability and Security of the IoT   149

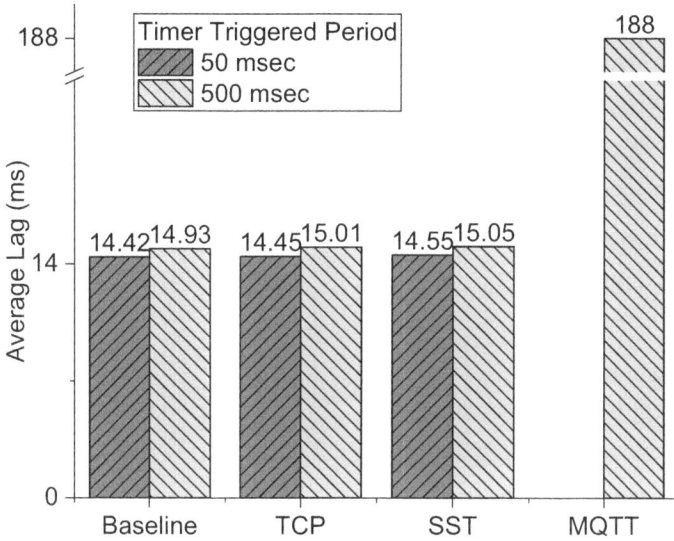

Fig. 8. Average lag compared with the baseline LF code. The federate Source sends messages when the timer is triggered as in line 10, with a period of 50 and 500 milliseconds.

time lapse between the system clock (physical time) and the semantic notion of intended global time agreed by all nodes (logical time) [31]. From the example above in Fig. 5b, the timer on line 10 triggers the Source reactor to send a message to the Destination reactor every 500 milliseconds. Due to the timeout in line 3, this will stop when the logical time reaches 500 s. We also set the timer with a period of 50 milliseconds, with a timeout of 50 s, to test sending messages in shorter periods. Consequently, in both cases, the message will be sent 1,000 times. We measure the average of each message's lag.

Figure 8 shows the average lag when sending messages in a period of 500 milliseconds and 50 milliseconds. In the case when the period is 500 milliseconds and the TCP model is compared with the baseline code, there is a 0.53% increase in lag which is 0.08 milliseconds. The SST model also has a similar lag of 0.80% lag increase. When we compare the SST model to the TCP model, there is only a 0.26% increase in lag when is security enabled for authentication of federates and confidentiality/integrity of messages. The 50 millisecond period test also had a similar tendency, showing a very small overhead.

MQTT has a longer lag due to the synchronous behavior. To ensure deterministic timing, all messages in LF must be sent in order. MQTT's underlying TCP sends messages in order, but the protocol itself does not guarantee it without QoS level 2. So, the *netdrv_write()* must call MQTTClient_waitForCompletion() after publishing the message to ensure the message is delivered. For synchronous receive of incoming messages, the *netdrv_read()* calls MQTTClient_receive().

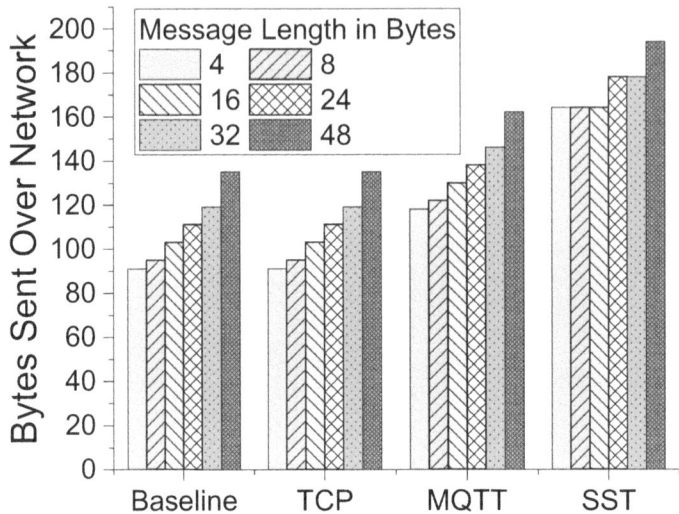

Fig. 9. Sizes of messages sent over the network in bytes for different communication types supported by the proposed API compared to the baseline implementation.

As the synchronous parts become the bottleneck, each netdrv_write() function call takes an average of 90 milliseconds to transfer a message. Due to the centralized coordination, passing a message via the RTI, the netdrv_write() is called twice, sending signals from the federate Source to the RTI, and RTI to the federate Destination. This explains the average latency of 188 milliseconds. However, when using MQTT, centralized coordination becomes inefficient because the RTI acts as a message broker while MQTT itself has its own broker. To remove this inefficiency, LF supports a different mode of runtime execution called *decentralized coordination*, where federates directly communicate with each other without the RTI. We plan to support this for MQTT in the future.

5.3 Message Size Overhead

We measure communication overhead in terms of the message size of each communication type using Wireshark [7]. To be specific, we measure the size of MSG_TYPE_TAGGED_MESSAGE, which includes a header with the destination information and a variable payload. Figure 9 illustrates the total message size when sending 4, 8, 16, 24, 32, and 48-byte long messages as payload. When this message is sent, a 21-byte header is attached, so when sending a 32-bit integer, a total of 25 bytes are sent.

Details are provided for when 4 bytes are sent. The baseline code and the client-server model is both based on TCP, so they have the same length of TCP/IP headers of 66 bytes, sending a total of 91 bytes. The message size of the baseline and client-server models both linearly increase in proportion to the message size. The MQTT model sent 118 bytes, including 66 bytes of TCP/IP headers and 27 bytes of MQTT headers. The MQTT header size varies depending on the topic name, which is 'MQTTTest_fed0_to_RTI'.

The SST model sent a total of 164 bytes, including the TCP/IP header, SST header, and encrypted message. SST employs the AES-128-CBC mode, which is a block cipher with a fixed block size of 16 bytes. As a result, the size of the encrypted output increases in 16-byte increments, demonstrating a stepwise pattern of growth.

Note that the block cipher encryption provided by SST also prevents side-channel attacks. As described in Sect. 4.5, LF has a message type with fixed lengths. Due to these fixed lengths, eavesdroppers can infer the message type from the message length. If the eavesdroppers can access to the trace of the messages, the message lengths can be a side channel, allowing them to determine which message types are being transmitted. By using a block cipher like AES-128-CBC, SST ensures that all encrypted messages are in block sizes, making it difficult to infer the message type.

5.4 Binary Size Overhead

To estimate the overhead in terms of binary sizes, we measure the binary size of the RTI, federate `Source`, and federate `Destination`. As illustrated in Fig. 10, the overhead introduced by the abstraction layer is minimal when compared with the baseline code. The increase in binary size for the RTI is 1.02%, 5.23%, and 1.44% for TCP, MQTT, and SST, respectively. For the same protocols, the Destination federate's binary size increases by 7.29%, 3.71%, and 7.67%, while the Source federate show a similar increase.

The increased binary size for both RTI and federates mostly comes from compilation of the network abstract layer as a separate library, making it more modular and flexible to extend. The RTI has a smaller overhead compared to the federates because it functions only as a *Listener*, whereas the federates have dual roles, serving as both *Connectors* and *Listeners* in *decentralized coordination* which has been briefly introduced in Sect. 5.2. In this case, the RTI does not relay messages to other federates; instead, the federates communicate directly with each other. However, the decentralized coordination of currently only supports TCP yet and supporting other protocols will be future work.

Among different modes of communication, MQTT shows the largest binary size for the RTI. However, for the federates, MQTT has smaller binary sizes than TCP and SST. Although we did not perform a detailed analysis of this difference, we speculate compiler optimizations and additional libraries caused this increase.

5.5 Software Design Analysis

Figure 11 shows the improved software design from the previous existing LF network module. In the previous software design, as shown in the left side of Fig. 11, each federate owns a TCP socket instance to call network functions

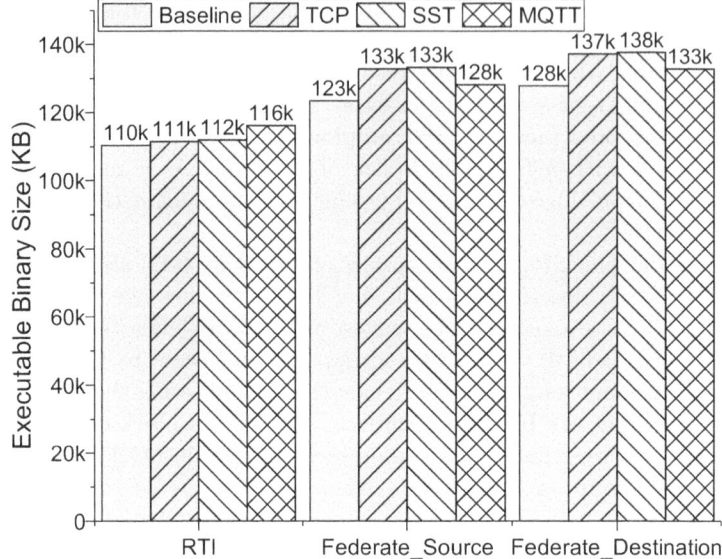

Fig. 10. Binary size in kilobytes compared with the baseline LF code.

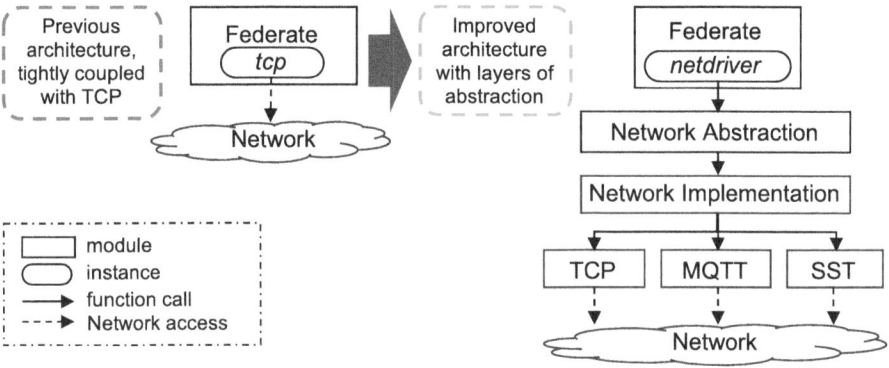

Fig. 11. Improved software architecture after applying the proposed API design to the previous LF network node.

to communicate with RTI or other federates. The direct inclusion of the TCP socket instance requires a large number of code modifications in the federate to add other types of communication protocols (e.g., publish-subscribe or secure communication mode) in addition to the existing TCP protocol. The right side of Fig. 11 denotes the improved software design after applying the proposed API design, which decouples a TCP instance from a federate.

To evaluate the effectiveness of the proposed API in terms of software design, we conduct a qualitative analysis focused on how the application of the proposed approach achieves the software design considerations in [50]. This paper applies only to qualitative parts of the design considerations, i.e., abstraction, modularity, and information hiding since quantitative evaluation using software engineering metrics is not the primary concern. Analyzing software design based on quantitative metrics such as cyclomatic complexity [34] is left to future work.

Abstraction. The goal of abstraction is to concentrate only on the essential features. To achieve this, two types of abstraction, i.e., *procedural abstraction* and *data abstraction* should be achieved. *Procedural abstraction* aims to find a hierarchy in the software's control and can be achieved by decomposing the software into sub-modules step-wisely so that the abstraction draws a hierarchical structure. In this structure, the top node denotes the problem to be solved through the software. As shown in the right side of Fig. 11, our design shows the APIs to solve the network access from federate as a top node, separates implementation and protocol types as next-level nodes, and forms a hierarchical structure. Our design also separates protocol types as another layer and hides data types from a federate.

Modularity. Coupling between modules should be considered to increase software modularity [50]. Six types of coupling, i.e., content coupling, common coupling, external coupling, control coupling, stamp coupling, and data coupling, need to be analyzed to evaluate the software modularity. Content coupling occurs when a module changes another module's data or control flows using jump indicators, and common coupling occurs when two modules share data. Since we separate *Network Abstraction* as the API layer and *Network Implementation* as its implementation, content and common coupling are avoided. External coupling occurs when modules communicate through an external medium, such as a file, and control coupling occurs when one module sets control flags that are reacted to by the dependent modules. Neither files nor control flags are used as arguments in our API design. Either stamp or data coupling occurs by passing complete data structures or simple data between modules. Our API design also does not pass any data between modules, avoiding all six types of coupling.

Information Hiding. The key to implementing information hiding is to hide design decisions and details from other modules. By designing each selection of a communication protocol as an option *comm-type* in Fig. 5b and implementing each protocol as a separate C source file, we can hide all design decisions and implementation details of each communication protocol into each source file. This design also enables developers to easily add other communication protocols with a simpler, easier-to-maintain, and more robust interface.

6 Conclusion and Future Work

In this paper, we perform a case study of the design of an API and its runtime to achieve network interoperability and security in the IoT and distributed CPS.

The proposed API encapsulates the underlying network implementation layer through seven key API functions. We implement our approach using open-source software, Lingua Franca, and SST as a case study. The evaluation assesses the proposed API design using our case-study implementation by measuring the communication time overhead, message size, and binary size, showing a minimal overhead. The proposed API and its implementation will be available on GitHub.

As future work, we plan to support more communication modes in our API. We also plan to allow distributed nodes with different communication modes to join a single federation. In this case, we envision that a centralized entity such as RTI will be able to handle multiple protocols in one process. Security options for communication among distributed nodes can be extended to enable fine-grained configurations in our proposed API.

Acknowledgment. This work was supported in part by the NSF I/UCRC for Intelligent, Distributed, Embedded Applications and Systems (IDEAS) and from NSF grant #2231620. This work was supported in part by ATTO Research.

References

1. MQTT Version 3.1.1. OASIS Standard (2014). https://docs.oasis-open.org/mqtt/mqtt/v3.1.1/os/mqtt-v3.1.1-os.html
2. Abounassar, E.M., El-Kafrawy, P., Abd El-Latif, A.A.: Security and interoperability issues with Internet of Things (IoT) in healthcare industry: a survey. In: Security and Privacy Preserving for IoT and 5G Networks: Techniques, Challenges, and New Directions, pp. 159–189 (2022)
3. Albouq, S.S., Sen, A.A.A., Almashf, N., Mohammad Yamin, A.A., Bahbouh, N.M.: A survey of interoperability challenges and solutions for dealing with them in IoT environment. IEEE Access **10**, 36416–36428 (2022)
4. Amjad, A., Azam, F., Anwar, M.W., Butt, W.H.: A systematic review on the data interoperability of application layer protocols in industrial IoT. IEEE Access **9**, 96528–96545 (2021)
5. Arduino: Arduino: Open-source electronics platform (2005). https://www.arduino.cc/
6. Bateni, S., et al.: Risk and mitigation of nondeterminism in distributed cyber-physical systems. In: Proceedings of the 21st ACM-IEEE International Conference on Formal Methods and Models for System Design, pp. 1–11 (2023)
7. Beale, J., Orebaugh, A., Ramirez, G.: Wireshark & Ethereal Network Protocol Analyzer Toolkit. Elsevier (2006)
8. Carvalho, G., Cabral, B., Pereira, V., Bernardino, J.: Edge computing: current trends, research challenges and future directions. Computing **103**(5), 993–1023 (2021). https://doi.org/10.1007/s00607-020-00896-5
9. Costin, A., Eastman, C.: Need for interoperability to enable seamless information exchanges in smart and sustainable urban systems. J. Comput. Civ. Eng. **33**(3), 04019008 (2019)
10. Eclipse Foundation: Eclipse Paho C Client Library (2009). https://github.com/eclipse/paho.mqtt.c

11. Giang, N.K., Lea, R., Blackstock, M., Leung, V.C.: Fog at the edge: experiences building an edge computing platform. In: International Conference on Edge Computing (EDGE), pp. 9–16. IEEE (2018)
12. Gomez, D.L., Montoya, G.A., Lozano-Garzon, C., Donoso, Y.: Strategies for assuring low latency, scalability and interoperability in edge computing and TSN networks for critical IIoT services. IEEE Access **11**, 42546–42577 (2023)
13. Gürdür, D., Asplund, F.: A systematic review to merge discourses: interoperability, integration and cyber-physical systems. J. Ind. Inf. Integr. **9**, 14–23 (2018)
14. Jellum, E.R., et al.: Beyond the threaded programming model on real-time operating systems. In: Fourth Workshop on Next Generation Real-Time Embedded Systems (NG-RES 2023). Schloss Dagstuhl-Leibniz-Zentrum für Informatik (2023)
15. Jo, Y., Cho, Y., Kim, H.: Secure and lightweight access control for highly decentralized and distributed file systems. In: Proceedings of the 1st International Workshop on Middleware for the Computing Continuum, pp. 1–6 (2023)
16. Kannan, B.M., Solainayagi, P., Azath, H., Murugan, S., Srinivasan, C.: Secure communication in IoT-enabled embedded systems for military applications using encryption. In: 2nd International Conference on Edge Computing and Applications (ICECAA), pp. 1385–1389. IEEE (2023)
17. Kim, D., Jo, Y., Kim, T., Kim, H.: SST v1.0.0 with C API: pluggable security solution for the Internet of Things. SoftwareX **22**, 101390 (2023). https://doi.org/10.1016/j.softx.2023.101390
18. Kim, D., Kim, H.: Poster abstract: securing edge-based real-time IoT systems. In: Proceedings of the 21st ACM Conference on Embedded Networked Sensor Systems, SenSys 2023, pp. 544–545 (2024). https://doi.org/10.1145/3625687.3628408
19. Kim, H.: Secure programming platform for edge-based IoT: wild-and-crazy-idea paper. In: 2023 Forum on Specification & Design Languages (FDL), pp. 1–4. IEEE (2023)
20. Kim, H., Kang, E., Broman, D., Lee, E.A.: Resilient authentication and authorization for the Internet of Things (IoT) using edge computing. ACM Trans. Internet Things **1**(1) (2020). https://doi.org/10.1145/3375837
21. Kim, H., Kang, E., Lee, E.A., Broman, D.: A toolkit for construction of authorization service infrastructure for the Internet of Things. In: The 2nd ACM/IEEE International Conference on Internet-of-Things Design and Implementation, Pittsburgh, PA, pp. 147–158 (2017)
22. Kim, H., Wasicek, A., Mehne, B., Lee, E.A.: A secure network architecture for the Internet of Things based on local authorization entities. In: The 4th IEEE International Conference on Future Internet of Things and Cloud (FiCloud), Vienna, Austria, pp. 114–122 (2016)
23. Koo, J., Kim, Y.G.: Interoperability requirements for a smart city. In: Proceedings of the 36th Annual ACM Symposium on Applied Computing, pp. 690–698 (2021)
24. Lai, X., Fan, L., Lei, X., Deng, Y., Karagiannidis, G.K., Nallanathan, A.: Secure mobile edge computing networks in the presence of multiple eavesdroppers. IEEE Trans. Commun. **70**(1), 500–513 (2021)
25. Lee, E., Seo, Y.D., Oh, S.R., Kim, Y.G.: A survey on standards for interoperability and security in the Internet of Things. IEEE Commun. Surv. Tutor. **23**(2), 1020–1047 (2021)
26. Li, B., Dong, W.: Edge-centric programming for IoT applications with automatic code partitioning. IEEE Trans. Comput. **71**(10), 2408–2422 (2021)
27. Light, R.A.: Mosquitto: server and client implementation of the MQTT protocol. J. Open Source Softw. **2**(13), 265 (2017)

28. Lin, H., Zeadally, S., Chen, Z., Labiod, H., Wang, L.: A survey on computation offloading modeling for edge computing. J. Netw. Comput. Appl. **169**, 102781 (2020)
29. The Linux Foundation: Zephyr RTOS (2016). https://www.zephyrproject.org/
30. Lohstroh, M., Menard, C., Bateni, S., Lee, E.A.: Toward a lingua franca for deterministic concurrent systems. ACM Trans. Embed. Comput. Syst. (TECS) **20**(4), 1–27 (2021)
31. Lohstroh, M., Menard, C., Schulz-Rosengarten, A., Weber, M., Castrillon, J., Lee, E.A.: A language for deterministic coordination across multiple timelines. In: 2020 Forum for Specification and Design Languages (FDL), pp. 1–8. IEEE (2020)
32. Lohstroh, M., et al.: Reactors: a deterministic model for composable reactive systems. In: Chamberlain, R., Edin Grimheden, M., Taha, W. (eds.) CyPhy/WESE -2019. LNCS, vol. 11971, pp. 59–85. Springer, Cham (2020). https://doi.org/10.1007/978-3-030-41131-2_4
33. Margaria, T., Chaudhary, H., Guevara, I., Ryan, S., Schieweck, A.: The interoperability challenge: building a model-driven digital thread platform for CPS. In: Margaria, T., Steffen, B. (eds.) ISoLA 2021. LNCS, vol. 13036, pp. 393–413. Springer, Cham (2021). https://doi.org/10.1007/978-3-030-89159-6_25
34. McCabe, T.J.: A complexity measure. IEEE Trans. Softw. Eng. **4**, 308–320 (1976)
35. Ning, H., Li, Y., Shi, F., Yang, L.T.: Heterogeneous edge computing open platforms and tools for Internet of Things. Futur. Gener. Comput. Syst. **106**, 67–76 (2020)
36. Noura, M., Atiquzzaman, M., Gaedke, M.: Interoperability in Internet of Things: taxonomies and open challenges. Mob. Netw. Appl. **24**, 796–809 (2019)
37. Object Management Group (OMG): Data Distribution Service (DDS) Version 1.4. OMG Specification formal/2015-04-10, Object Management Group (OMG) (2015). https://www.omg.org/spec/DDS/1.4/PDF
38. Oh, S.R., Koo, J., Kim, Y.G.: Security interoperability in heterogeneous IoT platforms: threat model of the interoperable OAuth 2.0 framework. In: Proceedings of the 37th ACM/SIGAPP Symposium on Applied Computing, pp. 22–31 (2022)
39. OPC Foundation: The OPC Unified Architecture (UA) (2008). https://opcfoundation.org/about/opc-technologies/opc-ua/
40. Pereira, P.H.M., Cainelli, G., Pereira, C.E., Costa, J.P.J.D., Freitas, E.P.D.: An interoperability middleware for IIoT. In: International Symposium on Industrial Electronics (ISIE), pp. 1–6. IEEE (2023)
41. Quadir, M.S.E., Chandy, J.A.: Embedded systems authentication and encryption using strong PUF modeling. In: 2020 IEEE International Conference on Consumer Electronics (ICCE), pp. 1–6. IEEE (2020)
42. Quigley, M., et al.: ROS: an open-source robot operating system. In: ICRA Workshop on Open Source Software, Kobe, Japan, vol. 3, p. 5 (2009)
43. Rana, B., Singh, Y., Singh, P.K.: A systematic survey on Internet of Things: energy efficiency and interoperability perspective. Trans. Emerg. Telecommun. Technol. **32**(8), e4166 (2021)
44. Raspberry Pi Foundation: RP2040 Microcontroller Datasheet (2021). https://datasheets.raspberrypi.com/rp2040/rp2040-datasheet.pdf
45. Saltzer, J.H., Reed, D.P., Clark, D.D.: End-to-end arguments in system design. ACM Trans. Comput. Syst. (TOCS) **2**(4), 277–288 (1984)
46. Shapsough, S., Aloul, F., Zualkernan, I.A.: Securing low-resource edge devices for IoT systems. In: International Symposium in Sensing and Instrumentation in IoT Era (ISSI), pp. 1–4. IEEE (2018)

47. Sigwele, T., Hu, Y.F., Ali, M., Hou, J., Susanto, M., Fitriawan, H.: An intelligent edge computing based semantic gateway for healthcare systems interoperability and collaboration. In: 2018 IEEE 6th International Conference on Future Internet of Things and Cloud (FiCloud), pp. 370–376. IEEE (2018)
48. Sánchez, S.M., et al.: Edge computing driven smart personal protective system deployed on NVIDIA Jetson and integrated with ROS. In: De La Prieta, F., et al. (eds.) PAAMS 2020. CCIS, vol. 1233, pp. 385–393. Springer, Cham (2020). https://doi.org/10.1007/978-3-030-51999-5_32
49. Tedeschi, P., Sciancalepore, S.: Edge and fog computing in critical infrastructures: analysis, security threats, and research challenges. In: 2019 IEEE European Symposium on Security and Privacy Workshops (EuroS&PW), pp. 1–10. IEEE (2019)
50. van Vliet, H.: Software Engineering: Principles and Practice. Wiley, Hoboken (2008)
51. Xiao, Y., Jia, Y., Liu, C., Cheng, X., Yu, J., Lv, W.: Edge computing security: state of the art and challenges. Proc. IEEE **107**(8), 1608–1631 (2019)
52. Yu, W., et al.: A survey on the edge computing for the Internet of Things. IEEE Access **6**, 6900–6919 (2017)
53. Zeyu, H., Geming, X., Zhaohang, W., Sen, Y.: Survey on edge computing security. In: 2020 International Conference on Big Data, Artificial Intelligence and Internet of Things Engineering (ICBAIE), pp. 96–105. IEEE (2020)
54. Zhang, J., Keramat, F., Yu, X., Hernández, D.M., Queralta, J.P., Westerlund, T.: Distributed robotic systems in the edge-cloud continuum with ROS 2: a review on novel architectures and technology readiness. In: 2022 Seventh International Conference on Fog and Mobile Edge Computing (FMEC), pp. 1–8. IEEE (2022)

ShadowConn: Breaking the Entanglement of Cross-Platform IoT Delegation in Multi-user Environments

Huan Bui(✉) and Chenglong Fu

The University of North Carolina at Charlotte, Charlotte, NC 28223, USA
{hbui11,chenglong.fu}@charlotte.edu

Abstract. Most popular smart home IoT platforms allow devices to be shared with other users or delegated to other platforms. We find that many IoT platforms have insufficient and vulnerable design of access control in the device sharing and delegation process. Existing literature has revealed non-negligible security and privacy risks caused by flawed access control of IoT platforms and proposed various solutions to enhance the device sharing process. However, they are focusing on securing device sharing within the same IoT platform and leave the access control of cross-platform IoT device sharing an open problem. In this work, we conduct the first systematic study of access control in cross-platform IoT device sharing and identify the issues of entangled associations among devices, users, and IoT platforms. Based on our study, we propose a practical solution named Shadow Connector (ShadowConn), which decouples IoT devices from users and platforms. ShadowConn creates virtual shadow device instances for each specific scenario of device sharing and delegation, synchronizing their states with the real devices. By regulating the state synchronizations between virtual and real devices, fine-grained and flexible access control policies can be enforced. Additionally, we design a Large Language Model (LLM) agent to assist users in specifying accurate and comprehensive access control policies that make use of shadow device instances. We implement the ShadowConn prototype on commercially available smart home platforms and conduct performance evaluations using four access control scenarios. ShadowConn successfully achieves a 100% satisfaction rate in establishing the access control policy based on users' requests and syncing the status of devices used by multiple users across different platforms.

Keywords: Smart Home · Internet of Things · Access Control

1 Introduction

The emergence of Internet of Things (IoT) automation platforms has revolutionized the smart home ecosystem by providing uniform device access and management interfaces and user customizable automation rules. IoT device manufacturers such as Philips Hue [11] and TP-Link [21] all provide options for the users to integrate their devices to IoT automation platforms such as SmartThings [13], and Alexa [1]. At the same time, IoT automation platforms also provide utilities for users to share their devices with other users.

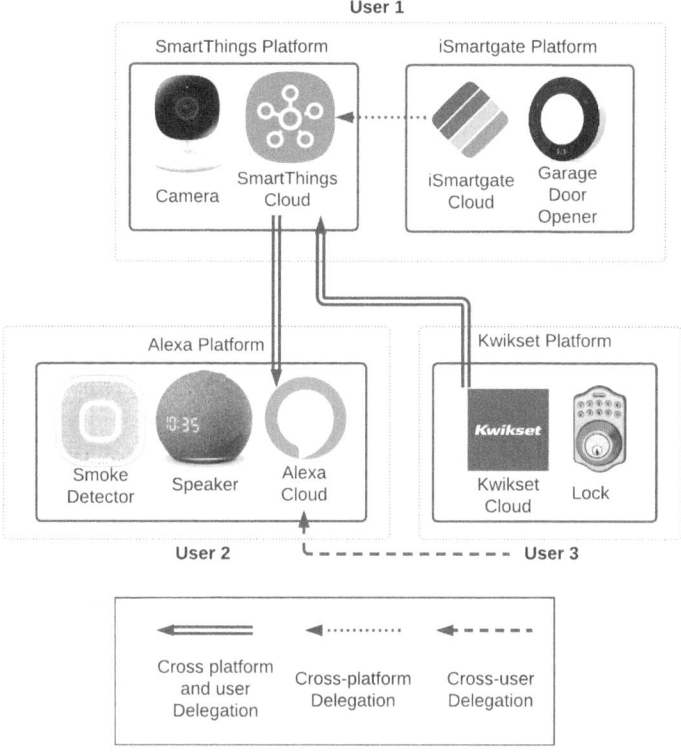

Fig. 1. Example of cross-user and platform IoT device delegation.

With the feature of device sharing, users can share their IoT devices with others in shared environments such as multi-resident apartments, rental homes, and offices [20,24], greatly facilitating collaboration and resource sharing. However, considering IoT devices' capabilities in interacting with the physical world and collecting sensitive information, it could be dangerous if some devices are exploited by malicious users. Existing literature has already revealed substantial risks of interpersonal abuse caused by overprivileged access to IoT devices [9,23]. This highlights the urgent need for fine-grained and context-aware access control solutions to regulate access to IoT devices in shared smart environments.

Mainstream IoT platforms like Alexa, SmartThings, and Google Home allow device sharing within the same platform. Previous works, such as Kratos [14] and [6], enhance access control for device sharing but assume all users are on the same platform. However, a survey by Chi et al. [3] reveals that users often prefer different IoT platforms, making cross-platform sharing common in shared environments. Although current platforms provide 'account linking' to delegate devices, they assume the same user controls both delegator and delegatee platforms, granting unrestricted access. For example, in Fig. 1, three users in a smart home environment (User1, User2, and User3) use different platforms (SmartThings, Alexa, and Kwikset) to share devices. This setup poses safety and privacy risks. The delegation of the smart lock from Kwikset to SmartThings lacks fine-grained control policies (e.g., restricting access when User3 is not home). Additionally, devices like User1's security camera may be unintentionally shared with User2, raising privacy concerns. Moreover, chained delegation complicates command tracking, making it difficult for User3 to distinguish between commands issued by User1 or User3.

To address the issue of flawed access control in cross-platform IoT device sharing, we present a systematic analysis of overprivileged access and introduce ShadowConn, a flexible and fine-grained access control framework. ShadowConn creates shadow device instances for each access endpoint, using virtual devices instead of real ones to enforce dynamic access control by synchronizing states between real and shadow devices. For example, separate shadow instances of a smart lock can be created for User1 and User2, allowing policies such as 'lock access only when User3 is not home' to be enforced by blocking state synchronization. This decouples the associations between users, devices, and platforms, allowing more precise control.

Although the new ShadowConn architecture provides a highly flexible and customizable access control framework, it also significantly complicates access control policies by introducing a large number of shadow devices. These policies must be tailored according to the IDs and attributes of the shadow devices, making the policy provision procedure overwhelming for common IoT users. To help users effectively and accurately create access control policies and implement them with shadow device instances, we also develop an LLM agent that can automatically generate accurate and comprehensive access control policies through interactive question-and-answer sessions with users. In addition to that, we also develop the tool to enable the LLM to automatically spawn the needed shadow device instances for fulfilling the created access control policies. This provides an end-to-end automatic solution that is accessible to common IoT users even with no technical backgrounds. ShadowConn is designed to integrate seamlessly with mainstream. The performance comparison between GPT-4-turbo-preview and GPT-3.5-turbo revealed significant differences. GPT-4-turbo-preview excelled in detecting missing information and accurately generating access control policies, achieving a perfect score and 100% accuracy. In contrast, GPT-3.5-turbo, while faster in initial processing, struggled with detecting missing details and produced incomplete policies, leading to a much lower accuracy of 35.67%. Additionally, in our assessment of synchronization between devices, the Aoetec and TP-Link switches recorded average response times of 343.7 ms and 256.7 ms, with maximum response times of 568.5 ms and 334 ms respectively. A total of 50 access control policies generated by the LLM (GPT-4-turbo-preview) were evaluated across 20 distinct test cases. In the acceptance scenarios, the allowed commands were synchronized, while in rejection cases, unauthorized actions were correctly blocked, highlighting the model's ability to manage dynamic access control requirements.

Our contributions can be summarized as below:

- We systematically analyze the issues of insufficient access control of existing smart home IoT delegation and sharing utilities.
- We propose ShadowConn as a novel framework that provides a universally applicable solution for fulfilling various cross-platform access control scenarios. The framework decouples the entangled associations among IoT devices, platforms, and users, making it possible to implement fine-grained access control policies.
- We designed an LLM-based intelligent agent to assist in the generation and deployment of access control policies. This agent allows effortless provision of the proposed framework through simple natural language descriptions of desired access control scenarios.
- We implement the prototype of ShadowConn on real-world testbed and conduct experiment with participants. The experiment results show that our solution achieves satisfying performance while incurring minimal overhead.

In Sect. 2, we provide an overview of IoT platforms and discuss relevant research on automation platforms, multi-user environments, and IoT delegation. In Sect. 3, we present the system model and enumerate the existing issues. Section 4 presents the design of ShadowConn. The implementation and evaluation are presented in Sect. 5. Section 6 covers related work, while Sect. 7 discusses the limitations and broader implications of our approach. Finally, Sect. 8 concludes the paper.

2 Background

2.1 Smart Home IoT Platforms

Smart home IoT platforms comprise collections of hardware, software, and services that enable IoT users to access and manage their IoT devices. An IoT platform typically includes IoT devices, IoT cloud services, and mobile applications, with optional hub or bridge devices. Among these components, cloud services play a pivotal role in the ecosystem. They allow users to remotely access their devices by forwarding messages between IoT devices and mobile applications, and they facilitate autonomous actions of IoT devices by hosting automation rules and routines.

Based on the features they provide, we can roughly classify IoT platforms into endpoint platforms and integration platforms. Endpoint platforms are hosted by IoT device manufacturers and are designed to connect and support their own devices. These platforms usually have cloud servers that communicate directly with the IoT devices. In contrast, integration platforms, such as SmartThings and Alexa, mainly integrate devices from other endpoint platforms. They focus on providing unified device access interfaces and automation services.

2.2 Multi-user Environment

As surveyed and discussed in existing literature [6,9,14,24], it is common to see IoT devices being shared among family members, roommates, and between tenants and landlords. Surveys have examined multi-user environments, such as the one conducted by Tabassum et al. [20], which surveyed 163 smart home IoT users and found that 47.8% of them have shared at least one IoT device with non-family members, driven by safety or convenience purposes.

In a shared smart home environment, IoT devices can be owned by different users and shared with each other. Since IoT devices can interact with the physical world (e.g., locks) and capture sensitive information (e.g., cameras), improper sharing of device access could lead to severe safety and privacy issues. In [9,19], authors reveal how IoT devices are being exploited for interpersonal abuse in shared smart home environments. Additionally, in [9,24], authors discuss the revealed and potential privacy issues caused by shared IoT devices. As a result, fine-grained access control that can be tailored according to different users' roles and needs becomes an urgent necessity for addressing these issues.

2.3 Device Delegation and Location Sharing

IoT device delegation and location sharing are two utilities that are provided by IoT platforms for users to share their devices across different IoT platforms and users.

Device Delegation refers to the process by which users delegate access to their devices from one IoT platform to another. This typically occurs between endpoint platforms and integration platforms, allowing users to access devices from different manufacturers through a unified interface and automate them using automation rules. For example, a TP-Link Kasa smart plug inherently operates on the Kasa platform as its endpoint platform, but users can choose to delegate its access to the Alexa platform. This enables users to control the plug along with other devices in the household via mobile applications, routines, or voice commands. In the process of cross-platform device delegation, "account linking" is usually employed to authorize the sharing of device access from the delegator platform to the delegatee platform. Although the account linking process uses the OAuth 2.0 protocol [15] for secure authorization, there is no common standard for the design of access control solutions. Most endpoint IoT platforms, such as Kasa and Philips Hue, assume that both the delegator and the delegatee platforms are owned by the same user and do not provide any access control options during the device delegation process. As a result, once the account linking is successfully performed, full access to all available devices in the user's account is granted to the delegatee platform.

Location Sharing allows the owner of IoT devices to share their devices with other users in a shared environment. Typically, IoT users register a location in their account on the IoT platform, which generally represents a household containing all the IoT devices the user owns. The owner of the IoT devices can share their devices with other users by inviting them to the same location. IoT platforms usually provide simple permission systems that allow the owner to assign different permission levels to the invited members for accessing the shared IoT devices. However, these permission systems mainly follow an "all-or-nothing" design and do not offer options for dynamic and fine-grained access control policies. For example, on the Alexa and SmartThings platforms, the owner can only specify whether the invited member has permission to add or delete devices. Once invited, the member gains unrestricted access to all devices in the owner's location.

3 System Model and Problem Statement

3.1 System Model

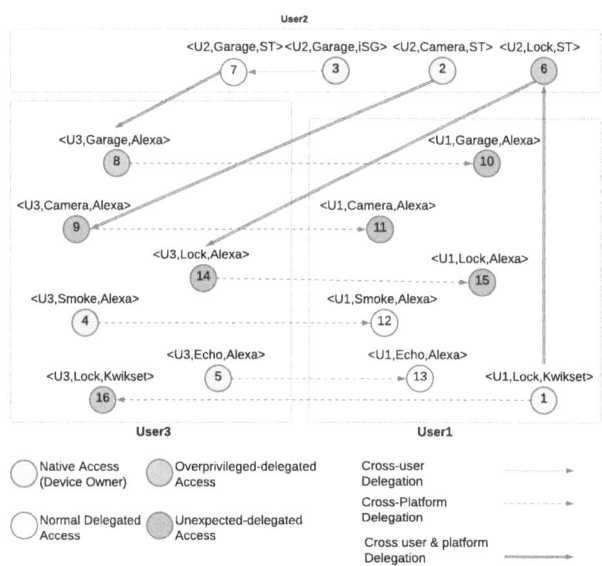

Fig. 2. The analysis of over-privilege issue of the example case in Fig. 1.

In this paper, we consider typical shared smart home environments as described in [6,14,24], which include dozens of IoT devices owned by multiple residents. Each resident may want to share some of their IoT devices with other residents while keeping others private. Additionally, according to a survey conducted by Chi et al. [3], it is common for users to use multiple IoT platforms simultaneously in a smart home to access their devices. Therefore, it is reasonable to assume that different users in a shared smart home may use different platforms to access IoT devices that are either owned by themselves or shared by other users.

As described in Sect. 2.2 and Sect. 2.3, users in a shared smart home can share their IoT devices with others through member invitations and device delegations across different platforms. To systematically study the access control issues involving these two device-sharing methods, we formalize the device access entry as triplets comprising three essential elements for IoT access control: the user who is granted access, the device being accessed, and the IoT platform used by the user to access the device. For example, the triplet $\langle U2, \text{Lock}, \text{ST} \rangle$ represents User2's access to the smart lock via the SmartThings platform. Using this formal representation, we can model the device-sharing relationship, as illustrated in the example scenario in Fig. 1, into the digraph shown in Fig. 2. In the graph, each access entry is represented as a node. Each device has one native node representing its owner and the manufacturer-hosted platform. For instance, the native access endpoint for a Kwikset lock is represented as $\langle U1, \text{Lock}, \text{Kwikset} \rangle$. All other nodes represent access that is either shared or delegated by other users or platforms. We use three types of edges to represent the device access sharing and delegation relationship: 1) Cross-platform delegation: this occurs when a device is delegated between two IoT platforms used by the same user; 2) Cross-user sharing: this represents a scenario where a device is shared between different users on the same platform; 3) Cross-platform and user sharing: this is a combination of the first two relationships, allowing a user to share their device access with another user on a different platform. In Fig. 2, we highlight three delegation channels in three different types of lines.

3.2 Problem Statement

Based on our formalization of device sharing and delegation, we reveal two critical issues of security and privacy in device sharing and delegation of existing smart home IoT platforms.

Insufficient Access Control. As introduced in Sect. 2, both account linking and location sharing utilities focus primarily on authentication but offer limited access control options. Device delegation is designed with the assumption that the devices are being delegated between platforms used by the same user, and therefore does not offer access control options. Some popular platforms, such as SmartThings, automatically delegate access to all devices in a location to other platforms like Google Home and Alexa. When this occurs between two different users, some devices might be unexpectedly shared. As illustrated in Fig. 2, User2 only wants to share access to his garage door with User3, but the account linking process mistakenly shares access to User2's camera and lock with User3. We highlight the nodes caused by this unintended device delegation in purple. More importantly, in a chained-device delegation scenario, when a command is forwarded by an intermediate platform, it inherits both the permissions and identity of that platform. Consequently, the subsequent platform receiving the forwarded command has no way to trace the original user's account or the originating platform that generated the command. As a result, enforcing access control policies becomes impossible. The node 14 as shown in Fig. 2 represents such a case. Since access to the lock is delegated twice, when User3 sends a command to the lock, the Kwikset platform cannot determine whether the command originated from User2 or User1. In both cases, the command will be sent via SmartThings with User2's identity, making it impossible to distinguish between the two scenarios.

Even for devices that are intentionally shared using device delegation and location sharing, there is still a lack of support for fine-grained access control policies. While some platforms attempt to differentiate device owners and other users by assigning different privileges (e.g., restricting non-owner users from adding or deleting devices), these static permissions fail to meet the dynamic and context-aware access control needs identified in previous studies [24]. We highlight the access nodes that are anticipated but potentially overprivileged in orange in Fig. 2.

Limited Information Flow Tracking. User studies indicate a strong preference for mechanisms that log IoT message flows for forensic and troubleshooting purposes, especially in multi-user environments [2,22]. Existing methods, however, can only monitor information flows within individual platforms and cannot trace the flow of IoT events and commands across different platforms. Chained device delegation further complicates this scenario by enabling IoT messages to be forwarded by multiple intermediate platforms, potentially owned by different users. As shown in Fig. 2, User3 has two access endpoints to the smart lock: one directly shared by User1 (node 16) and the other indirectly delegated by User2 (node 14). Consequently, when disputes arise among the three users regarding the lock being unexpectedly unlocked, there is no way to trace the origin of the commands.

3.3 Threat Model

ShadowConn aims to address issues of overprivileged and unexpected device access in shared IoT environments. We consider the potential adversaries to be legitimate but potentially malicious users within these shared environments, who may exploit their access to IoT devices to induce hazardous situations or compromise the privacy of other users. Attacks exploiting IoT system vulnerabilities, such as weak authentication and flawed API designs, are outside the scope of this paper. We also assume that mediator devices can be securely managed and operated within the local area network, which is a commonly assumed by some other recent literature about smart home IoT security [3,4,25]. We also discuss the alternative cloud-based deployment of ShadowConn in Sect. 7, which addresses the security concerns associated with relying on local mediator devices.

4 Design of ShadowConn

We propose the Shadow Connector (ShadowConn) as a practical and non-intrusive solution to the aforementioned security and privacy issues that are caused by entangled relationships among IoT devices, platforms, and users.

ShadowConn effectively decouples the complexities of IoT delegation and sharing channels by hosting shadow device instances for each delegated or shared access endpoint on a mediator service. Then, fine-grained access control policies can be enforced by the mediator device by regulating the state synchronization between real devices and their shadow instances. This design effectively separates a device subject to delegation into two distinct instances and ensures that access control can be tailored and adjusted with precision, addressing the inherent security risks without requiring modifications to the existing infrastructure of IoT platforms (Fig. 3).

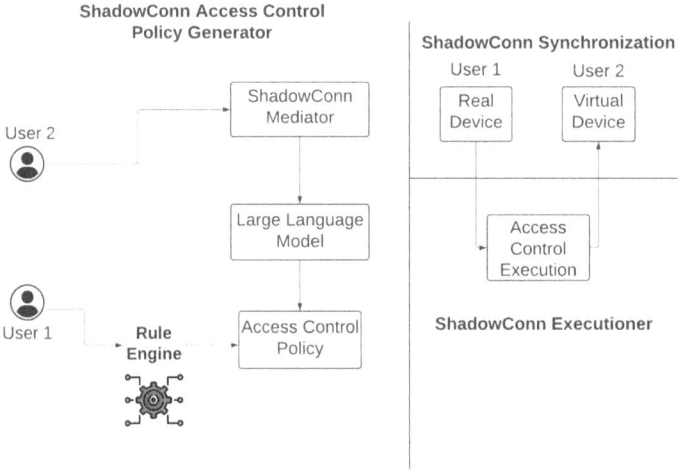

Fig. 3. The Architecture of ShadowConn.

4.1 Shadow Device Instances

The operation of ShadowConn is illustrated in Fig. 2, exemplified through the delegation of access to a smart lock between two users across distinct platforms. This process introduces a mediator platform (denoted as Pm), positioned intermediary to the delegator's and delegatee's platforms. Initially, both User1 (the delegator) and User2 (the delegatee) establish accounts on the mediator platform. Subsequently, User1 links their smart lock from its original platform ($P1$) to the mediator platform (Pm) through account linking, while User2's account remains initially devoid of devices.

It's critical to note that, although the account linking process might inadvertently extend delegation to all of User1's devices on platform $P1$ to Pm, this action does not lead to over-privilege or unintended delegation concerns. This is because the account on Pm is private, accessible solely by User1. Following this setup, User1 activates the mediator's smart application, which is designed to monitor the lock's events and commands. This application generates a pairing code, which User1 shares with User2.

Upon receiving the code, User2, assuming the role of delegatee, installs the same smart application and inputs the provided pairing code. This action triggers the creation of a shadow instance of the smart lock ($lock_s$) within the application, initiating synchronization of states between the actual lock ($lock_r$) and its shadow counterpart. This setup culminates in User2 being granted the ability to delegate access to the shadow smart lock instance to their preferred platform ($P2$), thus completing the secure delegation process.

The mediator smart application can be hosted on either third-party cloud services or locally on devices such as Raspberry Pis. It is equipped with a user interface that enables delegator users to define fine-grained access control policies. These policies can incorporate a wide variety of contextual information, including time, weather, and the states of other devices, and are enforced by a Rule Engine. This engine evaluates each state synchronization request between shadow and real devices against all loaded access control policies. All synchronization requests can also be logged for clear information flow tracking. We made ShadowConn publicly available on Google Drive[1].

4.2 Access Control Policy Generation

This research project is centered around the development of a novel framework for the synchronization of shadow devices and the formulation of dynamic access control policies using an LLM. The methodology is divided into two primary components: the creation and management of shadow devices and the development of an intelligent access control system.

Figure 4 models the interaction between two entities: User1 and User2. User1 operates a real smart device via SmartThings, while User2 controls a corresponding shadow device through Alexa. The process initiates when User2 sends a request to access the smart device under User1's account. Subsequently, this request prompts the generation of a JSON file by an LLM.

4.2.1 Access Control Policy Definition and Policy Formulation

A JSON file is generated based on user input, which serves as a command. Although the user input can be straightforward, it must contain sufficient details for the LLM to formulate the access control policy. This project requires collecting extensive information from users across various platforms to formulate the access control policy. The access control policy template, presented in the Appendix (Listing 1.2), defines a structured framework for managing interactions among users, platforms, and devices in a smart home environment". The access control policy offers a fine-grained, rule-based system composed of four key components:

---

[1] https://drive.google.com/drive/folders/1nZ15xjolYMcZCiBU3Kr37Zh-6Air9rAd?usp=sharing.

Algorithm 1. Device Synchronization Algorithm
───
1: User2 sends a request to access User1's smart device
2: LLM generates JSON data based on the request
3: Rule Engine evaluates the JSON to determine access approval
4: **if** request approved **then**
5:     ShadowConn synchronizes the real device with the shadow device
6: **else**
7:     Access is denied
8: **end if**
───

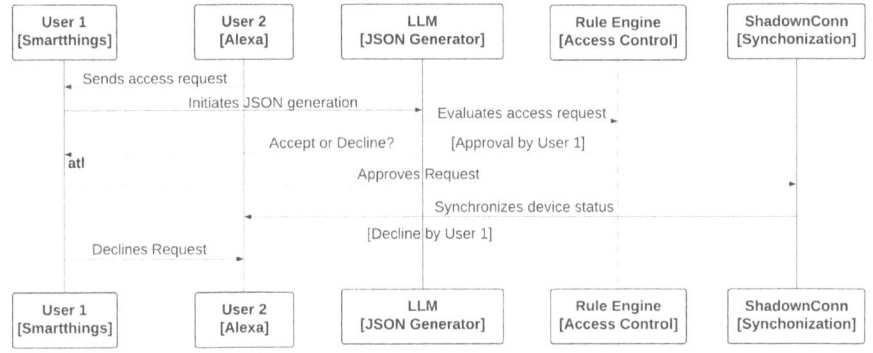

Fig. 4. Smart Device Access Control Process

1. **Trigger**: This section establishes the initial criteria for activating the policy. It identifies the source ('src') of the request, which includes 'user1' and the originating 'platform', and the destination ('dst'), specifying the target user and platform, along with the 'device_type' involved. These elements serve as the foundational context for the policy's decision-making process.

2. **Condition**: This component introduces contextual rules to refine the trigger. It includes 'location', indicating where the policy applies, and 'time', which sets temporal constraints. The time constraints can include specific ranges (such as "00:00" to "23:59") or distinguish between day and night, enhancing the flexibility of the policy.

3. **Decision**: This section outlines the process for handling the access request. It specifies the 'requester' who initiates the request, the 'approver' responsible for evaluating it, and the 'outcome', indicating whether the request is accepted or denied. This step is crucial in ensuring that only authorized interactions proceed.

4. **Actions**: This section defines the subsequent actions based on the decision outcome. It employs an "if-then-else" structure: if the decision is "accept", it specifies a set of commands to execute on specific devices, including device identifiers, component types, capabilities, and commands. If the decision is "deny", it provides an alternative set of commands to handle the rejection scenario, ensuring that both acceptance and denial paths are accounted for.

Within this framework, access control policies are articulated through a structured syntax designed to specify the interactions between users and devices within the context of the smart environment. The syntax is formulated as follows:

<user1> : <user2> : <devices> : <attributes> : <context> : <decision> : <actions>

In this syntax, <user1> and <user2> denote the roles of delegator and delegatee, respectively, capturing their usernames, platforms, and roles within the system. The <devices> field enumerates the devices that the delegator intends to share. <attributes> and <context> provide detailed descriptions of the shared devices' attributes and the environmental context (e.g., location, time), which are critical for decision-making processes. The <decision> parameter dictates the outcome of the access request, either "accept" or "reject". Lastly, <actions> delineates the procedures for request execution or denial, thereby defining the request/response mechanism for system synchronization.

4.2.2 LLM-Assisted Access Control Policy Generation

Large language models, such as ChatGPT, have demonstrated remarkable performance in responding intelligently to various questions. Exploring their application in assisting users with specifying accurate and comprehensive access control policies is a worthwhile endeavor. In ShadowConn, the LLM assists users in managing shadow devices. In a typical smart home environment, there may be dozens of smart devices. With ShadowConn, the number of devices increases further due to the addition of shadow devices across different smart home platforms. At this point, the LLM helps users with the access control policy and the synchronization of devices in the environment. For example, in Fig. 1, the cross-platform environment initially contains five different devices. After implementing ShadowConn, the total number of devices, including both real and shadow devices, can reach up to seventeen. However, the specification of access control policies heavily relies on the contextual information of the user, IoT devices, and platforms. Moreover, the generated access control policy must comply with the predefined format.

To overcome these challenges, we have designed a multi-step conversational prompt template using a few-shot learning architecture. This approach enables the language model to interactively collect all necessary contextual information from users through multi-round question-and-answer sessions and generate well-formatted access control policies (Algorithm 1).

To create an access control policy based on user input, the LLM first processes the user's request. Often, users may not provide sufficient information for the model to generate the access control policy. To address this, the LLM engages in a conversation with the user to gather the missing details, utilizing a technique known as few-shot learning. Few-shot learning is a machine learning approach that enables a model to perform well on a task with only a small number of training examples. Unlike traditional methods that require large datasets for generalization, few-shot learning allows the model to make accurate predictions or decisions with minimal data. By leveraging templates and examples, the LLM with few-shot learning can identify missing information and prompt the user to supply the necessary details, thereby facilitating the accurate creation of access control policies.

The deployment of the LLM facilitates the creation of access control policies in JSON format, thereby simplifying the processes associated with command issuance and device management across platforms such as Alexa and SmartThings. The newest model from OpenAI is employed as the LLM, with the subsequent prompt serving as an example in the Appendix (Listing 1.3) to illustrate the input to the model.

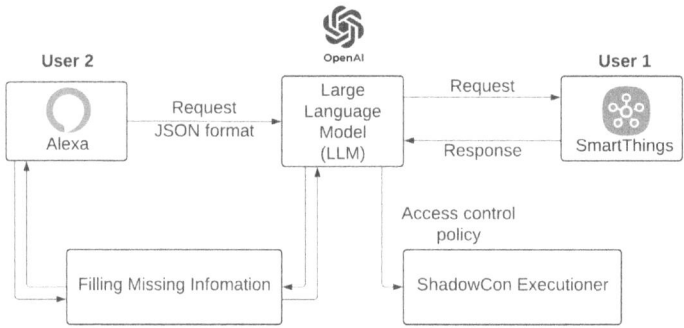

Fig. 5. Large Language Model Flow

Upon receiving the prompt, users can issue commands to the LLM, such as "U2 wants to use U1's smart light". Subsequently, the LLM generates a JSON file that serves as the access control policy. If the prompt lacks certain required details to finalize the access control policy, the LLM will prompt the user to provide the missing information as in Fig. 5. This approach ensures that users do not need technical skills to use or create access control policies, making the process accessible to a broader range of users and enhancing the overall usability of smart home systems.

4.3 Cross-Platform Access Control Enforcement

The initial phase involves the deployment of a server using Docker Compose to manage and synchronize shadow devices. Device information, including status, is fetched via the SmartThings API and persisted in a MongoDB database. For each shadow device, a dedicated Docker container is instantiated, providing endpoints for status retrieval and updates. This approach addresses the challenges in real-time device management and synchronization within a smart environment.

In the development of a shadow device, we establish a Cloud Connected Device leveraging Schema and webhooks. This virtual representation termed a shadow device, is integrated within the SmartThings application on the user's mobile device, functioning equivalently to a physical device with attributes such as device ID and status. Shadow devices refer to shadow device instances that mimic the behavior and state of physical IoT devices, enabling secure and manageable device delegation in smart home environments. Furthermore, the shadow device is responsible for maintaining the synchronization with the delegatee's device.

The policy framework adopts a syntax that encapsulates the relationships between users (delegators and delegatees), devices, and the context of access requests. This allows for the specification of permissions based on user roles, device attributes, and environmental conditions. For example, a delegatee's request to control a device is evaluated against the delegator's policies, demonstrating the system's capability to handle dynamic permissions. The Rule Engine evaluates these commands to determine whether User1 approves or declines the access request. ShadowConn retrieves the data required to synchronize the status of User1's real device and User2's shadow device. In one scenario, User2 requests User1's permission to access function A in Device1, and User1 accepts the request. If User2 then requests to execute function B on Device1, the mediator will reject the request without consulting User1. The Rule Engine then evaluates this JSON to determine whether User1 approves or declines the access request. ShadowConn retrieves the JSON data to synchronize the status between User1's real device and User2's shadow device.

From the access control policy, specific commands are derived to synchronize the device status, facilitating the user's intention to grant access control. Within this policy, the value of `command_section` serves as the trigger for the delegatee to synchronize the shadow device with the actual device.

To prevent conflicting policies deployed by different users, ShadowConn can review the content (`conditions` and `commands`) of the existing policies. This ensures that users cannot create policies that are already present in the list.

5 Implementation and Evaluation

5.1 Implementation

In this section, we describe the details of implementing the prototype of ShadowConn and evaluate its performance in terms of access control policy generation and synchronization.

We implement the Prototype of ShadowConn using a Raspberry Pi 4B and the SmartThings IoT platforms, which serves as the admin device and the intermediary platform, respectively. On the Raspberry Pi, we implement an IoT device server based on the SmartThings cloud-connected device SDK [18], which allows us to generate arbitrary shadow devices and integrate them into SmartThings. The Raspberry Pi is deployed in the user's home local area network for hosting and managing shadow device instances and synchronizing their states to different accounts on SmartThings. Unlike intermediary box-based approaches (e.g., PFirewall [4]), our strategy utilizes third-device integration interfaces on the intermediate platform instead of building them from scratch (Fig. 6).

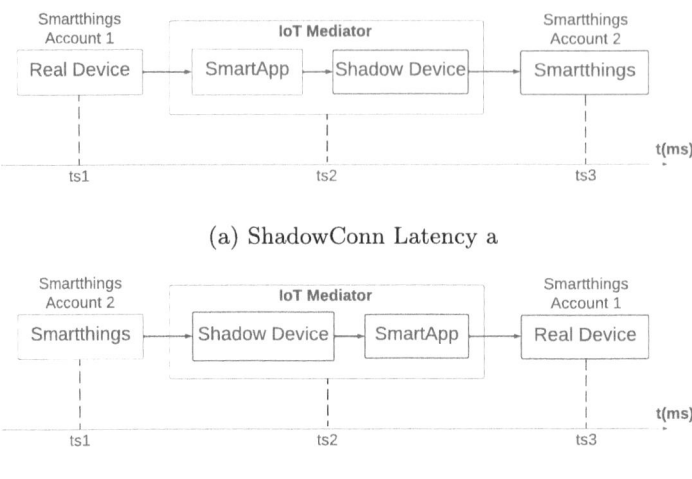

(a) ShadowConn Latency a

(b) ShadowConn Latency b

Fig. 6. ShadowConn Latency

The intermediary platform, represented by SmartThings in this particular instance, has a crucial function in the architecture of ShadowConn. It serves as an intermediary between the tangible gadgets and their digital counterparts. This platform must facilitate both inbound delegation (from other platforms to ShadowConn) and outbound delegation (from ShadowConn to other platforms). Within this approach, SmartThings functions not only as a passive conduit, but also as an active participant that oversees the allocation of device access. This ensures smooth integration and communication between different platforms.

5.2 Evaluation

To evaluate ShadowConn, we created a dataset of 50 commands generated by non-expert users, each designed to test the system across various smart home scenarios, including different platforms and threat models. The devices used, such as smart switches, were tested in an ideal environment. Each command serves as user input, which the LLM processes into distinct access control policies. Following this conversion, synchronization ensures that the shadow devices align with their real counterparts. An LLM-based evaluator then assesses whether the generated access control policy includes all expected details. We also generated random requests to test the 50 access control policies created by the LLM. By storing the access control policies in a CSV file and converting the requests to the same template, we can check if the requests match the policy.

5.2.1 Evaluation Metrics

The evaluation score is determined using a 6-point system, awarding one point for each of the following elements: source (src), destination (dst), device type (device_type), location, time, and decision. ShadowConn's efficiency was initially demonstrated through a use case illustrating its application in various smart home operations.

Further evaluation involved four different usage scenarios to verify the effectiveness of access control enforcement. ShadowConn formulates a new access control policy and establishes a shadow version of devices that remains synchronized with the actual devices. The synchronization evaluation also considers the latency between devices and platforms, which is particularly critical in a smart home setup. This setup includes a living room, bedroom, kitchen, and garage, operating under varying conditions and utilizing different smart devices throughout the day.

5.2.2 Evaluation of Access Control Policy Generator

We investigated four scenarios for the policy generation and negotiation processes as detailed below:

Scenario 1: Granting access control from User1's sharable devices. User2 (Alexa) requests control of specific devices in the living room, managed by User1 (SmartThings). The LLM uses a predefined list of devices shared by User1. Only three devices are shared: the smart switch, floor lamp, and TV. User2 can turn these devices on and off, but cannot adjust settings like thermostat temperature. The access control policy is auto-generated based on User1's permissions, testing selective sharing and command restrictions in a multi-user environment.

Scenario 2: Access control based on User1's decision. User2 requests control of devices in the bedroom through a voice assistant, with User1 managing them via a smart home hub. The LLM simulates interactions, and User2 is granted permission to operate the smart curtains, lamp, blinds, TV, and speaker. User2 can open curtains, turn on devices, and adjust blinds during designated times. This scenario tests User1's ability to set specific control over shared devices.

Scenario 3: User1-defined access based on triggers. User2 requests control of kitchen devices. The LLM processes the request and allows User2 to operate specific appliances like the smart speaker, kettle, coffee maker, and oven during designated times. User1 restricts access to devices like the microwave, thermostat, and stove. The policy enables granular control over which devices User2 can operate, based on time-based triggers.

Scenario 4: Secondary user management of access control. User1 delegates control over bathroom devices (e.g., mirror, lights) to User2, who manages access for User 3 (Kwikset). User3 is allowed to use the smart mirror at 7:30 AM but is denied control over the dehumidifier, heating system, and lights during other specified times. This scenario demonstrates the ShadowConn system's ability to handle fine-grained access control with multiple users.

In this section, we evaluate the performance of two models (GPT-4-turbo-preview and GPT-3.5-turbo) in generating access control policies for smart home devices. The temperature parameter was set for all LLMs at 0.5, with a maximum response length of 3000 tokens. The evaluation focuses on two main aspects: the LLM's ability to detect missing information and its overall performance in creating accurate policies. To assess the models, we designed four test scenarios that represent typical smart home interactions. These scenarios involved various users, devices, and actions, as in Table 1.

Table 1. Test Scenarios for LLM-Based Access Control Generation

Scenario	Users	Devices	Actions	LLM Role
1	User1, User2	Smart Light, TV	Turn On/Off	Grant access from pre-approved device list
2	User1, User2	Curtains, Lamp, TV	Open, Turn On/Off	Generate policy based on User1's decision
3	User1, User2	Kettle, Coffee Maker	Turn On	Time-limited control based on User1's preferences
4	User1, User2, User3	Smart Mirror, Lights	Adjust Settings, Turn On/Off	Multi-user delegation controlled by User2

Both LLMs were tested using identical prompts to generate access control policies. These prompts occasionally lacked critical information, such as device type, location, time, and decision, requiring the LLMs to gather additional data from the user.

GPT-4-turbo-preview: This model excelled in detecting missing information and engaged in a conversation with the user to gather it. After processing user commands, GPT-4-turbo-preview effectively generated comprehensive access control policies with all necessary details. The model achieved a score of **6/6**, demonstrating its ability to define the device type, location, time, and decision accurately.

GPT-3.5-turbo: While this model completed the initial steps more quickly, it struggled to detect missing information effectively. Before gathering additional data, GPT-3.5-turbo performed relatively well, scoring **5/6**. However, after attempting to collect missing details, it failed to define key elements such as device type and location, resulting in a final score of **2.14**. Its accuracy dropped significantly, achieving only **35.67%**, compared to GPT-4-turbo-preview's **100%** accuracy. Table 2 summarizes the models' performance in terms of their ability to detect missing information and generate accurate policies.

5.2.3 Evaluation of ShadowConn Execution

A comprehensive examination was undertaken to assess the synchronization mechanism between real devices and shadow devices, across various SmartThings accounts. The research employed Aoetec and TP-Link switches to analyze time intervals t1-t2, t2-t3, and t1-t3 in order to determine the synchronization events' mean, maximum, and standard deviation. In terms of transitions from RD to SD, the Aoetec Switch demonstrated mean synchronization times of 343.7 ms, 224.8 ms, and 568.5 ms during the corresponding time intervals, with the t1-t3 interval exhibiting the longest time of 735 ms. On the contrary, the average performance of the TP-Link Switch was marginally more consistent than that of

Table 2. LLM Performance Comparison in Access Control Policy Generation

LLM Model	Score Before Missing Information	Score After Missing Information	Accuracy (%)
GPT-4-turbo-preview	6/6	6.0	100%
GPT-3.5-turbo	5/6	2.14	35.67%

the Aoetec Switch. In terms of transitions from RD to SD, the TP-Link Switch demonstrated mean synchronization times of 256.7 ms during the corresponding time interval. The higher standard deviation values suggest that the synchronization times for the TP-Link devices are relatively more variable, which may be due to differences in network or device processing capabilities (Fig. 7).

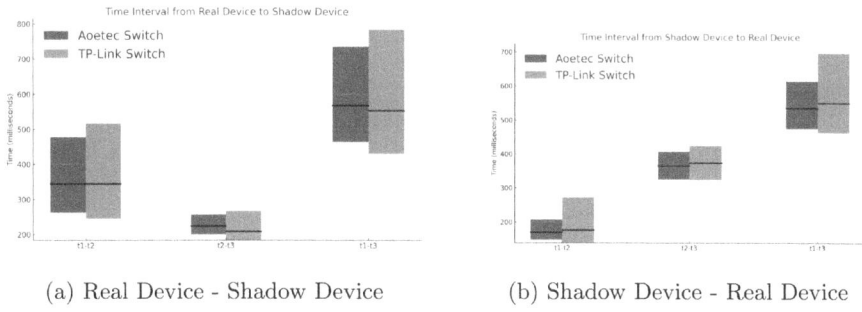

(a) Real Device - Shadow Device   (b) Shadow Device - Real Device

Fig. 7. Charts of ShadowConn Latency

In the initial t1-t2 interval, the synchronization times improved substantially when the reverse transition from SD to RD was considered; the Aoetec and TP-Link Switches recorded average times of 169.5 ms and 176.1 ms, respectively. The decrease in synchronization time highlights the effectiveness of the mechanism for updating virtual devices to physical ones. The observed maximal times and standard deviations for these transitions suggest that the synchronization process is relatively consistent and foreseeable, although it is possible that device-specific factors or network conditions introduce some degree of inherent variability. The results of this study indicate that although the synchronization mechanism between real devices and shadow devices, is successful, discernible variations in performance metrics may be due to the devices' inherent properties or the synchronization technique utilized.

A total of 50 access control policies generated by the LLM (GPT-4-turbo-preview) were tested using 20 distinct test cases, consisting of 10 acceptance and 10 rejection scenarios. The access control policies effectively detected the decisions in each case and synchronized the commands permitted by User1. In the acceptance scenarios, the policies accurately enforced User1's permissions, ensuring the correct execution of allowed commands, while in the rejection scenarios, unauthorized actions were successfully blocked.

6 Related Work

The advancement of smart home systems has created unique opportunities and challenges in terms of access control, security, and privacy. This section reviews the relevant literature, particularly focusing on smart home security, cross-platform delegation, and the integration of language models for access control. Zeng and Roesner [24] address the critical issue of multi-user security and privacy in smart homes, exploring conflicts between residents due to power imbalances between tech-savvy users and others. They highlight concerns about transparency, restricted access, and privacy violations through behavioral logs and monitoring applications. To mitigate these issues, they propose design principles for smart home platforms, emphasizing access control flexibility and transparency. Their prototype mobile app features location-based access controls and activity notifications. In a month-long in-home study, they found that positive household dynamics could sometimes reduce the need for technical solutions, though complex configurations could hinder usability. Rivkin et al. [12] introduce the SAGE system, which uses a sequential decision-making framework managed by an autonomous language model agent to provide fine-grained access control and dynamically synchronize states between devices and their virtual counterparts. SAGE's tools facilitate flexible device interaction, continuous monitoring of device states, and natural language comprehension, demonstrating significant improvements in flexibility and autonomy. Similarly, Sasha [8] employs LLMs to understand and respond to natural language commands, enhancing smart home functionality by interpreting user goals and generating action plans. Sasha effectively addresses under-specified commands by producing coherent action plans and managing automation routines. Kratos [14] offers a multi-user, multi-device-aware access control system with a user interaction module, back-end server, and policy manager to resolve conflicting demands and generate final policies for enforcement. Their experiments show a 100% success rate in resolving conflicting demands and preventing various threats in a multi-user smart home environment. Fernandes et al. [5] explore security risks in trigger-action platforms like IFTTT and introduce Decentralized Action Integrity to prevent misuse of compromised OAuth tokens. They present the Decentralized Trigger-Action Platform (DTAP), which decentralizes token management and enhances security with minimal latency and overhead. Geeng and Roesner [6] investigate the dynamics in multi-user smart homes, highlighting complexities during different phases of smart device use. They provide design recommendations for long-term smart home use and better support for non-expert users. Nandi and Ernst [10] present TrigGen, a tool that analyzes user-written rules in smart home systems to generate necessary event-based triggers, addressing the issue of incorrect or insufficient triggers. He [7] explores access control challenges in home IoT settings, emphasizing capability-centric models and suggesting default policies based on user relationships and contextual factors.

7 Limitations and Discussions

Currently, ShadowConn is implemented on a local device residing within the user's home area network. However, this approach introduces the risk that malicious users could gain physical access to the device, potentially manipulating the states of virtual devices or altering deployed access control policies. To address this limitation, we envision an alternative deployment option where ShadowConn is hosted as a third-party cloud service that integrates with various smart home IoT platforms. This service would be managed by independent service operators, trusted to securely maintain virtual device instances and access control policies.

In this model, delegating devices across different users and platforms requires both the delegator and the delegatee to sign up for the cloud service and establish access control policies. The device owner then grants access to their devices through automation APIs (e.g., Smart Apps APIs for SmartThings [17]). The cloud service would create dedicated virtual device instances, which would be integrated into the delegatee's account via cloud-to-cloud integration channels [16]. Although having the same architecture as the local-based deployment, this alternative design mitigates the risks associated with local deployment and reduces hardware costs, as the cloud service would feature an independent authentication system to prevent unauthorized access. We leave this exploration as future work.

8 Conclusion

We have introduced Shadow Connector (ShadowConn), a novel framework simplifying cross-platform access control by decoupling devices, users, and platforms. Using shadow devices, it streamlines delegation and enhances security. With LLM integration, access control policies are generated from natural language input. Our implementation on commercial smart home platforms demonstrates that ShadowConn achieves significant control enhancements with minimal overhead, maintaining system performance while reducing usability downgrades. ShadowConn represents a significant advancement in cross-platform IoT delegation, paving the way for more efficient and secure smart home environments.

A Access Control Policy Template

This appendix provides a comprehensive outline of the access control policy template and includes an illustrative example of the access control policy.

```
actions = json_string['actions']
then_commands = actions[0]['if']['then']
commands = then_commands[0]['command']['commands']

if_then_section = actions[0]['if']['then']
```

```
# Get to the 'command' key inside the first 'then' item
command_section = if_then_section[0]['command']

# Access 'devicesId2' inside 'command'
devices_id2 = command_section['devicesId2']

url = "https://api.smartthings.com/v1/devices/" + devices_id2 + "/
    commands"

payload = "{\n  \"commands\": " + str(commands) + "}"
headers = {
    'Content-Type': 'text/plain',
    'Authorization': '[SmartThings Bearer Token]]'
}

response = requests.request("POST", url, headers=headers, data=payload)
```

Listing 1.1. Synchronization using Smartthings API

```
{
  "trigger": {
    "src": {
      "user1": "",
      "platform": ""
    },
    "dst": {
      "user": "",
      "platform": ""
    },
    "device_type": ""
  },
  "condition": {
    "location": "",
    "time": ["00:00", ["day", "night"]]
  },
  "decision": {
    "requester": "",
    "approver": "",
    "outcome": ""
  },
  "actions": [
    {
      "if": {
        "decision": "accept",
        "equals": {
          "left": {
            "device": {
              "deviceId1": "",
              "component": "",
              "capability": "",
              "attribute": ""
            }
```

```
            },
            "right": {
              "string": "on"
            }
          },
          "then": [
            {
              "command": {
                "devicesId2": "",
                "commands": [
                  {
                    "component": "",
                    "capability": "",
                    "command": ""
                  }
                ]
              }
            }
          ],
          "else": [
            {
              "command": {
                "devicesId2": "",
                "commands": [
                  {
                    "component": "",
                    "capability": "",
                    "command": ""
                  }
                ]
              }
            }
          ]
        }
      }
    ]
}
```

Listing 1.2. Access Control Policy Template

```
{
  "trigger": {
    "src": {
      "user1": "U1",
      "platform": "Smartthings"
    },
    "dst": {
      "user": "U2",
      "platform": "Alexa"
    },
    "device_type": "smart_light"
  },
```

```
"condition": {
  "location": ["living room"],
  "time": ["08:00", ["day"]]
},
"decision": {
  "requester": "U1",
  "approver": "U2",
  "outcome": "accept"
},
"actions": [
  {
    "if": {
      "decision": "accept",
      "equals": {
        "left": {
          "device": {
            "deviceId1": "Light_1",
            "component": "main",
            "capability": "switch",
            "attribute": "switch"
          }
        },
        "right": {
          "string": "on"
        }
      },
      "then": [
        {
          "command": {
            "devicesId2": "Light_2",
            "commands": [
              {
                "component": "main",
                "capability": "switch",
                "command": "on"
              }
            ]
          }
        }
      ],
      "else": [
        {
          "command": {
            "devicesId2": "Light123",
            "commands": [
              {
                "component": "main",
                "capability": "switch",
                "command": "off"
              }
            ]
          }
```

```
65        }
66      ]
67    }
68  }
69 ]
70 }
```

Listing 1.3. Example of Access Control Policy

B Access Control Scenarios and Policies

We present a detailed list of access control scenarios and policies used during the implementation and evaluation of ShadowConn in Table 3.

Table 3. Access Control Scenarios and Policies in a Multi-User, Cross-Platform Smart Home Environment

ID	User 1	User 2	User 3	Owned Devices	Shared Devices/Actions	Access Control Policy
1	Smart Things	Alexa	N/A	Smart switch, smart floor lamp, air conditioner, smart wall plug, smart vacuum, air purifier, smart thermostat, smart TV, smart sound system	Smart switch, smart floor lamp, smart TV	Granting access control from the list of sharable devices of User 1
2	Smart Things	Alexa	N/A	Smart curtains, ceiling fan, humidifier, smart lamp, smart blinds, alarm clock, smart TV, speaker	Smart curtains, smart lamp, smart blinds, smart TV, speaker	Granting access control from User 1's decision
3	Smart Things	Alexa	N/A	Smart curtains, ceiling fan, humidifier, smart lamp, smart blinds, alarm clock, smart TV, speaker	Opening the smart curtains, turning on the smart lamp and TV, adjusting the smart blinds	User 1 decides what User 2 can do with the shared devices based on the trigger
4	Smart Things	Alexa	Kwikset	Light, Smart workbench light, Smart garage door opener, Smart security camera, Smart security alarm, Smart mirror, Smart dehumidifier, Smart heating system, Smart lights, Smart irrigation system, Smart pool cover, Smart lamp, Smart TV (study room), Computer light, Smart lock (front door), Smart lock (garage door), Smart light strips (hallway)	Light, Smart workbench light, Smart garage door opener, Smart mirror, Smart irrigation system, Smart lamp, Smart lock (front door), Smart light strips (hallway)	User 3 grants access control of User 1's devices through User 2

C LLM Prompt Construction and Response

We provide an example of ShadowConn's Prompt Construction and Response.

[**Task Info**]
You are an AI that helps the user ask for access control and synchronize the smart devices.
[**Requirements**]
- Device type (input: light common name)
- Location (input: light uuid)
- deviceId1 (input: light uuid)
- deviceId2
- Access control decision
- Time
[**Format Info**]
Use the access control policy template (Listing 1.1)
[a] **User input 1**: Turn on the light
[**History Info**]
LLM response:
To generate an access control policy as a JSON file based on the user request described, we'll follow a step-by-step approach. Please provide the missing information as we go through each step. Let's start with the information provided and what's missing:
1. Trigger Section:
- src (source): We know the source user is 'U1' and the platform is 'Smartthings'.
- dst (destination): The destination user is 'U2' and the platform is 'Alexa'.
- device_type: Since U2 wants to use U1's smart light, the device type is "smart light".
2. Condition Section:
- You mentioned that the user needs to specify the time. Please provide the "time" during which this policy should be active. Also, specify the "location" if applicable.
3. Decision Section:
- requester: Since U2 is the one who wants to use the device, U2 is the requester.
- approver: U1, being the owner of the device, would be the approver.
- The outcome is set to be either "accept" or "decline", as per your instructions.
4. Actions Section:
- This section is predefined based on the decision outcome. If the decision is "accept", the smart light (device) will turn on; if "decline", it will remain off. The device IDs need to be specified.
Given these points, we need you to fill in the following information to complete the JSON file:
- Time during which this policy should be active (e.g., "18:00-22:00"):
- Location (if specific location conditions apply, otherwise leave as is):
- Device ID for the smart light referred to as 'deviceId1' and 'devicesId2':
Once you provide these details, I can generate the complete JSON access control policy for you.
[b] **User input 2**:
- The device type is "smart_light".
- The location is "living room".
- The deviceId1 is "Light_1". The deviceId2 is "Light_2"
- Decision will be taken from the delegatee.

References

1. Alexa, A.: Amazon Alexa official site (2024). https://alexa.amazon.com/. Accessed 26 Apr 2024
2. Celik, Z.B., et al.: Sensitive information tracking in commodity IoT. In: 27th {USENIX} Security Symposium ({USENIX} Security 2018), pp. 1687–1704 (2018)
3. Chi, H., Zeng, Q., Du, X.: Detecting and handling {IoT} interaction threats in {Multi-Platform}{Multi-Control-Channel} smart homes. In: 32nd USENIX Security Symposium (USENIX Security 2023), pp. 1559–1576 (2023)
4. Chi, H., Zeng, Q., Du, X., Luo, L.: Pfirewall: semantics-aware customizable data flow control for home automation systems (2019)
5. Fernandes, E., Rahmati, A., Jung, J., Prakash, A.: Decentralized action integrity for trigger-action IoT platforms. In: Proceedings 2018 Network and Distributed System Security Symposium (2018)
6. Geeng, C., Roesner, F.: Who's in control? Interactions in multi-user smart homes. In: Proceedings of the 2019 CHI Conference on Human Factors in Computing Systems, pp. 1–13 (2019)
7. He, W., et al.: Rethinking access control and authentication for the home internet of things (*IoT*). In: 27th USENIX Security Symposium (USENIX Security 2018), pp. 255–272 (2018)
8. King, E., Yu, H., Lee, S., Julien, C.: Sasha: creative goal-oriented reasoning in smart homes with large language models. Proc. ACM Interact. Mob. Wearable Ubiquit. Technol. **8**(1), 1–38 (2024)
9. Mare, S., Roesner, F., Kohno, T.: Smart devices in airbnbs: considering privacy and security for both guests and hosts. In: Proceedings on Privacy Enhancing Technologies (2020)
10. Nandi, C., Ernst, M.D.: Automatic trigger generation for rule-based smart homes. In: Proceedings of the 2016 ACM Workshop on Programming Languages and Analysis for Security, pp. 97–102 (2016)
11. Philips Hue: Philips hue smart lighting (2024). https://www.philips-hue.com/en-us. Accessed 26 Apr 2024
12. Rivkin, D., et al.: Sage: smart home agent with grounded execution. arXiv preprint arXiv:2311.00772 (2023)
13. Samsung: Samsung smartthings (2024). https://www.samsung.com/us/smartthings/. Accessed 26 Apr 2024
14. Sikder, A.K., et al.: Kratos: multi-user multi-device-aware access control system for the smart home. In: Proceedings of the 13th ACM Conference on Security and Privacy in Wireless and Mobile Networks, pp. 1–12 (2020)
15. SmartThings: Oauth integrations (2024). https://developer.smartthings.com/docs/connected-services/oauth-integrations. Accessed 24 May 2024
16. SmartThings Developer Documentation: Bring your cloud connected devices to smartthings (2024). https://developer.smartthings.com/docs/devices/cloud-connected/get-started. Accessed 06 Sept 2024
17. SmartThings Developer Documentation: Create a smartapp (2024). https://developer.smartthings.com/docs/connected-services/create-a-smartapp. Accessed 06 Sept 2024
18. SmartThings Developer Documentation: Smartthings core SDK documentation (2024). https://developer.smartthings.com/docs/sdks/core. Accessed 11 July 2024
19. Stephenson, S., Almansoori, M., Emami-Naeini, P., Chatterjee, R.: " It's the equivalent of feeling like you're in {Jail"}: lessons from firsthand and secondhand

accounts of {IoT-Enabled} intimate partner abuse. In: 32nd USENIX Security Symposium (USENIX Security 2023), pp. 105–122 (2023)
20. Tabassum, M., Kropczynski, J., Wisniewski, P., Lipford, H.R.: Smart home beyond the home: a case for community-based access control. In: Proceedings of the 2020 CHI Conference on Human Factors in Computing Systems, pp. 1–12 (2020)
21. TP-Link: Tp-link official website (2024). https://www.tp-link.com/us/. Accessed 26 Apr 2024
22. Wang, Q., Hassan, W.U., Bates, A., Gunter, C.: Fear and logging in the Internet of Things. In: Network and Distributed Systems Symposium (2018)
23. Yuan, B., Jia, Y., Xing, L., Zhao, D., Wang, X., Zhang, Y.: Shattered chain of trust: understanding security risks in {Cross-Cloud}{IoT} access delegation. In: 29th USENIX Security Symposium (USENIX Security 2020), pp. 1183–1200 (2020)
24. Zeng, E., Roesner, F.: Understanding and improving security and privacy in multi-user smart homes: a design exploration and in-home user study. In: 28th {USENIX} Security Symposium ({USENIX} Security 2019), pp. 159–176 (2019)
25. Zhang, W., Meng, Y., Liu, Y., Zhang, X., Zhang, Y., Zhu, H.: Homonit: monitoring smart home apps from encrypted traffic. In: Proceedings of the 2018 ACM SIGSAC Conference on Computer and Communications Security, pp. 1074–1088 (2018)

Hardware and Firmware Security

Hardware-Assisted Runtime In-vehicle ECU Firmware Self-attestation and Self-repair

Josh Dafoe, Job Siy, Niusen Chen, and Bo Chen[✉]

Department of Computer Science, Michigan Technological University,
Houghton, MI, USA
bchen@mtu.edu

Abstract. Modern vehicles are largely controlled by many embedded computers, known as Electronic Control Units (ECUs). The increased use of ECUs has brought many in-vehicle security concerns. Specifically, injection of malware into ECUs poses a significant risk to vehicle operation. Indeed, many ECU malware injection attacks have been performed, and much work has been introduced towards mitigating these vulnerabilities. A main defense is for ECUs to perform a self-attestation over their firmware state. However, most current self-attestation solutions do not enable runtime checking due to their high computational cost. Additionally, existing solutions mostly do not incorporate any ECU self-repairing in coordination with the attestation mechanisms.

In this work, we have designed FSAVER, a highly efficient self-attestation and self-repair framework for in-vehicle ECUs. For the self-attestation, we adapt highly efficient spot-checking techniques, so that the firmware can be checked periodically at runtime. To perform these attestations, we rely on the TEE already equipped within each ECU. For self-repair, we take advantage of the isolated flash memory controller (FMC) in the storage device. Specifically, we coordinate it with the update mechanism and self-attestations to guarantee that the latest benign firmware version can always be restored. To realize this while malware is running, a special mechanism has been carefully developed to notify the FMC of the malicious presence.

Keywords: Autonomous Vehicles · Firmware Attestation · Self Repair · Trusted Execution Environment · Flash Memory Controller · Flash Translation Layer

1 Introduction

Modern vehicles increasingly contain many specialized computers, mostly known as Electronic Control Units (ECUs). These embedded real-time ECUs perform specialized tasks, many of which are safety critical vehicle operations such as controlling the brakes, throttle, or steering. Vehicles today can contain up to

150 ECUs [17]. With the advent of autonomous vehicles, these ECUs are gaining more responsibilities and so require more sophisticated software.

To facilitate the transmission of control and communication signals between ECUs, the original Controller Area Network (CAN) protocol was introduced in 1986. Additionally, to provide users with diagnostic information and the ability to connect to the CAN bus, an OBD-II port has been introduced. This has been mandated on all new vehicles in the US since 2009 [51]. The control and communication signals sent over the CAN are essential information, used by many ECUs to initiate their safety-critical behavior. For example, the airbag control module receives information from deceleration sensors, yaw rate sensors, and seat occupancy sensors, all of which need to communicate properly in order to activate the airbags [49]. Upon airbag activation, the airbag control module communicates with the engine control module to cut off fuel supply which prevents fire [47]. If *any* one of the mentioned ECUs are compromised with malware, then the *entire* airbag deployment system may fail during a collision, or even be initiated during normal vehicle operation.

To introduce such malware to an ECU, an attacker would need to perform ECU reprogramming. Usually, reprogramming of ECUs is performed over the CAN bus. To do this, a diagnostic tool connected to the bus will initiate a reprogramming mode. During this mode, new software is sent to the target ECU. A widely deployed protocol which enables such a reprogramming mode is Unified Diagnostic Services (UDS). Before initiating the reprogramming mode, (and so allowing arbitrary manipulation of the ECU code) UDS performs an authentication of the diagnostic tool. This allows anyone with a valid diagnostic tool to enter programming mode [29]. This is the only authentication that UDS performs. Consequently, anyone who can pass this authentication can typically write arbitrary code to any in-vehicle ECU. Unfortunately, this authentication can usually be passed even without access to a valid diagnostic tool, enabling malware to be written directly to a target ECU [19,29,31]. Therefore, any compromised device connected to the CAN bus is able to reprogram any target ECU. These compromised devices include mobile phone apps, OBD-II dongles, or even other ECUs [7,15,28,53].

To mitigate such attacks, one main strategy has been to employ a passive defense [50] solution which aims to detect malicious *behavior*, then perform some isolation and restoration. Towards accurate detection of malicious behavior, many intrusion detection systems have been implemented [16,18,20,57]. These systems aim to identify malicious messages being sent over the CAN bus by a compromised device (e.g. an ECU, OBD-II dongle, etc.), and localize that device [9,37,56]. Upon this detection and localization, [30] proposes several potential mitigation measures by placing the target ECU in a safety mode, or disabling the attack messages. However, they do not provide explicit methods to achieve this, and have no evaluation of their method. To provide such explicit restoration means, [23] proposes enabling a roll back to the original firmware before resetting the ECU, which allows a restoration of the compromised ECUs.

Another approach to detecting malware on an ECU has been to explicitly check the firmware contents, rather than performing detection based on adversarial CAN messages. This approach has several advantages. For example, ECU malware does not always cause malicious behavior on the CAN bus, meaning that such malware may go undetected by CAN-based intrusion detection methods. However, checking of the firmware contents will immediately reveal any malicious code injections. One such firmware content based detection mechanism is to perform a self-attestation, where the entire firmware content is usually checked against a signature provided by the OEM [13,35,38,41,42,52]. A few of these self-attestation solutions check this signature *only* immediately after writing it to the flash memory [35,38,52]. However, performing this check only once may not be sufficient [25,38,41]. To resolve this, several solutions rely on isolated software (i.e. bootloader, hardware security modules, etc.) which performs an attestation upon each reboot before loading the firmware into memory [13,25,41,42]. However, there is still need for a good *runtime* self-attestation solution [25,38]. A challenge in ECU self attestation schemes has been the strict real time requirement that ECUs have [8,25], making it difficult to implement a runtime self-attestation. In particular, since ECUs are usually low power embedded systems, the operations required to verify a cryptographic signature over the entire firmware are very time consuming. Towards providing such an efficient runtime self-attestation scheme, Kaster et al. [25] propose organizing the firmware into d blocks with b cells per block, then "slicing" the blocks so that one cell from each block is in a slice. Each slice can be checked individually. This allows a probabilistic self-attestation which is much more efficient than other designs. However, this solution still requires expensive cryptographic operations (CBC-MAC) for the attestation, and requires the use of additional secure hardware which is expensive.

In addition to the need for an efficient runtime self-attestation scheme which is adapted to the in-vehicle real time requirements, there is a need for quickly restoring the old firmware upon malware detection [8,35,38]. Mansor, et al. [35] proposes enabling such restoration by *temporarily* storing old firmware on a "central communication unit". However, their design allows only one firmware backup at a time, so that restoration is possible only after a single initial attestation. When performing runtime self-attestation, a persistently available restoration mechanism is desirable. Dafoe et al. [23] adapted a firmware rollback mechanism which was built into the flash memory controller software. This allows an automatic implicit backup of the old firmware version, taking advantage of the out-of-place updates nature of NAND flash memory, and the isolation of the flash memory controller. However, this rollback mechanism was adapted for a intrusion detection approach rather than a self-attestation.

In this work, we introduce FSAVER, which is a **F**irmware **S**elf **A**ttestation and in-**V**ehicle **E**CU **R**epair design. FSAVER is highly scalable (as demonstrated in Sect. 6), and conforms to the real-time requirement of a runtime attestation. We achieve this design by utilizing existing hardware components to develop two functions that will run on each ECU: self-attestation, and self-repair.

Self-attestation. In order to regularly check the firmware contents for malware, we need a trusted entity to perform regular attestations (i.e. a trusted *auditor*). Typically, ECUs are equipped with ARM Cortex-A or Cortex-M [43–45,55] CPUs which have TrustZone capabilities [32,33]. The TrustZone is a hardware-level security feature built into the processor which enables a trusted execution environment (TEE, introduced in Sect. 2.3) isolated from the normal insecure execution environment. Even if the ECU firmware is compromised, the execution running in the TEE secure world remains uncompromised. Thus, we can establish the TEE secure world within each ECU as the trusted auditor. As mentioned above, typical in-vehicle self-attestation solutions [13,35,38,41,42,52] will hash the entire firmware image to check its integrity against a signature provided by the OEM. This solution is usually viable when checking a *small* firmware image during boot (i.e. not runtime). However performing runtime attestation in this way over large firmware images will compromise the real-time requirement for in-vehicle ECUs [8,25]. In cloud storage applications, however, a highly efficient data checking solution known as spot checking has been used for some time [5,24] which challenges a small subset of the data periodically (Introduced in Sect. 2.1). Therefore, we can adapt the spot checking technique to the in-vehicle scenario, using the TEE as the auditor.

Self-repair. Once compromised firmware is detected, the next step is to quickly restore it to the latest benign firmware. Our insight is that the ECU firmware is typically stored on a flash memory medium [11,21] managed by an isolated Flash Memory Controller (FMC, introduced in Sect. 2.2). This isolation ensures that even if the ECU firmware is compromised, the FMC flash memory management software remains intact. Our key observation is that the software running on the FMC will usually perform out-of-place updates, conforming to the unique hardware nature of flash memory. This implies that any updates to the ECU firmware will only invalidate rather than delete the original code. Therefore, by manipulating the garbage collection strategy implemented by the FMC, this invalidated code can always remain on the flash memory medium. Then, we can efficiently restore the latest benign firmware by simply rolling back the invalidated code. To perform this repair, the FMC would need to be aware of the firmware corruption. Therefore, we develop a secure notification mechanism between the TEE and FMC.

2 Background

2.1 Remote Data Integrity Checking (RDIC)

Remote data integrity checking enables a *client* to check the integrity of data outsourced to any *storage provider*. Such RDIC schemes were originally introduced primarily for cloud storage applications, and the primary RDIC schemes are Provable Data Possession, (PDP) [4,5] or Proof of Retrievability (PoR) [24,46]. The essential idea is that rather than checking the entire data, the data are viewed as a collection of blocks, with a small random subset periodically selected for integrity checks. This approach is known as spot checking, and it is able to detect data corruptions with an arbitrarily high probability for a given amount of corruption over the entire data [4,5]. To enable such spot checking, verification tags are computed over each block. These tags may be constructed in either a privately or publicly verifiable manner [5]. In the privately verifiable schemes, the same private key which generated the tags is used in data integrity verification. In the publicly verifiable schemes, the verification relies on a public key associated with the private key originally used in tag generation. In general, the publicly verifiable schemes are much more expensive in terms of the computations performed for both generating and verifying the integrity proof due to the use of asymmetric cryptographic primitives. After generating the tags, both these and the data are outsourced to the storage provider. The "setup phase" in an RDIC scheme involves both generation of the tags, and distribution of the tags and data. After the setup phase, the "verification phase" (in our adaptation, we call this the "attestation phase") starts, during which a client can issue challenges to the server, requesting a proof that a random subset of the data are stored correctly. In response, the storage provider will use the challenged data and tags to compute a proof. Importantly, the size of this proof is independent of the number of blocks challenged, significantly improving the efficiency of verification, and enabling the ability to aggregate proofs over any number of arbitrarily selected blocks. The proof is returned to the client, who can check its validity based on the maintained keys.

2.2 Flash Memory

Typically, in-vehicle computers are equipped with NAND flash memory to store their firmware image [11,21]. This is used in vehicles due to its very high I/O throughput, which is necessary due to the real-time requirements. However, NAND flash has some unique hardware properties. NAND flash is organized into many contiguous storage chunks known as blocks, each containing a certain number of pages. The read and program (write) operations always occur over pages, while the erasure operation is over entire blocks. In order to perform a program operation, the encompassing block data should be erased. However, each block can only sustain a finite number of program/erase cycles before it becomes unreliable. Therefore, to manage this unique hardware nature, it is more economical to perform writes to a new page rather than the original data location. This is known as an *out-of-place updates* strategy. Upon a programming operation, this strategy results in the data for a constant logical location to end up in a different physical location. To manage this, the NAND flash also performs *address translation* by maintaining a dynamic mapping between the logical and physical locations. Another mechanism for relocating data is known as *wear leveling* which regularly swaps blocks to evenly distribute programming/erasure operations. Upon moving to a new physical location, the previous physical page is marked as invalid. Once a block is full of invalid pages, it can be removed, and so the *garbage collection* mechanism will eventually perform an erasure over it. These sophisticated NAND flash management operations require a firmware layer on top of the memory hardware. This firmware layer is usually either a flash translation layer or a flash file system. This firmware layer is implemented on a dedicated embedded processor known as the Flash Memory Controller (FMC). Essentially, the FMC is isolated from the host computer, such that if the host OS is compromised, any software running in the FMC will remain intact. Additionally, the FMC presents a very limited read/write interface to the host OS, and so has a very limited attack surface.

2.3 Trusted Execution Environment (TEE)

Many processors today are equipped with a Trusted Execution Environment (TEE) such as Intel SGX, AMD SEV/SME/TSME, and ARM TrustZone. The TEE allows a sensitive application to run in a secure memory area, where the application code and data can be isolated at the hardware level. The secure memory area and execution state are known as the secure world, while all other memory and computation is known as the untrusted world. Typically an operating system is running in the untrusted world so we refer to this as the host OS. Typically a TEE enabled application has two components, a trusted application (TApp) running in the secure world, and an untrusted application (UApp) running on the Host OS. Therefore, the TApp component is protected even if the host OS is compromised. Additionally, there is an interface for UApp to call predefined functions on the TApp. Further, TEE enables a mechanism called sealing, where any data which needs to be persistently stored, such as keys, can be securely stored on the host device. Typically, in-vehicle ECUs use embedded systems processors, many of which are ARM based. Specifically, many ECUs [43–45,55] are equipped with ARM Cortex-A or Cortex-M CPUs, which have TrustZone capabilities enabled [32,33].

2.4 In-vehicle Network Architecture

Today's in-vehicle electronics systems are typically organized into a domain based architecture [2], in which the Electronic Control Units (ECUs) are organized into *functional* domains, which are collections of functionally related ECUs. The ECUs in each domain are connected to each other via a shared CAN FD bus. CAN FD was introduced in 2012 as an extension of the original CAN. CAN is a protocol for communications between many nodes connected by two wires, where each message is broadcast to all other connected nodes. Compared to the original CAN protocol, CAN FD supports 64 byte messages in each data frame, and an increased throughput up to 5MB/s (CAN is capped at 1MB/s). Due to the broadcast nature of the CAN FD bus, all ECUs in a domain can freely broadcast messages to all other ECUs in their domain. At the "head" of each domain is a dedicated *gateway* unit [22]. Essentially, gateways are special ECUs which, rather than containing low-power embedded processors, are equipped with much more powerful processors. Gateways act as an intermediary between domains, providing the ability for inter-domain communications [22]. Additionally, gateways are equipped with capabilities to communicate with external networks for functions such as over-the-air updates. As mentioned in Sect. 1, an OBD-II port has been mandated in the US since 2009 [51], providing direct external physical access to the CAN FD bus. Due to the domain-based architecture, however, when communicating with a particular ECU through the OBD-II port (e.g. during reprogramming), all messages are sent through a gateway before being forwarded to the target ECU. Recently, a zonal architecture has been introduced, where rather than being organized by functional domains, ECUs will be organized by geographical zones. In this model, the ECUs in each zone still share a CAN FD bus, and at the head of each zone is a zonal gateway. In this paper, when we refer to a "domain", it may be applied to either domains or zones, depending on the vehicle architecture (Fig. 1).

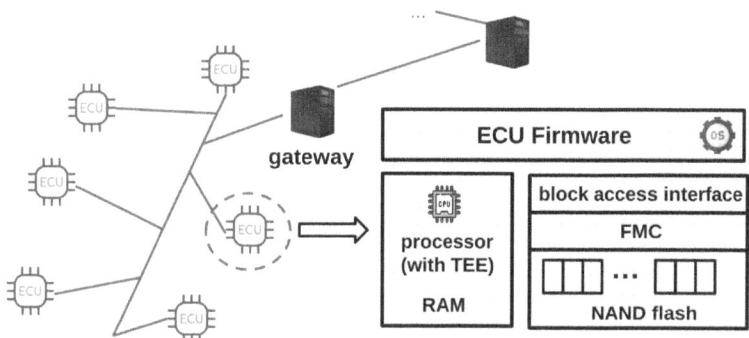

Fig. 1. System Model.

3 Models and Assumptions

3.1 System Model

We consider a typical vehicle consisting of multiple ECUs connected to a CAN FD bus (later in the paper, we refer to this simply as CAN). The ECUs are separated into a few domains, where each ECU in a domain may communicate freely within that domain. Without loss of generality, our solution focuses mainly on interactions within a single domain. Our prototypical domain contains one gateway unit, which acts as an intermediary between domains. The typical ECUs are low power embedded systems, while the gateway contains far more powerful processing capabilities. Each ECU is equipped with a processor, RAM, and flash memory. The flash memory is managed by a flash memory controller (FMC) which runs software isolated from the host OS (Sect. 2.2). This software performs out-of-place updates to manage the unique hardware nature of flash memory. Additionally, the processor is equipped with a TEE on which a secure world is enabled. In the gateway, this can be on TrustZone, SGX, or

SME/TSME, depending on the gateway CPU, while in the other ECUs, this will usually be TrustZone. In the secure world, we run a trusted application (TApp), which can directly communicate only with an untrusted application, (UApp) running in the normal world. UApp has kernel level privileges, which is typical in real-time operating systems [36]. Thus, UApp can directly access the read/write interface provided by the FMC. Additionally, UApp can send communication messages over the CAN FD bus (via communication with the CAN controller). Note that a valid firmware update will always pass through the gateway before ECU reprogramming (Sect. 2.4).

In the gateway, TApp is equipped with a certificate from the OEM which can be established securely during vehicle manufacturing. Further, we assume that a valid firmware image is always equipped with a signature signed by the OEM private key, and that an adversary will not obtain this key, an assumption which is standard in self-attestation schemes [35,41,42,52]. The existence of a shared secret key (K_d) between TApp in all ECUs in a domain is also assumed. Within each ECU, there is a key (K_s) and counter (γ) shared between TApp and FMC. The length of all keys is κ. All the shared secrets can be established securely during manufacturing.

3.2 Adversarial Model

The firmware of any in-vehicle computer may be compromised and misbehave. Specifically, an adversary may cause an ECU to enter into reprogramming mode via any means of connecting to the in-vehicle network (e.g. OTA-updates, diagnostic tools, remote OBD-II connection, etc.). During this time, the code of UApp which is stored in the flash memory will be modified. This may result in malicious behaviour such as disruptions to ECU operation or sending malicious CAN messages (*Adversary I*). Additionally, the malware is in between TApp and FMC, and so can arbitrarily manipulate the communications between them (*Adversary II*). The Adversary II may perform any of the following attacks which manipulate the communications between TApp and FMC: 1) Manipulation of message content during transmission. 2) Generating arbitrary message responses on the fly (without the knowledge of any secrets). 3) Attempt to imitate one of the trusted components (a spoofing attack). 4) Delay the communications. 5) Block the communications completely (a DoS attack). We do not consider an adversary which will shut down the TEE completely. Both Adversary I and II are computationally bounded.

3.3 Assumptions

Our design relies on a few assumptions: 1) the TEE is secure (Sect. 2.3), a common assumption for any TEE-based applications [12]. 2) The FMC is secure (Sect. 2.2). This assumption is also reasonable as the FMC is isolated from the OS by the storage hardware [54]. 3) The initial firmware on a given ECU is benign. 4) If there is no malware in the flash memory, then there is no malware in the RAM. This implies that if there is malware in RAM, then necessarily there is malware in the flash memory (this is the contrapositive). Note that the converse statement is not necessarily true. An implication of this assumption is that misbehaviour will always indicate malware is present in flash memory.

4 FSAVER

4.1 Design Overview

To ensure that the in-vehicle ECUs are always malware-free, we take an approach of explicitly checking the firmware contents. To do this, we use a self-attestation. Our self-attestation design improves on existing solutions by incorporating an efficient runtime checking by using a spot checking based RDIC scheme (Sect. 2.1). In particular, we employ UApp as the storage provider, and TApp as the client analogue, performing regular checks over the committed firmware "data". To enable an efficient RDIC scheme in the in-vehicle scenario, we carefully adapt the setup and attestation phases.

In addition to checking the firmware contents and so detecting malicious code, ensuring that the ECUs are malware-free requires some active response. To enable this, we use self-repair mechanisms within FMC. The first step towards self repairing is for FMC to become aware of a malicious presence through notification. Next, FMC will rollback to the latest benign firmware, which is enabled by ensuring that the old firmware data are maintained and locations are known. Next, a reboot is initiated so that the malware is replaced in memory by the benign firmware (Fig. 2).

4.2 Self-attestation

Setup Phase. As established above (Sect. 1), a typical ECU update strategy is to initiate a reprogramming mode by authenticating some flashing device [29] over CAN via a protocol like UDS[1]. When the flashing device is a physical diagnostic device, it is attached to the OBD-II port. With the rise of over-the-air updates [27], updates can now be sent directly to a gateway, which then acts as the flashing device, performing authentication with the target ECU.

Once authenticated, the flashing device can send arbitrary code to the ECU. This code may be malicious, so it is essential to check whether it was provided by the OEM. Towards achieving this, the most natural solution which incorporates RDIC is to use a publicly verifiable RDIC scheme. The OEM would generate publicly verifiable tags, and each TApp would store an OEM certificate to verify RDIC proofs. However, using a publicly verifiable scheme is much less efficient than a privately verifiable scheme (Sect. 2.1). Consequently, our design adapts a privately verifiable scheme. In order to do this, the key used in tag generation must also be used during attestation. However, if the OEM generates these tags (as in the publicly verifiable case), then its private key must be shared with all the ECUs. This would involve complex key management and potential security

---
[1] FSAVER does not modify UDS.

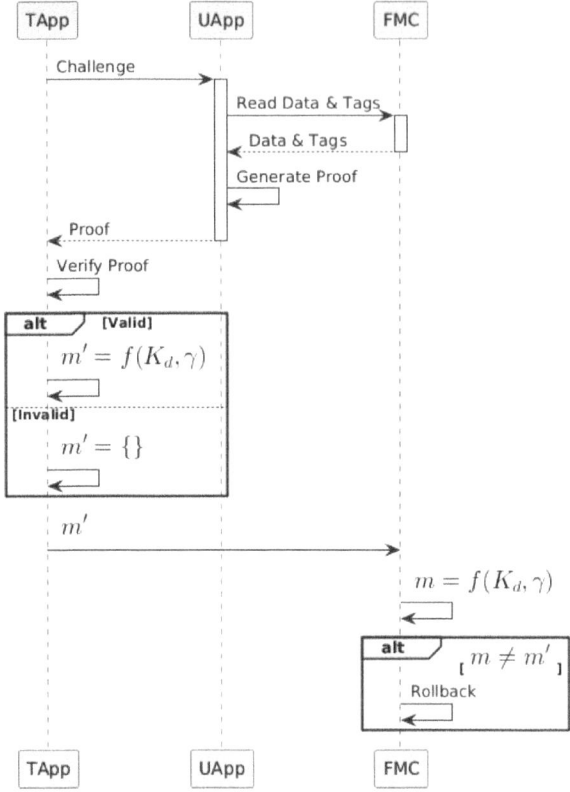

Fig. 2. A Sequence diagram for *attestation*, *notification* and *rollback*. Valid indicates no malware detected while invalid indicates malware. Note that the diagram uses "alt" (alternative) fragments to show different paths of execution based on conditions.

vulnerabilities. FSAVER addresses these challenges by leveraging the gateway generate tags before the firmware reaches the target ECU.

We observe first that the benign new firmware contents will always pass through the gateway before it is received by the target ECU (Sect. 2.4). In FSAVER, as in the existing solutions [13,35,38,41,42,52], the OEM will initially provide a signature over the new firmware. As the new firmware is passed through the gateway, TApp will verify the signature using its OEM certificate. In this way, the gateway TApp checks whether the received firmware is from the OEM. Next, the privately verifiable tags must be generated. The gateway TApp, having verified the firmware source, will generate these tags using K_d, the private key which is shared between all TApp in the domain. Next, the firmware, along with the generated tags, are written to the target ECU. Since the tags were generated using K_d, which is shared between TApp in all domain ECUs, the target ECU TApp is able to perform a self-attestation (as described below).

Sometimes, the target ECU may be the gateway itself. In this case, we observe that there are multiple gateways which communicate with one another via their network interfaces. Therefore, another connected gateway can perform the role typically designated to the domain gateway (i.e. considering the gateways together as a domain).

Attestation Phase. Once the new firmware and privately verifiable tags are stored in the target ECU, the attestation phase will begin. We establish an epoch of duration T_{epoch} so that within each epoch, one spot checking self-attestation will occur. To perform such a attestation, TApp will act as the client in an RDIC scheme, randomly selecting which blocks will be challenged. This challenge will be generated close to the beginning of the epoch, as it is a very quick operation, and can typically be scheduled at any time. This quick challenge allows a duration close to T_{epoch} for the remainder of the attestation phase. This consists of UApp receiving the challenge and, acting as the storage provider, generating a proof that the challenged blocks are stored properly. Upon receiving the proof, TApp will perform verification. The verification results will either be valid or invalid, indicating that the firmware is benign or malicious respectively. This approach provides a wide time window within each epoch to perform proof generation and verification, allowing these processes to be flexibly scheduled without compromising the real-time requirements of the system.

4.3 Self-repair

Notification. Upon malware entering the ECU, FMC, which performs the self-repair, should be aware as soon as possible. To become aware of this, we have designed a notification checking mechanism which is activated each epoch in the FMC. To receive a notification, we rely on the key length κ, a security parameter l, as well as the key (K_s) and counter (γ) which is shared between TApp and FMC.

First, we define a pseudo random function (PRF) as $f : \{0,1\}^\kappa \times \{0,1\}^* \to \{0,1\}^l$. Upon computing self-attestation results, TApp will compute $m = f(K_s, \gamma)$ then increment γ. Next, if the attestation results are *valid*, TApp will write m to the flash storage at a location agreed upon with FMC (we call this location the "notification address"). If the results are *invalid*, TApp will write an arbitrary, unrelated value to the same location.

To receive this notification, FMC will have a built in hardware timer which triggers a timer interrupt at the end of each epoch. To handle this interrupt, FMC will also compute $m = f(K_s, \gamma)$ then increment γ. FMC then reads the value m' from the notification address. If m and m' match, then there is no malware present. Otherwise, FMC will know malware is present and so can initiate self-repair.

Rollback. Once FMC is aware of a malicious presence, FMC should be able to rollback to the previous benign firmware state. To enable this rollback, there are two challenges: 1) The old firmware should be present on the flash storage device, and 2) this data should be easily located.

Essentially, solving *challenge 1* requires modifying the garbage collection policy, which typically would erase the old firmware data soon after it is invalidated (Sect. 2.2). Our solution is to coordinate our garbage collection strategy in time with ECU reprogramming, so that we can ensure any activated rollback will always be *from* malware *to* the latest benign firmware. In FSAVER, garbage collection is disabled when the original firmware is invalidated, after a new firmware image is written. This preserves the old firmware for rollback. Garbage collection is re-enabled for the invalidated firmware only when the new firmware is invalidated and replaced by another update, allowing old firmware versions to be replaced in sequence.

Before GC is re-enabled on the invalidated firmware, however, it is crucial to ensure that it does not erase the latest benign firmware. To do this, we need to determine whether the active firmware or the invalidated firmware (i.e., the previous version) is the latest benign version before proceeding with reprogramming. This is verified through a self-attestation over the active firmware. There are two possible self-attestation results:

1. **Valid Result:** If the self-attestation confirms that the current firmware is benign, garbage collection is enabled before reprogramming. This allows the old firmware data to be reclaimed.
2. **Invalid Result:** If the self-attestation indicates that the active firmware is compromised, a self-repair is initiated to restore the previous firmware, which is ensured to be the latest benign version (Sect. 5.2). After restoring the firmware, garbage collection is enabled to remove the identified malware before reprogramming.

In both scenarios, the latest benign firmware is overwritten during subsequent reprogramming. This ensures that the firmware to be restored upon activating rollback is always the latest verified benign version. A proof of this is given is Sect. 5.2.

Solving *challenge 2* requires extending the address translation policy, which upon writing the new firmware would typically *replace* the old logical to physical mapping with the new one (Sect. 2.2). In order to restore the old firmware, the old mapping simply needs to be maintained. To ensure this, we reserve some space as a backup mapping table.

Recall that prior to reprogramming, we ensured (in solving challenge 1) that the active firmware is the latest benign version (i.e., the version to which we want to roll back). Thus we can simply save the mapping table in this state prior to reprogramming to maintain its location. Since the corresponding data is maintained as established in challenge 1, these mappings will continue to point to the correct firmware. Therefore, upon restoring this mapping table, the latest benign firmware will also be restored.

Reboot. Upon rolling back to the latest benign firmware in the flash storage, malware will still be contained in the ECU RAM and be running on the CPU. To remove the malware from RAM and run the benign firmware again, the target ECU should simply reboot. Fortunately, this reboot can easily be initiated by

TApp. However, in order to always initiate this reboot, TApp should always be aware of when rollback is completed by FMC. When self-attestation verification returns a invalid result, TApp can always be assured that rollback will be performed. Thus, in this case, TApp can initiate a reboot after the epoch expires, since by this time, FMC would have checked the results and performed rollback. However, there is a case where self-attestation verification returns a valid result, yet there is still malware present, and rollback is subsequently performed by FMC (see Sect. 5) while TApp is unaware. To make TApp aware of this, TApp will always read from the notification address at then end of each epoch. TApp will then check whether m', the value read, matches m, the previously generated shared random value. If it does match, then the special case did not occur. Contrarily, if it does not match, then this case did occur and a reboot should be performed. A safety-focused discussion on when this reboot should be performed is given in Sect. 7.2.

5 Security Analysis

5.1 Self-attestation

Setup. This is the phase during which an adversary would write malicious firmware to the target ECU. The aim would be to perform this writing so that malware is not detected by the self-attestation. To perform such an attack, tags would need to be generated so that verification of a self-attestation proof using K_d would always return valid results. We show this to be infeasible through the two possible attacker strategies:

1. The attacker obtains the OEM private key (with key length κ) and uses this to sign the malware. Subsequently, the attacker will enter reprogramming mode with the target ECU and transmit the malware through the gateway. Then, the gateway will verify the signature and generate tags using K_d. Thus, if the adversary can obtain the OEM private key, then this attack can be successfully performed. Assuming that the OEM will properly secure this key, the best strategy is for the attacker to randomly guess it with probability $\frac{1}{2^\kappa}$.
2. The attacker obtains the key K_d (with key length κ) and uses this to generate tags for the malware. Subsequently, the attacker will enter reprogramming mode either 1) directly with the target ECU, bypassing the gateway which is typically involved in the ECU reprogramming, or 2) through a compromised gateway, which will forward malware to the ECU regardless of signature verification results. Thus, if the adversary can obtain K_d and bypass the gateway, then this attack can be successfully performed. Since the TEE sealing mechanism (Sect. 2.3) will secure the storage of K_d on the target ECU, so that it is inaccessible to the adversary, the best strategy is for the attacker to randomly guess K_d with probability $\frac{1}{2^\kappa}$.

Attestation. Periodically, TApp will challenge a random subset of the firmware blocks. For each attestation, the probability of successfully selecting "corrupted" blocks to challenge is at least $P_m = 1 - (1-\beta)^c$ [5], where c is the number of blocks being challenged, and β is the proportion of blocks containing at least one bit of corruption. For α attestations, this probability is $P_{\text{detect}} = 1 - (1-P_m)^\alpha$. As an example, when $c = 100$, $\beta = .01$, and $\alpha = 5$, $P_{\text{detect}} = .993$[2].

Upon challenging "corrupted" blocks, several adversarial responses by UApp are possible. Specifically, five different methods may be used by the Adversary II, as outlined in Sect. 3:

1. For method 1, UApp may attempt to modify the challenged data so that it can pass the challenge. However, UApp does not have access to K_d, required to generate valid tags on the fly. Additionally, UApp does *not* have immediate access to the previous firmware content or tags, which may be used to pass an attestation. Therefore, the best strategy is to guess K_d with probability $\frac{1}{2^\kappa}$, then generate tags for the malware. This guessing is infeasible for a computationally bounded adversary, especially within duration T_{epoch}.
2. For method 2, UApp will attempt to generate some data or tags which will result in computing a valid proof. This is again not feasible for a computationally bounded adversary, which cannot access K_d.
3. For method 3 (spoofing attack), there is no sensible attack.
4. For method 4 and 5, a response will be delayed greater than duration T_{epoch}. This will result in TApp generating a invalid verification result.

Due to the infeasible nature of performing these attacks within the capabilities of the Adversary II, challenging "corrupted" blocks will always result in a invalid verification result within TApp. In all these cases, TApp is assured that there is malware present, since misbehaviour will always indicate malware is present in flash memory (Assumption 4 in Sect. 3).

5.2 Self-repair

Notification. When the timer interrupt is triggered in FMC, it will compute $m = f(K_s, \gamma)$ and compare this with the value m' contained at the notification address. If m does not match m', then FMC will automatically trigger rollback. The security goal is that upon a invalid verification result from TApp, rollback will always occur. Upon a valid verification result, we ensure that rollback is not performed *unless malware is present in the ECU*. There are three cases to consider:

Case 1: Invalid verification result, with malware present. In this case, TApp will not reveal the value m to UApp. We consider the Adversary II, which

---
[2] Here, there is a design trade-off between the number of blocks and the number of audits performed to detect a given β proportion of malware. For smaller c, it will take more attestations (a higher α) to detect malware, but each attestation will be more efficient, so that the real-time requirement is met.

may employ five different methods, as outlined in Sect. 3. Using method 1, 2, or 3, UApp would need to generate m. The best strategy is to guess both K_s and γ, then compute m using f. The probability of correctly guessing K_s, however is $\frac{1}{2^\kappa}$, which is negligibly small for sufficiently large κ. Using methods 4 and 5, the timer interrupt in FMC would trigger before any value is written to the notification address. Therefore the value m' which is read will not match m.

Case 2: Valid verification result, with malware present. In this case, TApp will reveal the value m to an adversarial UApp. We again consider the Adversary II. If the adversary does not manipulate the writing of m to the notification address, then FMC will not perform rollback. However, UApp does not always know whether TApp computes valid or invalid verification results, and so may still manipulate the communications between TApp and FMC. Using methods 1, 2, and 3, the value m' written to the notification address will very likely be different than the value m provided by TApp. Therefore, FMC will perform rollback. Using methods 4 or 5, the timer interrupt in FMC would again trigger before any value is written to the notification address. Therefore the value m' which is read will not match m, and rollback will be triggered.

Case 3: Valid verification result, no malware present. In this case there is no adversary present, so the correct notification $m' = m = f(K_s, \gamma)$, generated by TApp, will always be written to the correct location. Therefore, a rollback will never be triggered in this case.

Rollback. To demonstrate the security of our firmware rollback mechanism, we prove the following theorem:

Theorem 1. *For any firmware version P_{n+1}, if P_{n+1} is identified as malicious, then the rollback mechanism will restore the system to P_n, which is the latest benign version prior to P_{n+1}.*

Proof. We have established in the above analysis that under our adversarial model, if P_{n+1} is identified as malicious, then the rollback mechanism will be triggered. First, we show that the last active firmware (malicious or benign) P_n can always be restored to. That is, that through our garbage collection management and address translation management, the data and mappings for the previous firmware version P_n can always be restored (i.e. the data and mappings for P_n are always present when P_{n+1} is active).

We can prove this by induction. The initial firmware update is from P_0 to P_1. Our garbage collection polity is that before this reprogramming GC is enabled. However, there is no previous firmware version to reclaim, and so nothing occurs. GC is disabled for any new writes, and so P_0 is maintained. Additionally, by copying the mapping table prior to the update, the locations for the firmware P_0 are maintained. Therefore, P_0 can always be restored while P_1 is active.

Now, suppose that for P_k such that $1 < k \leq n$, the firmware can be rolled back to P_{k-1}. Prior to updating from firmware version P_n to P_{n+1}, GC is enabled on firmware version P_{n-1}, and so it may be reclaimed. Again, as GC is disabled for new writes, the data for P_n is maintained when P_{n+1} is written. Additionally, as the mapping table for P_n is saved prior to writing P_{n+1}, the locations can be restored.

Now, we have to show that P_n, the prior firmware version, which is rolled back to, is always the latest benign version. This proof is by induction. By our assumption 3 (Sect. 3), the initial firmware version P_0 is always benign. Now, assume that for any firmware version P_k, where $0 < k \leq n$, that if P_k is identified as malicious, then P_{k-1} is the latest benign version prior to P_k. Now, suppose that the firmware version P_{n+1} is identified as malicious and so rollback is initiated. Note that before writing P_{n+1} to the flash memory, a self-attestation was performed over P_n. If the verification results were invalid, then P_{n-1} was restored to replace P_n, which, by the inductive hypothesis, is the latest benign firmware. If the results were valid, then P_n was already the latest benign firmware. Therefore, in both cases, P_n, at the time of rollback, is the latest benign firmware version. □

Reboot. Essentially, in order to effectively perform the rebooting after rollback, this will be triggered by TApp. Since we have established that rollback will occur only when there is malware present, it is simply required that TApp is always aware when rollback is performed. There are two cases when rollback may be performed:

1. **When a self-attestation verification returns invalid results:** In this case, TApp is aware of the invalid results because it is directly performing the self-attestation verification. TApp can initiate a reboot after the epoch expires, knowing that rollback has occurred.
2. **When a self-attestation verification returns valid results:** In this case, rollback can still occur due to other issues, such as message manipulation by an adversary (Adversary II) which is not *yet* detected via spot-checking. In this scenario, recall that TApp checks the value m' read from the notification address at the end of each epoch. If m' differs from the previously generated value m, it indicates that FMC has detected a problem and performed rollback. Therefore, TApp can still initiate a reboot even when the verification results were valid, but rollback has been triggered due to external manipulation.

This demonstrates that FSAVER ensures that TApp is aware of all rollback scenarios and can initiate a reboot as needed. For a discussion on the timing and implications of this reboot, refer to Sect. 7.2.

6 Implementation and Evaluation

Implementation. We have implemented the self-attestation component (including setup and attestation phases) in real-world hardware. As an ECU, we used a Raspberry Pi 3B+ [1] (With 1.4GHz 64-bit quad-core ARM Cortex-A53 CPU, and 1GB LPDDR2 SDRAM) with TEE enabled. We implemented UApp, which will receive the challenge from TApp and compute an RDIC proof, under the host OS Raspbian Stretch. TApp, which will generate the challenge and verify the received RDIC proof, was implemented by porting OP-TEE [39] (Open Portable Trusted Execution Environment) to the Raspberry Pi via the Raspbian-TEE open source project [6]. For our RDIC scheme, we used the privately verifiable scheme from Compact PoR [46]. Additionally, we have implemented a rollback mechanism on the USB header development prototype board LPC-H3131 [34] (with ARM9 32-bit ARM926EJ-S, 180Mhz, 32MB of SDRAM, and 512MB NAND flash). We have ported [48] the open source NAND flash manager OpenNFM [10] to the LPC-H3131, and verified the feasibility of a rollback based self-repair mechanism.

Evaluation. To evaluate the efficiency of the self-attestation component in both TApp and UApp, we timed the setup and attestation phases. For evaluating self-repair, we timed the rollback phase. We present our results in Table 1 as throughput rates. This table demonstrates the amount of data which can be attested, and mapping content which can be rolled back in given time.

For the setup phase, we timed the generation of privately verifiable tags in the gateway TApp. Note that this is a "one-time" operation, occurring only when a new firmware update is received. Additionally, while our implementation uses the Raspberry Pi 3B+, gateway units typically contain much more powerful processing capabilities. For the attestation phase, we have timed three distinct phases: 1) The challenge generation (TApp), 2) the proof generation (UApp), and 3) the proof verification (TApp). As anticipated (Sect. 4), we observe that the challenge generation throughput is significantly higher than the other procedures. This is because the computation is limited to generating a few random numbers. Additionally, we observe that the proof verification is slower than the proof generation despite the proof verification being a more lightweight computation. We suspect this is because the proof generation is run on the host OS, while verification is run in the TrustZone secure world. The TrustZone TEE is generally much slower than the host OS, and the Raspberry Pi is not optimized for the use of TrustZone. In general, our attestation results demonstrate that a small spot checking based self-attestation would be quickly performed, since the amount of data being checked is small.

For the rollback phase of self-repair, we have timed the restoration of the backup mapping table. The observed mapping restoration throughput of 1724KiB/s results in an associated *data* restoration rate of 836MiB/s. This is very quick, considering that the typical ECU firmware size is much less than 20MB [14].

These results also effectively demonstrate the scalability of FSAVER. Specifically, we have demonstrated above (Sect. 5) that by using the spot checking-technique, checking a small constant number of blocks will have the same probability to detect a given proportion of firmware corruption, regardless of the overall firmware size. Therefore, the attestation cost can be constant for *any* firmware size. In contrast, the rollback phase scales linearly with the firmware size. However, by only restoring *mappings*, we scale the *data* restoration speeds significantly, enabling us to restore any reasonably sized ECU firmware very quickly.

Table 1. Average throughput (KiB/s) for setup, attestation, and rollback.

Procedure	Component	Throughput (KiB/s)
Setup (one time)	Gateway TApp	22.04
Challenge Generation	ECU TApp	363.64
Proof Generation	ECU UApp	28.52
Proof Verification	ECU TApp	15.43
Rollback (Mapping Restoration)	ECU FMC	1724

7 Discussion

7.1 Limitations

During the *setup* phase for self-attestation, FSAVER first requires the gateway to compute a signature over a hash of the entire firmware image. However, doing this requires the full firmware image to be contained in gateway memory at one time. Due to memory constraints, this may not always be feasible. To mitigate this, this authentication may use a hash chain rather than a single hash over the complete data. As each block is processed, its tags can immediately be generated and stored in gateway memory while the block's contents are sent to the target ECU. Only if the authentication passes will the complete set of tags be released.

Another limitation of FSAVER is that the adversary may find some way to shut down the TEE, and thus disable TApp. If this occurs, FMC will still be aware (since an invalid m' will be present at the notification address), but TApp will not be able to perform the *reboot* step upon self-repair. Thus, the only way to clear the ECU memory and restore functionality would be to manually reboot the ECU. This is why we cannot consider the adversary which can shut down the TEE (Sect. 3).

It may be beneficial to notify either other ECUs or the driver of failed attestations prior to repair (Sect. 7.2), so that they can respond and be aware of the full vehicle state. However, under our solution, the Adversary II may disrupt any communication between TApp and the CAN bus. Therefore it is in general not feasible to perform this communication *before* self-repair is completed.

Due to these limitations inherent to most self-attestation approaches, we may in the future investigate a decentralized firmware attestation solution [26] which can establish external awareness of attestation results, and is resilient to the adversary which can shut down the TEE.

7.2 Real-Time vs Delayed Rebooting

While FSAVER technically enables almost immediate ECU restoration, it may be unsafe to do this during vehicle operation. This is because rebooting the ECU may cause undefined behaviour while driving, as usually the vehicle is idle when ECUs are powered on. This undefined behaviour may cause safety issues such as crashing the vehicle or stopping it in the middle of the road. Due to these safety concerns, Andréasson et al. [3] has suggested a comprehensive strategy: 1) always notifying the user upon malware detection, so that *in certain cases*, they can stop the vehicle while repair proceeds. 2) Upon detecting malware in non safety-critical ECUs, immediately performing repair while still informing the driver. 3) For safety-critical ECUs, wait until the vehicle is completely stopped before performing repair. In this case, the user would initiate repair when he/she feels it is safe. We believe this is a reasonable approach, though as noted above, an inherent limitation of self-attestation schemes is that they cannot notify other ECUs of the detected malware under the Adversary II. This makes it infeasible to reliably notify the user of malware detection. In the future, we may investigate a solution which enables such external communication.

8 Related Works

8.1 In-vehicle Firmware Self-attestation

Secure Firmware Flashing. In 2008, Weimerscirch [52] developed a secure software flashing strategy for in-vehicle ECUs. In their strategy, the bootloader will receive a certificate, then the new firmware will be flashed. During flashing, a hash is incrementally computed over each block and a signature provided by the OEM is checked. Nilsson et al. [38] introduce a distinction between control and functional systems within an ECU. The control system will manage the flashing procedure and will check that the firmware contents were provided by a trusted "portal". This attestation is accomplished via computing a hash chain with an incorporated challenge value, which is checked against a provided "verification code".

Secure Boot. Towards checking the firmware contents after the initial flashing, Gui et al. [13] developed a hardware based root of trust for ECUs, which includes secure boot component. This secure boot component largely relies on the maintenance of "golden measurements" of the trusted code. The ECU firmware is checked against these golden measurements before booting to ensure that the trusted code will run. Indeed, such secure boot solutions, where *upon each boot* a bootloader which will check the entire ECU firmware using cryptographic signatures is in widespread use by the automotive industry today [40, 42].

Runtime Self-attestation. The above solutions [13,35,40,42,52] all rely on checking a digest over the entire firmware image, and so are not efficient enough to meet the in-vehicle real time requirements for a runtime attestation [8,25]. However, an efficient runtime attestation solution is desired [13,52]. In 2023,

Kaster et al. introduced sliced secure boot [25], designed for in-vehicle ECUs. This solution uses a Hardware Security Module (HSM) to compute "re-usable fingerprints" over the firmware image, after the download has been authenticated. Using these "fingerprints", the HSM can challenge "slices" of the firmware image at runtime, using a Cipher Block Chaining Message Authentication Code (CBC-MAC) for authentication. Unlike the bootloader, the HSM is always available at runtime, allowing for runtime checking. The efficiency of their solution relies on a similar technique as spot-checking; by checking one "slice" at a time, the verification can be made much more efficient. Compared to FSAVER, sliced secure boot introduces the HSM as an added hardware component, while our solution uses the TEE, an existing trusted hardware component within ECUs native CPU (Sect. 2.3). Additionally, by using a publicly verifiable spot-checking RDIC based attestation (Sect. 2.1), FSAVER replaces the expensive cryptographic operations used by a CBC-MAC with simple linear combinations [46]. Further, sliced secure boot provides no explicit ECU restoration mechanism.

8.2 In-vehicle ECU Restoration

In cyber-physical systems such as vehicles, a quick repair mechanism is desirable [8,35,38], which the above solutions do not provide. In 2015, Mansor et al. [35] developed a mechanism which, in addition to verifying an OEM signature upon writing new firmware to the ECU, enabled temporarily storing the old firmware version on a "central communication unit". This only allows restoration immediately after flashing and is not adaptable to a secure boot or runtime attestation solution. Towards providing explicit ECU restoration mechanisms in conjunction with CAN intrusion detection [16,18,20,57] mechanisms, Kwon et al. (2018) introduced a design to reconfigure ECUs and limit their behaviour upon detection of malware [30]. The solution was to send a CAN message to a suspected ECU which initiates a generic "safe mode" and reboot. Additionally, all benign ECUs would ignore certain messages sent from the ECU to cause harm. In 2023, Dafoe et al. extended this idea by relying on the unique nature of flash storage to coordinate a repair between the TrustZone and flash firmware [23]. Specifically, upon detection of a malicious ECU via intrusion detection, a CAN message would be sent to the target ECU and firmware rollback would be initiated. Similar to FSAVER, this firmware rollback was enabled by altering the garbage collection strategy within the flash memory. Different from [23], FSAVER does not rely on CAN based intrusion detection, which can be unreliable, and usually assumes a trusted detection ECU. FSAVER is based upon a highly efficient runtime self-attestation, which is far more reliable. Additionally, [23] fails to protect against the Adversary II and so cannot guarantee successful ECU restoration.

9 Conclusion

In this work, we have designed FSAVER, a self-attestation and self-repair scheme for the in-vehicle scenario which is highly efficient, adapting to the real-time requirement of any cyber-physical system. Due to the increased efficiency, we are able to perform a runtime attestation using the TEE as a trusted auditor. For self-repair, we enable immediate rollback to the latest benign firmware version by incorporating the isolated flash memory controller into our design. We have implemented a prototype of the self-attestation and rollback mechanisms, and have effectively demonstrated its viability.

Acknowledgments. This work was supported by US National Science Foundation under grant number 2225424-CNS and 2043022-DGE.

References

1. Raspberry pi 3 model b+. https://www.raspberrypi.com/products/raspberry-pi-3-model-b-plus/
2. Aberl, P.: How a zone architecture paves the way to a fully software-defined vehicle (2023). https://api.semanticscholar.org/CorpusID:260712942
3. Andréasson, E., Lyesnukhin, I.: Device attestation for in-vehicle network (2022)
4. Ateniese, G., et al.: Remote data checking using provable data possession. ACM Trans. Inf. Syst. Secur. (TISSEC) **14**(1), 12 (2011)
5. Ateniese, G., et al.: Provable data possession at untrusted stores. Cryptology ePrint Archive, Paper 2007/202 (2007). https://eprint.iacr.org/2007/202
6. benhaz1024: Raspbian with op-tee support. https://github.com/benhaz1024/raspbian-tee. Accessed 15 July 2024
7. Bielawski, R., Gaynier, R., Ma, D., Lauzon, S., Weimerskirch, A.: Cybersecurity of firmware updates. Technical Report DOT HS 812 807, University of Michigan. Transportation Research Institute and University of Michigan, Dearborn and Volkswagen Group of America (Herndon, VA) (2020). https://doi.org/10.21949/1530213. Corporate contributor: United States. Department of Transportation. National Highway Traffic Safety Administration. Office of Vehicle Safety Research
8. Cárdenas, A.A., Amin, S., Sinopoli, B., Giani, A., Perrig, A., Sastry, S.: Challenges for securing cyber physical systems (2009). https://api.semanticscholar.org/CorpusID:13643850
9. Choi, W., Jo, H.J., Woo, S., Chun, J.Y., Park, J., Lee, D.H.: Identifying ECUs using inimitable characteristics of signals in controller area networks. IEEE Trans. Veh. Technol. **67**(6), 4757–4770 (2018). https://doi.org/10.1109/TVT.2018.2810232
10. Code, G.: Opennfm. https://code.google.com/p/opennfm/
11. Electronic Products: Memory use in automotive (2024). https://www.electronicproducts.com/memory-use-in-automotive/. Accessed 23 June 2024
12. Guan, L., et al.: Supporting transparent snapshot for bare-metal malware analysis on mobile devices. In: Proceedings of the 33rd Annual Computer Security Applications Conference, pp. 339–349 (2017)
13. Gui, Y., Siddiqui, A.S., Saqib, F.: Hardware based root of trust for electronic control units. In: SoutheastCon 2018, pp. 1–7 (2018). https://doi.org/10.1109/SECON.2018.8479266
14. Gupta, S.: The role of phase-change memory in automotive OTA firmware upgrades. Embedded.com (2023). https://www.embedded.com/the-role-of-phase-change-memory-in-automotive-ota-firmware-upgrades/. Accessed 8 Sept 2024

15. Hackenberg, R., Weiss, N., Renner, S., Pozzobon, E.: Extending vehicle attack surface through smart devices (2017)
16. Hamada, Y., Inoue, M., Ueda, H., Miyashita, Y., Hata, Y.: Anomaly-based intrusion detection using the density estimation of reception cycle periods for in-vehicle networks. SAE Int. J. Transp. Cybersecur. Priv. **1** (2018). https://doi.org/10.4271/11-01-01-0003
17. Hammerschmidt, C.: Number of automotive ECUs continues to rise. eeNews Europe (2019). https://www.eenewseurope.com/en/number-of-automotive-ecus-continues-to-rise/. Accessed 20 June 2024
18. Han, M.L., Kwak, B.I., Kim, H.K.: Anomaly intrusion detection method for vehicular networks based on survival analysis. Veh. Commun. **14**, 52–63 (2018). https://doi.org/10.1016/j.vehcom.2018.09.004. https://www.sciencedirect.com/science/article/pii/S2214209618301189
19. Van den Herrewegen, J., Garcia, F.D.: Beneath the bonnet: a breakdown of diagnostic security. In: Lopez, J., Zhou, J., Soriano, M. (eds.) Computer Security, pp. 305–324. Springer, Cham (2018)
20. Hoppe, T., Kiltz, S., Dittmann, J.: Applying intrusion detection to automotive it-early insights and remaining challenges. J. Inf. Assur. Secur. (JIAS) **4**, 226–235 (2009)
21. Premio Inc.: Autonomous vehicle data storage (2024). https://premioinc.com/pages/autonomous-vehicle-data-storage. Accessed 23 June 2024
22. Texas Instruments: Processing the advantages of zone architecture in automotive. Technical report (2023). https://www.ti.com/document-viewer/lit/html/SSZT211#:~:text=A%20zone%20architecture%20organizes%20the,module%20to%20manage%20network%20traffic. Accessed 04 July 2024
23. Josh, D., Harsh, S., Niusen, C., Bo, C.: Enabling real-time restoration of compromised ECU firmware in connected and autonomous vehicles. In: Yu, C., Chung-Wei, L., Bo, C., Qi, Z. (eds.) Security and Privacy in Cyber-Physical Systems and Smart Vehicles, pp. 15–33. Springer, Cham (2024)
24. Juels, A., Kaliski, B.S.: PORs: proofs of retrievability for large files. In: Proceedings of the 14th ACM Conference on Computer and Communications Security, CCS 2007, pp. 584–597. Association for Computing Machinery, New York (2007). https://doi.org/10.1145/1315245.1315317
25. Kaster, R., Ma, D., Behl, A., Bakalarczyk, B.: Sliced secure boot. In: 19th escar Europe: The World's Leading Automotive Cyber Security Conference (Konferenzveröffentlichung). sampled secure boot with re-usable fingerprints (2021). https://doi.org/10.13154/294-8354
26. Khodari, M., Rawat, A., Asplund, M., Gurtov, A.: Decentralized firmware attestation for in-vehicle networks. In: Proceedings of the 5th on Cyber-Physical System Security Workshop, CPSS 2019, pp. 47–56. Association for Computing Machinery, New York (2019). https://doi.org/10.1145/3327961.3329529
27. Kim, B., Park, S.: ECU software updating scenario using OTA technology through mobile communication network. In: 2018 IEEE 3rd International Conference on Communication and Information Systems (ICCIS), pp. 67–72 (2018). https://doi.org/10.1109/ICOMIS.2018.8645019
28. Klinedinst, D.: On board diagnostics: risks and vulnerabilities of the connected vehicle. Carnegie Mellon University, Software Engineering Institute's Insights (blog) (2016). https://insights.sei.cmu.edu/blog/board-diagnostics-risks-and-vulnerabilities-connected-vehicle/. Accessed 21 June 2024

29. Kulandaivel, S., Jain, S., Guajardo, J., Sekar, V.: Candid: a stealthy stepping-stone attack to bypass authentication on ECUs. ACM J. Auton. Transport. Syst. (2024). https://doi.org/10.1145/3657645
30. Kwon, H., Lee, S., Choi, J., Chung, B.H.: Mitigation mechanism against in-vehicle network intrusion by reconfiguring ECU and disabling attack packet. In: 2018 International Conference on Information Technology (InCIT), pp. 1–5 (2018). https://doi.org/10.23919/INCIT.2018.8584882
31. Lauser, T., Krauß, C.: Formal security analysis of vehicle diagnostic protocols. In: Proceedings of the 18th International Conference on Availability, Reliability and Security. ARES 2023. Association for Computing Machinery, New York (2023). https://doi.org/10.1145/3600160.3600184
32. Arm Ltd.: Trustzone for cortex-a (2024). https://www.arm.com/technologies/trustzone-for-cortex-a. Accessed 23 June 2024
33. Arm Ltd.: Trustzone for cortex-m (2024). https://www.arm.com/technologies/trustzone-for-cortex-m. Accessed 23 June 2024
34. Olimex Ltd.: LPC-h3131. https://www.olimex.com/Products/ARM/NXP/LPC-H3131/. Accessed 30 June 2023
35. Mansor, H., Markantonakis, K., Akram, R.N., Mayes, K.: Don't brick your car: firmware confidentiality and rollback for vehicles. In: 2015 10th International Conference on Availability, Reliability and Security, pp. 139–148 (2015). https://doi.org/10.1109/ARES.2015.58
36. u/Head Measurement1200: I am coming from developing in linux. i was wondering if my analogy is right that bare-metal programming is like operating in 'kernel' mode the whole time? (2022). https://www.reddit.com/r/embedded/comments/ug3kau/i_am_coming_from_developing_in_linux_i_was/. Accessed 14 July 2024
37. Murvay, P.S., Groza, B.: Source identification using signal characteristics in controller area networks. IEEE Signal Process. Lett. **21**(4), 395–399 (2014). https://doi.org/10.1109/LSP.2014.2304139
38. Nilsson, D.K., Sun, L., Nakajima, T.: A framework for self-verification of firmware updates over the air in vehicle ECUs. In: 2008 IEEE Globecom Workshops, pp. 1–5 (2008). https://doi.org/10.1109/GLOCOMW.2008.ECP.56
39. OP-TEE: Op-tee documentation. https://optee.readthedocs.io/en/latest/general/about.html. Accessed 14 July 2024
40. Ring, M., Frkat, D., Schmiedecker, M.: Cybersecurity evaluation of automotive e/e architectures. In: ACM Computer Science In Cars Symposium (CSCS 2018), vol. 92 (2018)
41. Sanwald, S., Kaneti, L., Stöttinger, M., Böhner, M.: Secure boot revisited: challenges for secure implementations in the automotive domain. SAE Int. J. Transp. Cybersecur. Priv. **2**(2), 69–81 (2019). https://doi.org/10.4271/11-02-02-0008. Also appears in SAE International Journal of Transportation Cybersecurity and Privacy-V128-11EJ
42. Schrotter, M.: Understanding and designing an automotive-like secure bootloader. In: Regensburg Applied Research Conference (2020)
43. NXP Semiconductors: S32G3 vehicle networking reference design (2024). https://www.nxp.com/design/designs/s32g3-vehicle-networking-reference-design:S32G-VNP-RDB3. Accessed 23 June 2024
44. NXP Semiconductors: S32K3 automotive telematics box (t-box) reference design board (2024). https://www.nxp.com/design/designs/s32k3-automotive-telematics-box-t-box-reference-design-board:S32K3-T-BOX. Accessed 23 June 2024

45. NXP Semiconductors: S32Z and S32E real-time processors (2024). https://www.nxp.com/products/processors-and-microcontrollers/s32-automotive-platform/s32z-and-s32e-real-time-processors:S32Z-E-REAL-TIME-PROCESSORS. Accessed 23 June 2024
46. Shacham, H., Waters, B.: Compact proofs of retrievability. In: Pieprzyk, J. (ed.) ASIACRYPT 2008. LNCS, vol. 5350, pp. 90–107. Springer, Heidelberg (2008). https://doi.org/10.1007/978-3-540-89255-7_7
47. Bosch Mobility Solutions: Airbag control unit (2024). https://www.bosch-mobility.com/en/solutions/control-units/airbag-control-unit/. Accessed 20 June 2024
48. Tankasala, D., Chen, N., Chen, B.: Creating a testbed for flash memory research via LPC-h3131 and opennfm-linux version. Technical report, Department of Computer Science, Michigan Tech (2022)
49. Team-BHP: Technically understanding airbag systems & SRS (2017). https://www.team-bhp.com/forum/road-safety/189259-technically-understanding-airbag-systems-srs.html. Accessed 20 June 2024
50. Thing, V.L., Wu, J.: Autonomous vehicle security: a taxonomy of attacks and defences. In: 2016 IEEE International Conference on Internet of Things (iThings) and IEEE Green Computing and Communications (GreenCom) and IEEE Cyber, Physical and Social Computing (CPSCom) and IEEE Smart Data (SmartData), pp. 164–170 (2016). https://doi.org/10.1109/iThings-GreenCom-CPSCom-SmartData.2016.52
51. U.S. Environmental Protection Agency: Final Rule for Control of Air Pollution from New Motor Vehicles and New Motor Vehicle Engines. EPA Regulation: Emissions from Vehicles and Engines (2009). https://www.epa.gov/regulations-emissions-vehicles-and-engines/final-rule-control-air-pollution-new-motor-vehicles-and
52. Weimerskirch, A.: Secure software flashing. SAE Int. J. Passeng. Cars Electron. Electr. Syst. **2**(1), 83–86 (2009). https://doi.org/10.4271/2009-01-0272. Also in: SAE International Journal of Passenger Cars - Electronic and Electrical Systems-V118-7, SAE International Journal of Passenger Cars - Electronic and Electrical Systems-V118-7EJ
53. Wen, H., Chen, Q.A., Lin, Z.: Plug-N-Pwned: comprehensive vulnerability analysis of OBD-II dongles as a new Over-the-Air attack surface in automotive IoT. In: 29th USENIX Security Symposium (USENIX Security 2020), pp. 949–965. USENIX Association (2020). https://www.usenix.org/conference/usenixsecurity20/presentation/wen
54. Xie, W., Chen, N., Chen, B.: Enabling accurate data recovery for mobile devices against malware attacks. In: International Conference on Security and Privacy in Communication Systems, pp. 431–449. Springer (2022)
55. Xilinx: Zynq ultrascale+ mpsoc zcu104 evaluation kit (2024). https://www.xilinx.com/products/boards-and-kits/zcu104.html. Accessed 23 June 2024
56. Yang, Y., Duan, Z., Tehranipoor, M.: Identify a spoofing attack on an in-vehicle can bus based on the deep features of an ECU fingerprint signal. Smart Cities **3**(1), 17–30 (2020). https://doi.org/10.3390/smartcities3010002. https://www.mdpi.com/2624-6511/3/1/2
57. Ying, X., Sagong, S.U., Clark, A., Bushnell, L., Poovendran, R.: Shape of the cloak: formal analysis of clock skew-based intrusion detection system in controller area networks. CoRR abs/1807.09432 (2018). http://arxiv.org/abs/1807.09432

Unveiling the Operation and Configuration of a Real-World Bulk Substation Network

Keerthi Koneru[1]([✉]), Juan Lozano[1], John Castellanos[2], Emmanuele Zambon[3], and Alvaro Cardenas[1]

[1] University of California, Santa Cruz, USA
{kekoneru,juclozan,alacarde}@ucsc.edu
[2] Hitachi Energy, Mannheim, Germany
john.castellanos@hitachienergy.com
[3] Eindhoven University of Technology, Eindhoven, Netherlands
e.zambon.n.mazzocato@tue.nl

Abstract. Electrical substations are distributed points where operators can monitor and control the power grid. In this paper, we perform the first in-depth study of the operation of a large (500 KV) real-world substation automation network. We provide a view of how these critical networks operate using packet captures and a Substation Configuration Description (SCD) file. We discuss the challenges we overcame to reconstruct a network with redundant paths, gateways, serial legacy devices, and sophisticated intelligent electronic devices (IEDs). Our work provides a deep-dive discussion of these critical networks in a real-world system and sheds light on their operation, configuration and security.

Keywords: Communication · GOOSE · IEC 61850 · IEC 62439-3 · Network measurement · Security · Passwords · Network Monitoring · Network architectures · PRP · Substation Automation · Substation · Substation Configuration Description (SCD) file

1 Introduction

In the last two decades, substations have undergone an automation revolution, changing from analog communications to modern networks and computers, including Intelligent Electronic Devices (IEDs). Furthermore, new standards focus on the fully automatic configuration of a substation. A Substation Automation System (SAS) comprises hardware and software components that monitor and control an electrical system locally and remotely. SAS replaces repetitive, tedious, and error-prone tasks with automated processes, enhancing system efficiency and productivity.

Despite the growing number of automated substations, the academic community has limited visibility into how these critical networks operate in the real world. This paper addresses this blind spot in our network measurement community. In particular, we analyze a pcap network capture and a substation configuration description (SCD) file from a large 500 kV substation. From these passive measurements, we aim to learn as much as possible about the operation of these networks, the devices, their configuration, and their security.

Finally, automating network reconstruction is an important security step. In Operational Technology (OT) networks, the most common first step for securing a network is to perform asset inventory: systematically collecting, cataloging, and managing all technological devices in the OT network. Many legacy OT operators do not even know what is on their network. However, reconstructing a substation network presents challenges for their unique use of proxies, legacy serial devices, multicast messages, and redundant parallel networks. These complexities require us to develop a new framework to analyze substation datasets.

Our contributions include the following:

- We create a framework for analyzing substation networks through packet captures and substation configuration description (SCD) files, enabling a systematic approach to understand their operation.
- We reconstruct the substation network using passive measurements, providing insights into the network's structure, protocols, and devices. As far as we know, we are the first to give this detailed view of the operation of a real-world substation.
- During our study, we had to overcome several technical challenges to uncover the presence of proxies, redundant networks, legacy serial devices, devices subscribed to multicast messages, etc. Our results shed light on unexplored aspects of real-world substation networks.

2 Related Work

Table 1. List of papers showing the analysis of previous research on substation datasets.

	[15]	[7]	[24]	[14]	[20]	[37]	[36]	[5]	[19]	[8]	[10]	[18]	Our Work
Substation Network	●	●	●	●	●	●	●	●	●	●	●	●	●
DPI	-	-	-	-	-	-	●	●	●	-	-	●	●
Use of SCD Files	-	-	-	-	-	-	-	-	-	●	-	●	●
Real World data	-	-	-	-	-	-	-	-	-	-	●	-	●

Legend: ●: Feature considered by authors. DPI: Deep Packet Inspection. SCD: Substation Configuration Description

There is a growing literature studying the networks in electric substations. As we can see in Table 1, while researchers have an interest in understanding substation networks, most studies use synthetic data, either from testbeds or simulations.

We are only aware of three previous papers looking at real-world power grid networks [4,10,23,25]. Out of these papers, only Formby et al. [10] focuses on substation networks (and therefore, it is the only paper in our table). Formby et al. focused primarily on traffic analysis (timing and packet sizes) of various devices. In contrast, in this paper, we use Deep Packet Inspection (DPI) and a substation configuration file to discover all the devices in the network and their specific roles and functions.

The remaining papers in the table use synthetic data from testbeds or simulations. Research efforts that look at the payload of the packets (DPI) include Konka et al. [17], which studies the headers and PDUs of Sample Values packets. Similarly, [5] and [18] analyzed Generic Object Oriented Substation Events (GOOSE) traffic, exploring statistical parameters and proposing anomaly detection methods. As far as we know, no previous work has used DPI to uncover legacy serial devices and proxies within substations.

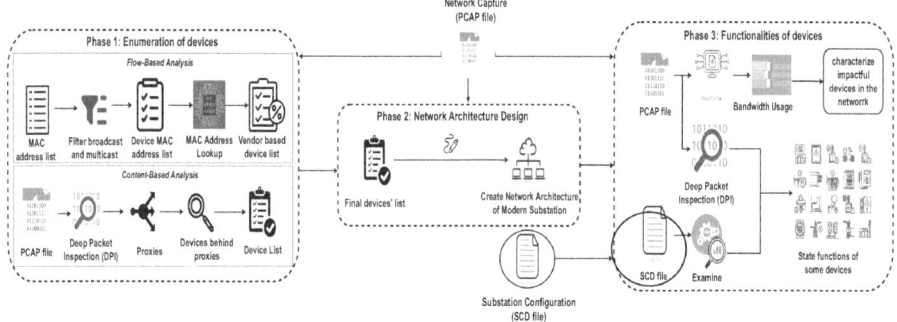

Fig. 1. Proposed framework to analyze substation networks.

Furthermore, we discovered several studies that actively utilized Substation Configuration Language (SCL) files for their research. These files are used for substation automation. For example, [8] developed a program capable of parsing SCL files and generating detailed reports on the identified components. Similarly, [13] proposed a methodology for estimating GOOSE and Sample Values (SV) traffic by extracting relevant information from SCL files. Also, [18] employed the SCL language to evaluate the domain of exchanged fields, incorporating it as supplementary data for their deep packet inspection analysis. We are unaware of any previous work using SCL files from real-world systems.

This paper uses a network trace from a real-world transmission substation and a configuration file for some of its devices. We then inspect the payloads to understand these networks' devices, services, and operations. In addition, the configuration files allow us to understand the producers and consumers of multicast information being sent in this network. Our final result is a more detailed description of an operational substation network in an academic publication. Our work fills a gap in the academic community in understanding real-world substation networks.

3 Substation Automation

Early substations had to convert several analog-to-digital signals at the local control room, requiring individual copper wires for each signal to traverse from the switchyard to the control room. This arrangement led to frequent failures and incurred significant maintenance and operation costs.

In the 1990s, the transition from conventional protection and control systems to digital Substation Automation Systems (SASs) brought significant changes. The distinction between the conventional and digital substations lies in the location of the information digitization process. This conversion happens at the source in digital substations, allowing the transmission of digital data to the local control room via more reliable channels like optical fiber. Figure 2 illustrates the difference between conventional and digital substations.

Digital substations offer significant advantages, including increased safety by eliminating the electrical connection between high voltage equipment and protection/control panels, reduced cabling, maintenance, and installation time, utilization of advanced communication protocols (such as IEC 61850 and IEC 62351), and high reliability.

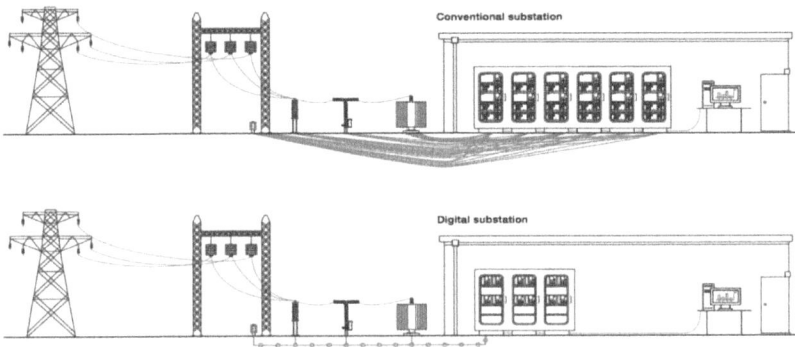

Fig. 2. Conventional Substation vs. Digital Substation: Digital Substations replace many copper wires with a single fiber-optic bus.

3.1 Substation Architecture

Substations are situated at various stages of the Power Grid, each with specific functions and equipment tailored to their location, such as generation, transmission, or distribution. Despite that diversity, a standardized substation architecture typically includes three levels based on function and location.

The three levels are: **process**, **bay**, and **station**-level.

Figure 3 shows different substation levels connected by process bus and station bus.

Process Level: The process level includes primary equipment like switching devices, circuit breakers, and instrument transformers. Information from measuring devices is transmitted through the process bus using the Sampled Values (SV) protocol, facilitating communication between Merging Units (MUs) and Intelligent Electronic Devices (IEDs) over Ethernet. IEDs at this level serve as sensors and actuators connected to the process bus through LAN technology. MUs synchronize time by receiving inputs from instrument transformers and circuit breaker auxiliary contacts.

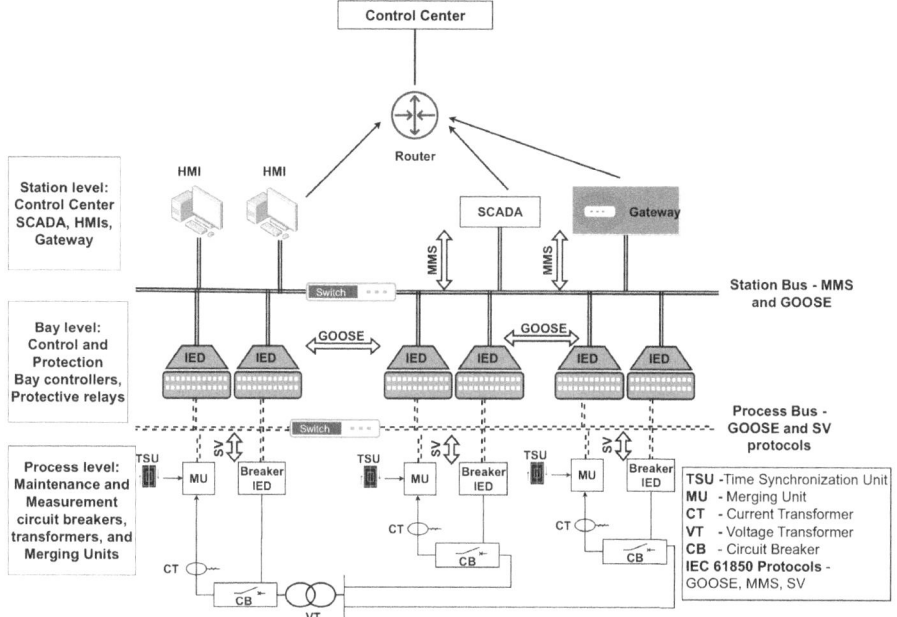

Fig. 3. Levels of a Substation Automation System.

Bay Level: The bay level consists of IEDs, which are microprocessor-based power system equipment with control and automation functions. These devices process sensor data and issue control commands to prevent failures. They utilize the GOOSE protocol for communication. Common IEDs include circuit breakers, protective relays, and PLCs.

Station Level: The station level in the control room includes Human-Machine Interfaces (HMI), SCADA systems, Fault Recorders, Alarms, and the Gateway to the remote control center. The client can read/write data, read configuration, and exchange files at this level.

Our network capture is at the interface between the Station level and the Bay level of the substation, so in the remainder of this paper, we focus on this supervisory network.

3.2 IEC 61850 in Substations

Improvements in automation and reliability need interoperability. The IEC 61850 standard by the Technical Committee Number 57 (TC57) Working Group 10 (WG10) has three main principles:

1. Defining a **unified information model.**
2. Usage of standard file (**substation configuration description**) with specific rules.
3. Defining a system-wide **communication protocol** to exchange data.

Information Model: The information model or data model of IEC 61850 includes a naming hierarchy (described in Appendix A.1) and specific data structures for any compliant device. Vendors should use the same concepts with the same name and a standard format to build their information. This feature reduces errors and format conversions.

Substation Configuration Description (SCD)
Integrating substation equipment in automation requires a standardized format to configure networking capabilities and improve performance. IEC 61850 Part 6 defines the Substation Configuration Description Language (SCL/SCD), using the XML schema, to describe IED functions and communication networks [13]. The detailed structure of the SCD file is described in Appendix A.2.

Communication Protocols
The IEC 61850 protocol stack ensures reliable communication between devices, promoting interoperability. It includes sub-protocols like Manufacturing Message Specification (MMS) over TCP/IP, while GOOSE and SV utilize Ethernet multicast for real-time data exchange. Ethernet multicast enables fast and prioritized data distribution to subscribers with low delays (within 3 ms or 20 ms).

GOOSE: We focus on GOOSE because our network capture includes this industrial protocol. GOOSE is a publish-subscribe communication service between IEDs [12]. It ensures fast and reliable distribution of binary status information through multicast services, enabling high performance and availability [11].

3.3 Parallel Redundancy Protocol (PRP)

Substations utilize redundancy measures to ensure uninterrupted operation and minimize downtime during failures or disruptions. In particular, the IEC released specifications for industrial-Ethernet communications under the IEC 62439 standard mandated by IEC 61850 [3].

PRP is a part of IEC 62439. PRP protocol provides redundancy by connecting devices to two independent Ethernet networks (LAN_A and LAN_B). This implementation allows Ethernet frames to reach their destination through an alternate port if one network fails. PRP also allows the connection of devices that do not support PRP (these devices will only see one Ethernet network). After we obtained our network capture, we first realized that all frames in the network were encapsulated in this PRP protocol. We will describe our analysis next.

4 Enumeration of Devices

Our data consists of a network capture (pcap file) of 14 continuous hours and a configuration file (SCD) from a large (500 KV) real-world substation automation network. Figure 1 shows our proposed framework for analyzing the pcap and configuration files from a substation.

The goal in the initial phase is to enumerate all the devices in the network. Our first attempt used flow-based analysis to discover unique MAC addresses. However, we later found that several devices in these networks are behind proxies, and to discover those, we need to perform a content-based analysis by looking at the payload of the packets.

Table 2. Taxonomy of devices based on vendors

Protocol	Device Vendor Name	# of Devices	Percentage
Ethernet Devices	ABB AUTOM	33	31.73%
	SIEMENS	10	9.61%
	IPCAS	10	9.61%
	ABB SWITZ	3	2.88%
	HP	5	4.8%
	AMETEK	4	3.84%
	RuggedCom	3	2.88%
	MegaSystem	3	2.88%
	Moxa Technologies Corp.	2	1.92%
	Advantech	2	1.92%
	CADAC	1	0.96%
	Reason Technologia SA	1	0.96%
	Dlink	1	0.96%
Serial Devices	SEL IED	9	8.65%
	Unknown Devices	16	15.38%
	Modem	1	0.96%

In the second phase, we build the network architecture by integrating the initial findings and the guidance of the IEC standard. In the final stage, we combine deep-packet inspection and SCD files to identify the functionality of most devices in the network.

This section focuses on the first phase, where we enumerate devices by network flow analysis and then by content-based analysis (to identify devices behind proxies).

4.1 Flow-Based Analysis

We found 97 unique MAC addresses in the network capture. Of them, 19 are broadcast/multicast destinations for different Ethernet and IP protocols such as PRP, GOOSE, IPv4/IPv6 multicast, and broadcast. For the remaining 78 Ethernet devices, we used a MAC address look-up [32] to classify them based on their vendor name as shown in Table 2. From this table, we can see that most of the Ethernet devices are from just three vendors: ABB AUTOM (which refers to the industrial company ABB), SIEMENS (referring to the industrial company Siemens), and IPCAS (which as we will show later, is also from Siemens).

4.2 Content-Based Analysis

During this step, we examined the payload of network traffic from each identified device in the flow-based analysis. Using deep packet inspection, we found the presence of two types of proxies in the network: RedBoxes (referred to as AS1, AS2, AS3) from ABB SWITZ and Serial to Ethernet Gateways (referred to as M1, M2) from MoxaTech Corp.

Redundancy Boxes (RedBoxes): RedBoxes serve proxies that facilitate the connection of devices that do not support the PRP protocol (we will discuss this protocol in the next section).

Table 3 shows the summary of the devices behind each RedBox.

Serial to Ethernet (S2E) Gateways: After identifying the MAC address of these gateways (which we refer to as M1 and M2), we figured they connected legacy serial devices to the substation network. We analyzed the traffic through these gateways and discovered 26 additional devices behind M1 and M2. We will describe more details in Sect. 6.3.

We are now ready to switch to Phase 2: identifying how these devices are connected to the network.

5 Network Architecture

The first surprise we found when analyzing the pcap file was that all Ethernet frames were encapsulated (PRP enclosure trailer 0x88fb) with a protocol we hadn't heard of before: PRP.

After finding the documentation about PRP, we discovered that this protocol is used for high reliability and availability. PRP provides full redundancy by connecting two Ethernet networks and using both to send all information. This level of redundancy makes sense in highly critical networks, where an alert about an unsafe condition (e.g., an overcurrent alert) needs to be propagated reliably to devices that can respond to this condition (e.g., opening a circuit breaker to prevent a fire).

The IEC standard classifies each device connected to the PRP network into three categories [8]:

- ***Doubly-Attached Nodes (DAN)*** - devices with two network interfaces and connected to both the LANs (LAN_A and LAN_B).
- ***Singly-Attached Nodes (SAN)*** - non-critical devices having a single network interface and only connect to one of the networks (either LAN_A or LAN_B) without requiring redundant communication [8].
- ***Redundancy Box (RedBox)*** - gateways for SAN devices without PRP support to connect to both networks. These devices, called VDANs, use RedBox messages that include the associated SAN's source MAC address. Figure 4 differentiates the RedBox Message from the normal DAN message.

From the above, devices that speak PRP are either DANs or RedBoxes.

Table 3. Devices connected to RedBoxes

RedBox	Devices Connected
AS1	No Devices
AS2	2 HP computers and 1 Advantech Industrial PC
AS3	1 HP computer and 1 Advantech Industrial PC

5.1 Network Architecture

We identified more than 50% of the devices as DANs, based on the PRP packets they send. Additionally, 2.8% of the devices incorporate RedBox Messages in their PRP packets. Behind the RedBoxes are five devices, accounting for 4.8% of the total devices. The network also includes two Serial to Ethernet Gateways, facilitating the connection of approximately 25% of the devices. And, 14.4% of the other SANs directly connect to the Ethernet switch without requiring a RedBox intermediary.

Figure 5 shows our inferred network architecture. In short, the critical devices in the network leverage the redundant PRP network, while less critical devices are only connected to one network interface. As we will see in the next phase, these critical devices are mostly IEDs, which are the devices that monitor and protect the substation equipment from accidents and failures.

6 Identifying Device Functionality

To identify device functionalities, we analyzed the protocols on the network.

Most (95%) of the network traffic consists of PRP, GOOSE, and TCP transmissions. PRP supervision packets comprise over 62% of the packets and 50% of the bandwidth, GOOSE contributes to more than 13% of packets and 31% of bandwidth, while TCP occupies 19% of packets and 15% of bandwidth.

This section details devices utilizing PRP, GOOSE, and TCP protocols.

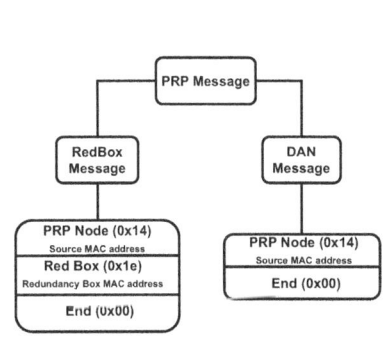

Fig. 4. Types of PRP packets

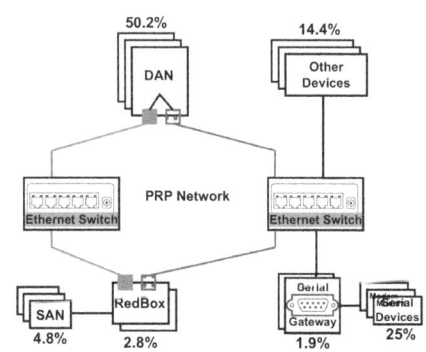

Fig. 5. Real-world substation network architecture

6.1 PRP

Most of the traffic in our network corresponds to heartbeat messages from the PRP protocol, so we are checking that the network is available. In more detail, DANs and RedBoxes regularly transmit PRP messages to notify other devices or network infrastructure that the originating device is still active and reachable, often known as "**heartbeat**" or "**keep-alive**" packets. These packets are small and multicast to a MAC address (`01:15:4e:00:01:00`), defined as PRP multicast address by an IEC standard [3].

The network capture shows that ABB AUTOM, SIEMENS, and IPCAS devices are DANs, and ABB SWITZ devices are RedBoxes with RedBox messages. All these devices are responsible for the PRP traffic.

Therefore, we confirm that all devices sending PRP traffic (connected to the redundant network) are industrial equipment that needs highly reliable communications [35]

6.2 GOOSE

After PRP traffic, GOOSE traffic is the second most prevalent in our network. GOOSE, a publish-subscribe protocol used by IEDs, facilitates the exchange of alarms, measurements, device statuses, or control commands via multicast frames. The IED standard [22] assigned the multicast addresses from `01:0c:cd:01:00:00` to `01:0c:cd:01:01:FF` to support this architecture.

We identify 28 devices publishing GOOSE messages to four multicast addresses in the network capture, as shown in Fig. 6. Among the 28 devices, 22 are manufactured by ABB AUTOM, four are SIEMENS, and two are IPCAS.

While we have identified a minimum of 28 IEDs in the network, we do not know what type of IEDs they are (e.g., a control switch, a protective relay, etc.) We also do not know if they communicate with one another: since this is a broadcast protocol, we cannot learn which devices consume the information sent by others. We need to analyze the SCD file we obtained to answer all these questions.

The goal of IEC 61850 is to facilitate substation automation, and SCD files are used to configure IEDs. Our SCD file contained information on all SIEMENS devices.

SCD File Analysis:

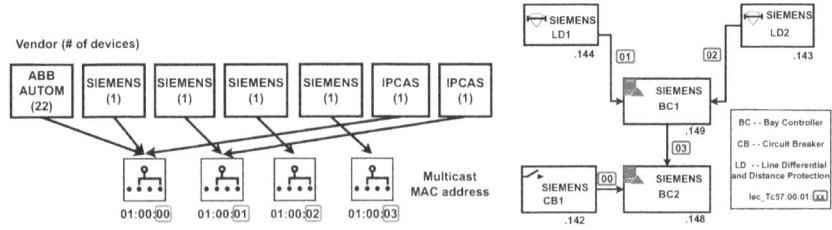

Fig. 6. PCAP analysis   **Fig. 7.** SCD file analysis

The SCD file configures Fig. 10 SIEMENS devices, which correlates with the number of SIEMENS devices observed in our pcap file.

We identify IED types using the vendor's website [29]. For example, 6MD85 (shown in Listing 1.1) is listed as a SIEMENS Bay Controller. In the SCD file, we find three Bay Controllers (BC), two Circuit Breakers (CB), two Line Differential and Distance Protection (LD) devices, two Digital Fault Recorders (DFR), and a Busbar Protection device.

In the remainder of this paper, these devices are named BC1, BC2, BC3, CB1, CB2, LD1, LD2, DFR1, DFR2, and BB.

```
<IED iedName="BC2" type="6MD85" manufacturer="SIEMENS" ....... >
    <LDevice desc="Signals" ....>
        <LN lnClass="GGIO" inst="2" desc="Goose">
            <Inputs>
                <ExtRef desc="Pos" iedName="BC1" ldInst="..."
                    lnClass="XCBR" lnInst="1" ... />
                <ExtRef desc="SPS4" iedName="CB1" ldInst="UD1"
                    lnClass="USER" lnInst="4" ... />
            </Inputs>
        </LN>
    </LDevice>
</IED>
```

Listing 1.1. Sample of Input signal. Subscriber: BC2. Publishers: BC1 and CB1

In addition to identifying devices, the SCD file provides information on which devices subscribe to GOOSE messages. While the pcap file can only indicate the devices that publish GOOSE messages, it does not provide information on devices that read them. We can determine which devices read GOOSE messages by examining the <Inputs> label in the SCD file, as shown in Listing 1.1. For example, we identify that the label inside the bay controller (BC2) instructs it to read messages from another bay controller (BC1) and circuit breaker (CB1) using the iedName. Through this analysis, we discovered that out of the ten devices, only two devices, namely the two Bay Controllers (BC1 and BC2), subscribe to GOOSE messages.

– **BC1**: subscribes to two Line Differentials (***LD1 and LD2***).
– **BC2**: subscribes to a bay controller and Circuit Breaker (***BC1 and CB1***).

This indicates that these BCs actively receive and process the GOOSE messages published by other devices within the network, as shown in Fig. 7.

Furthermore, suppose we want to learn the physical meaning of the messages they subscribe to. In that case, we can look at lnclass, which means Logical node class (defines the functionality of configured device), and desc, which means the description of the message configured.

The configuration of devices in the SCD file can have two types of Logical Nodes (LNs): (1) Standard and (2) User-Defined. The standard LNs define well-known functionalities such as Circuit Breaker (XCBR), Control Switch (CSWI), etc. The user-defined LNs (Generic Process I/O, GGIO) declare a functionality that the user gives and uses only when the functionality does not exist in standard or under catastrophic conditions.

Initially, we anticipated that all SCD files would adhere to well-known device configurations. However, we discovered that most functionalities in our SCD file were user-defined, which added complexity to our analysis as we had to establish connections between different device descriptions.

The standard configuration allows for the direct identification of input messages. For instance, in Listing 1.1, the input message from BC1 corresponds to the "Circuit breaker position" (denoted by XCBR for circuit breaker and Pos for Position). Conversely, to identify the message from CB1 (which utilizes a user-defined LN), we had to delve into CB1's configuration, search through all logical nodes for user-defined entries, and correlate them with the provided values in `Inputs` label to determine that it represents a "Busbar Interlocking Signal."

Unfortunately, the configuration of LD1 and LD2 also utilizes user-defined LNs. After an extensive search, we found BC1 subscribes to the "Busbar Interlocking Signal" and "Ground Interlocking Signal" messages from the Line Differentials. Table 4 lists the publisher and subscriber IEDs, the subscribed messages, and the corresponding Logical nodes. Below, we highlight the importance of these subscribed messages:

1. **Busbar Interlocking** is a control mechanism ensuring the safe operation of electrical busbars by coordinating device actions and maintaining network integrity and stability. It prevents faults, voltage transients, and excessive currents using predetermined rules or logic.
2. **Ground Interlocking** is a special case of busbar interlocking, where the busbar connects to the ground as an essential safety measure to protect personnel, equipment, etc.
3. **Position of Circuit Breaker** provides the status of the circuit breaker, whether ON/OFF or Transient. These inputs allow devices within the network to take appropriate actions, such as initiating protective measures or signaling other devices.

Using user-defined LNs complicates the configuration and poses challenges for modification during failures or outages. The SCD file analysis revealed unnecessary activation of user-defined LNs without utilizing their equivalent standard LNs, further adding complexity to the configuration.

Table 4. GOOSE input messages between IEDs

Subscriber	Publisher	Input	LN used
BC2	CB1	Busbar Interlocking (ON/OFF)	User Defined
BC2	BC1	Circuit break. Position (ON/OFF/Transient)	XCBR
BC1	LD1	Busbar and Ground Interlocking (ON/OFF)	User Defined
BC1	LD2	Busbar and Ground Interlocking (ON/OFF)	User Defined

In summary, our 14-hour GOOSE traffic analysis yields three key insights:
- The unchanged payload and timing of GOOSE messages indicate no critical events, affirming stable power grid conditions.

- All SIEMENS devices utilizing GOOSE are IEDs, suggesting that devices from vendors like ABB AUTOM and IPCAS could also be IEDs, totaling 53 in the network.
- The substation's configuration deviates from typical standards, likely due to vendors promoting proprietary setups that do not conform to a unified SCD file or standard LNs.

6.3 TCP

We finalize the analysis of this section by looking at TCP flows in our network, as they represent the third most popular traffic in the capture.

Almost all TCP traffic (99.95% of TCP flows) results from encapsulating serial communications to the serial devices behind the Serial to Ethernet proxy. The remaining TCP traffic includes 0.034% of TELNET packets and 0.011% of HTTP packets.

Serial to Ethernet (S2E): From the TCP traffic header, the S2E communication appears to be only with two devices, Moxatech Serial to Ethernet Gateways (M1 and M2). But, these devices expose multiple serial links [1].

Looking at the payload of the packets sending data as serial transmission (a stream of data bits sequentially transmitted), we find they are point-to-point `RS-232` serial links. It implies only one device connects to the gateway per serial link or ***one device per TCP destination port***.

The list of communicating ports for M1 are ***950-956; 959-960; 966-972; 975-976***, and M2 are ***950-953; 966-969***.

Looking at the different port numbers, we discover that behind the two S2E gateways are 26 devices, specifically, 18 connected to M1 and eight to M2.

We analyzed the payload of traffic sent to each port to identify the devices behind these ports. We find ASCII characters sent to 11 different ports.

For example, TCP packets sent from H1 to one of the M1 ports contained ASCII characters such as `Q`, `U`, `IT`, and `..` Analyzing these characters, we identified keywords like QUIT, ACCESS, OTTER, SHOWSET, etc. Looking online for these keywords, we find that they correspond to commands in the Schweitzer Engineering Laboratories (SEL) protocol [27], which is utilized by SEL IEDs. We find that nine devices were receiving these commands, indicating the presence of 9 SEL IEDs behind M1.

Two additional devices are receiving ASCII characters. The other device we could identify is a modem, as it receives `AT` commands that control the operation of a modem [31]. We could not identify the remaining device receiving ASCII characters, which include non-word characters like `@416DCF05S8A\r\n`.

We analyzed traffic from 11 devices out of the 26 S2E-connected devices. Among the remaining devices, 13 received data in an unidentified binary protocol, and two received only Acknowledgment (ACK) packets from H1.

We also notice that this substation employs **default passwords** (`OTTER`) to access these SEL IEDs [27] (detailed in Appendix A.3). This shows that the operators assume that any device inside the substation is trusted. This might be a reasonable assumption if the network is strongly segmented (or air-gapped) from external networks. Still, it does raise some security concerns, as any device in the network can connect to these SEL IEDs and may be able to change their configuration.

In summary, our analysis of the serial communications reveals nine SEL IEDs, one modem, and 16 unknown devices behind M1 and M2.

TELNET: We will now finish our analysis of TCP flows by studying the other application layer protocols. In the network capture, we find 167 TELNET packets sent to four AMETEK devices (AME 1-4). TELNET is a text-based communication protocol, and in the pcap, we find the TELNET commands sent sequentially to each of the AMETEK devices in the order shown below:

→ 1 command of SHOW DATE
→ 30 commands of SHOW TIME
→ 1 command of LIST RECORD
→ (*if record exists*) sends a single packet with a group of commands for each record (BATCH EXECUTE; SHOW RECORD <X>; SHOW MCCONFIG <X>; SHOW PROFILE <X>; SHOW FAULTDATA <X>), where <X> represents record
→ 1 packet of LIST RECORD

We infer from these commands that the AMETEK devices may be fault recorders or a type of historian or database used to keep records of faults.

HTTP: 54 HTTP packets sent from H1 to a CADAC device include a combination of GET requests and POST status messages. Listing 1.2 shows the sample of a GET request:

```
GET /data/fault/....,....,...,FLN,RPV311,AlstomGrid,....,fault.
    zic HTTP/1.1\r\n
```

Listing 1.2. GET Sample

We find the model number RPV311, AlstomGrid from the GET request. Looking online, we find that it is a Digital Fault Recorder **DFR** (a type of IED) with Fault Location and Phase Measurement Unit (PMU) [30]. The Alstom-Grid RPV311 DFR captures and stores high-speed electrical waveform data during fault events.

6.4 Final Network Characterization

Besides the mentioned core devices, miscellaneous devices in the network play smaller roles. We now summarize our findings.

HP Computers (H2, H3, H4, H5): We discovered four additional HP computers in the network, apart from H1. These computers perform network broadcasts such as NetBIOS, LLMNR, DHCPv6, SSDP, IGMP, and Sentinel LDK UDP broadcasts. Sentinel LDK packets indicate the usage of proprietary software.

The payloads of NetBIOS host announcements reveal that these computers use Windows 7 or Windows Server 2008 operating systems, suggesting their roles as workstations. These workstations serve multiple purposes, including HMIs, alarm systems, control and monitoring devices, and enabling remote management and supervision of substation processes and equipment.

During analysis, H4 consistently exhibited packets with bad checksums due to checksum offloading (the network card allows hardware to calculate the checksum, causing random values in captured traffic). This behavior only affects the host capturing the network traffic and does not impact the transmitted packets, which identifies H4 as the computer responsible for capturing the traffic.

Industrial PC (IPC): We find two Industrial PCs from vendor Advantech. IPC is a specialized computer system designed to operate in harsh and demanding industrial environments. We only see NetBIOS and Sentinel LDK UDP broadcasts sent from this host. They use the Windows XP operating system from the NetBIOS host announcements.

MegaSystems Devices: We find three devices from the vendor MegaSystem Technologies in the network. The online information [2] reveals that these devices are UPS management cards that offer remote monitoring and control of a UPS by connecting it to the network.

RuggedComm Devices: In the network traffic, we find three RuggedComm Devices. They are robust industrial communication equipment designed for harsh environments and include networking components for reliable communication. One of these devices only receives NTP from a RedBox, and by observing the payload, we find it acts as an NTP client. The second one broadcasts STP, and the other sends ARP. We cannot identify the other two even after inspecting their payloads.

7 Timing Analysis

The preceding research [21], the GOOSE protocol's timing mechanisms for the same network capture was examined. This analysis shed light on the distribution and timing variations of data packets by focusing on the inter-arrival times and bandwidth utilization. The main findings include:

- Identifying a network behavior characterized by traffic segmentation into four clusters according to their packet inter-arrival spans - intervals of 2, 3.3, 5, and 10 s, respectively.
- Our clustering approach demonstrated a network exhibiting periodicity and latency, reinforcing the network's capability to maintain stability and communication patterns among connected devices over durations, as evidenced across a 14-hour monitoring timeframe.

Expanding upon these findings, in this research, we investigated the timing analysis of TCP traffic, contrasting it with the periodic nature of the previously studied PRP and GOOSE traffic, which covered the entirety of the 14-hour observation period. During the TCP traffic analysis, we observed packet dissections occurring within brief time frames, suggesting a different operational dynamic compared to the transmissions of PRP and GOOSE packets.

To analyze this hypothesis, we implemented a "change detection" statistic on the amount of TCP packets we saw over a period of time. In particular, we implement the nonparametric CUmulative SUM (CUSUM) on the number of packets transmitted per second in the network traffic. We used Algorithm 1 to analyze this data, and Table 5 describes parameters used in the CUSUM algorithm.

Algorithm 1. Nonparametric CUSUM

1: **procedure** CUSUM($\Delta P_t, t, \tau$)
2:     $S_0 \leftarrow 0$
3:     **for** t in ΔP_t **do**
4:         **if** $t > 0$ **then**
5:             $S_t = S_t + S_{t-1} - \delta$ ▷ true $\Longrightarrow \delta = 1$
6:         **else**
7:             $S_t = \Delta P_t$
         end
8:     **if** $S_t \geq \tau$ **then**   ▷ $\tau = median\ of\ S_t\ series$
9:         The traffic includes data packets.
10:    **else**
11:        The traffic includes no data packets.
12: **end**

Table 5. Parameters of CUSUM algorithm

Parameter	Description
t	Time in seconds
S_0, S_t	CUSUM value at $t = 0$ and $t = t$, respectively
ΔP_t	Difference in the number of transmitted packets between t and $t + 1$
τ	Threshold to classify packets transmitting data (constant)

The results, as shown in Fig. 8, indicate a clear storm of TCP activity in a limited time window. In particular, we find that this flurry of TCP activity started at around 7600 s (nearly 2 h) after the capture began and lasted for a maximum of 428 s (7.13 min). We also found that the time gap between the commands sent to the same device is less than 30 ms, which suggests that all the TCP traffic we analyzed was automated or pre-configured execution of scheduled jobs or tasks.

In addition to the big flurry of activity, we also see some smaller activity spread throughout the data capture. These smaller peaks correspond to a single connection to

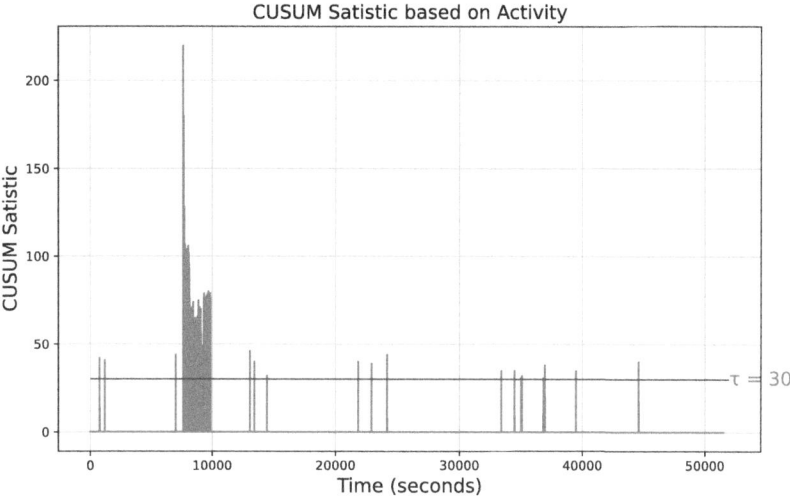

Fig. 8. Timing analysis of TCP protocol - CUSUM statistic

port 969 of gateway M2. Unfortunately, this is one of the few devices we could not identify.

8 Discussion

To finalize our analysis, we now illustrate the full substation network in Figure 9.
In summary, the real-world substation network analysis reveals:

- The vast majority of packets in the network correspond to keep alive (heartbeat) messages from two protocols: PRP and GOOSE. This shows that most of the traffic is used to check the network's status and to ensure that all devices are ready to act in case of an emergency event (e.g., an overcurrent).
- This leads us to the second observation: having long periods without events inside a real-world substation appears to be common.
- GOOSE traffic is broadcasted in the network, so classical flow analysis cannot identify destinations (or consumers of information). To obtain that information, we need SCD files.
- The configuration of the substation network does not follow standard procedures, such as using a single SCD file for all IEDs in the network or using standard LNs.
- The substation operates under a "trusted insider" assumption. This assumption may make sense if the network is fully segmented from the operational network and uses a private network. However, the modern trend in security is to assume "zero trust" networks (which means eliminating default trust assumptions, requiring verification for every access request regardless of location). So, we believe substations need to start following this zero trust configuration in the future to make them more resilient to potential future attacks.

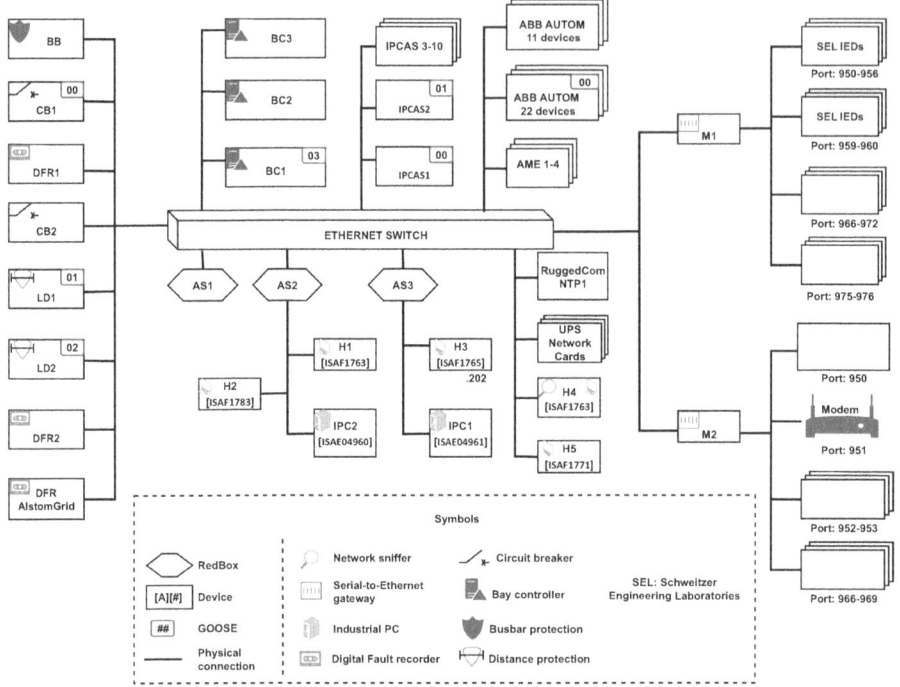

Fig. 9. Reconstructed substation network.

Our network analysis offers a unique perspective by investigating the coexistence of legacy devices and cutting-edge IEDs, setting us apart from existing literature. While previous research tends to focus exclusively on either IEC 61850 protocols [5, 17, 18] or legacy serial communication [6], our study examines both simultaneously. For instance, Chang et al. highlight the presence of IEDs and an HMI in station bus configuration, stating that *"The IEC61850 station bus interconnects 17 multifunction IEDs and the station HMI gateway unit through two independent LANs in a double-star configuration"* [8]. Similarly, other researchers acknowledge that *"IEDs connect the process bus and station bus"* [7, 18, 36]. However, our analysis uncovers that the real-world substation is more intricate, involving additional devices.

We draw attention to the existence of proxies and serial devices, which have not been discussed in previous research studies about substation networks.

We also highlight our effort to use data from actual measurements and perform a detailed architecture of the substation. As far as we are aware, Formby et al. [10] is the only other research paper discussing data from an operational substation; however, Formby provides only traffic flow analysis and does not attempt to enumerate all devices and protocols used in these networks, giving us only a glimpse of what their substations look like.

Our inspection of GOOSE packets revealed individual timestamp values for each IED, highlighting the reliance on external synchronization sources. Time synchronization is crucial in substation networks, as failures can lead to artificial phase shifts and potential tripping [28]. The IEC 61850-7-5 standard emphasizes the need for careful consideration of time synchronization [26] to prevent undesired consequences.

Furthermore, our analysis of the SCD file provided valuable insights into the substation's configuration. We discovered that the substation relied on segmented SCD files, which can challenge managing and updating the design. Previous research, such as Wang et al. [34], has proposed management and control modules for more accessible configuration file updates. Still, segmented files add another layer of complexity to this task.

Additionally, our analysis of the SCD file revealed the use of generic user-defined Logical Nodes (GGIO) in the substation. While GGIOs provide flexibility in cases where the IEC 61850 standard lacks a standardized structure, they come at the cost of losing functional descriptions. This deviation from specialized logical nodes and reliance on GGIOs can impact the interoperability and traceability of information over time, as noted by Kaneda et al. [16].

It is also worth mentioning that besides industrial protocols, some devices, such as SEL IEDs, utilize proprietary protocols over serial communications. We are not aware of other papers discussing this use in substations. The existence of these devices and unencrypted protocols may also raise many security concerns. One such finding is the use of default passwords, like OTTER, to access these IEDs, despite the option for users to modify them according to the manuals [27]. This neglect of changing default passwords can allow unauthorized personnel to access these devices. The attacks against the power grid in Ukraine are a stark reminder that sophisticated adversaries are now targeting the power grid [9].

Impacts of Findings and Future Design Strategies for Substation Networks: From our analysis, we have summarized some of the implications and suggestions for future design strategies below:

1. *Redundant Paths on Network Reliability and Efficiency:* The study's findings on redundant paths, mainly through protocols like PRP, significantly boost network reliability and fault tolerance. With advanced monitoring and management systems, careful planning and implementation of redundant paths can optimize reliability while minimizing efficiency overhead.
2. *IEDs and their evolution in future substation designs*: The study highlights the role of IEDs in substation automation and the critical need for cybersecurity in safeguarding vital infrastructure. As substation automation progresses, IEDs become essential for enhancing system intelligence, efficiency, and resilience. Future IEDs may integrate advanced cybersecurity features, such as secure communication protocols and encryption, to counter cyber threats effectively.
3. *Identified vulnerabilities and recommendations for cybersecurity:* The analysis revealed several vulncrabilities in the substation network's cybersecurity posture, including:
    - Lack of encryption for sensitive communication channels, leaving data vulnerable to interception and tampering.
    - Inadequate access control measures, allowing access to critical devices and systems multiple times unlimitedly.

To address these vulnerabilities and enhance cybersecurity:
- Implement end-to-end encryption for critical communication channels to protect data confidentiality and integrity.
- Enforce strong access control policies, including restricting any unauthorized access.
- Regularly update firmware and software on network devices to patch known vulnerabilities and mitigate security risks.
- Deploy intrusion detection and prevention systems to monitor network traffic for anomalies.

4. ***Enhancement of Network Understanding through SCD File Analysis and Challenges Encountered:*** By analyzing the SCD file, we can:
- Identify network devices, functionalities, and interconnections based on the standardized IEC 61850 data model.
- Validate device configurations and communication mappings against the intended network design and operational requirements.
- Document network configurations and settings for maintenance, troubleshooting, and future upgrades.

However, challenges encountered in using the SCD file include:
- Complexity in parsing and interpreting the XML-based format of the SCD file, requiring specialized tools or scripts for analysis.
- Inconsistencies or discrepancies between the SCD file and the network configuration necessitate manual verification and validation.
- Limited support for proprietary extensions or vendor-specific configurations, leading to incomplete or inaccurate network representations.

Despite these challenges, the SCD file remains valuable for understanding substation networks and facilitating effective network management and optimization.

9 Conclusion

As far as we know, in this paper, we provide the most detailed view of an operational substation network. We also present a novel framework for analyzing and understanding such networks by analyzing packet captures and utilizing an SCD file.

Our study found topics not commonly discussed in the industrial control network literature, such as the use of redundant Ethernet networks to guarantee reliability, the existence of several proxies (Red Boxes and Serial to Ethernet Converters), and the coexistence of legacy serial IEDs with modern IEDs following the GOOSE protocol. Most of the devices in the network were IEDs (63), emphasizing the critical role of these devices in substation automation.

We hope this study motivates future work to characterize and understand these critical networks, and perform ongoing asset inventory to identfy devices that need protection, as well as to identify any new malicious endpoint added to the network.

Acknowledgments. This work was partially supported by the Air Force Office of Scientific Research under award number FA9550-24-1-0015, by the NSF CPS program under CNS-1929410 and by the INTERSECT project, Grant No. NWA.1162.18.301, funded by the Netherlands Organisation for Scientific Research (NWO).

A Concepts of Substation

A.1 Information Model

The IEC 61850 standard defines a hierarchical information model as shown in Fig. 10 that includes:

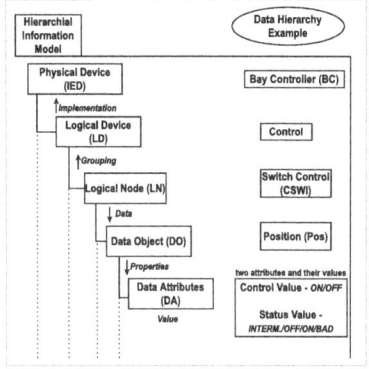

Fig. 10. Hierarchial Information Model

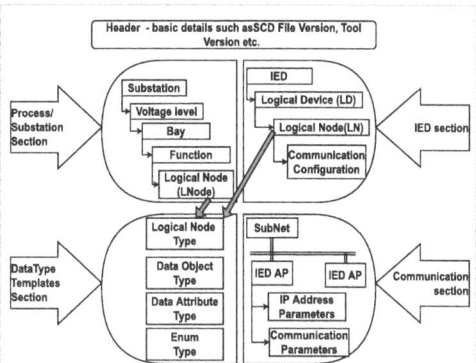

Fig. 11. SCD File Template

IED (or) Physical Device: This is the device connected to the network directly.

Logical Device (LD): The LDs are logical containers organizing the information of an IED, splitting it into different categories. Vendors categorize their information using logical device names such as PROT(protection), CTRL (control), REC (recorder), etc.

Logical Node (LN): The LNs are the functions or components that automate the system.
They are named by four letters; the first indicates their category. For example, LN CSWI stands for Control Switch.

Data Object (DO): Each LN includes a set of mandatory and optional data objects to fulfill their actions. The data objects represent status information, position, measurements, set points, controllable points, or descriptive information.

Data Attribute (DA): Each Data Object includes a group of data attributes that define the properties of the object. For example, properties such as Control Value, Status Value, timestamp, etc.

A.2 Substation Configuration Description

Integration of substation equipment is a complex task in the automation process. Using a standardized format to configure various IEDs' networking capabilities, equipment functionality, is vital in improving the performance Hence, Part 6 of the IEC 61850 standard defines the Substation Configuration description Language (SCL/SCD) that describes functions of IED and its communication network using extensive markup language (XML) schema [13]. Figure 11 shows a substation's template of an SCD file. Every SCD file template contains five main sections [33], defined as follows:

- **Header:** It helps to identify the version of the IED configurator used to create the SCD file.
- **Substation:** It defines the substation's name and its different entities, including various devices, interconnections, and other functionalities that help identify electrical connections and functions.
- **Communication:** This section describes the communication network of IEDs and the protocols they can use. For example, if an IED uses GOOSE, it is configured using a sub-element GSE.
- **IED:** This section describes the complete configuration (functions, access points, logical devices, logical nodes, data objects, inputs, and other information) of the connected IED. This section also helps to identify the interconnection between the logical nodes of multiple IEDs.
- **DataType Templates:** It provides a standardized structure for defining various communication data formats and characteristics. These templates ensure consistency and interoperability between devices and systems.

A.3 SEL Device Functionality in the Real-World Network

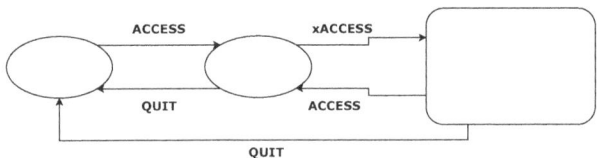

Fig. 12. Transition between different access levels in SEL devices

There are seven access levels in SEL devices. The ACCESS command is sent to enter any access level. If password protection is enabled, then it prompts for the password. The default password for access level 1 is OTTER, and access level 2 is TAIL. In access level 1, we can read the data and status information; in access level 2, we can also write the data. Access level B performs breaker control and shows breaker data. Access levels P, A, and O show protection settings, automation settings, and output settings, respectively, along with the functions of access level B. In the deep packet inspection, we only find commands from three access levels (access level 0, access level 1, and a single command from access level 2). These commands provide read access to different information. If an access level 2 command is used, we can write data to these devices from host machines [27]. Figure 12 shows the transition between different access levels in SEL devices.

The QUIT command sends the system to Access level 0 from any given level. Users utilize this command after communicating with the device to prevent unauthorized access.

The HISTORY command summarizes up to 40 previous events. Each summary shows the date, time, event type, fault location, active setting group, and targets.

Unveiling the Operation and Configuration of a Real-World Bulk Substation Network 233

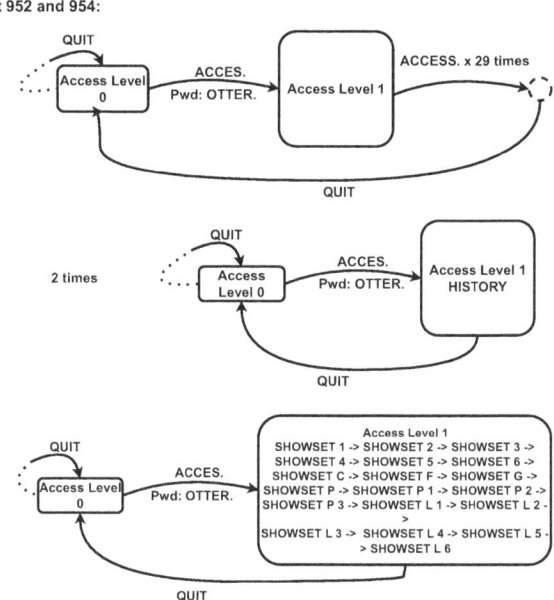

Fig. 13. State machine of commands to ports M1:952 and M1:954

The SHOWSET command displays the device settings for the selected group. One should use the command SET at access level 2 to perform update/write operations on these settings. Table 6 shows different SHOWSET commands. SHOWSET with 'A' displays all the inactive settings.

The STATUS command allows inspection of self-test status. The device executes the STATUS command whenever the self-test software enters a warning or failure state.

The TIME command displays and sets the internal clock.

We now summarize the commands we see and their destination:

- Commands to SEL Devices:
  - *M1:950*: QUIT.
  - *M1:951*: QUIT, ACCESS/OTTER, ACCESS, SHOWSET, HISTORY
  - *M1:952, 954*: QUIT, ACCESS/OTTER, ACCESS(multiples times in a row), HISTORY, SHOWSET 1-6, SHOWSET C, SHOWSET F, SHOWSET G,
    SHOWSET P, SHOWSET P 1-3, SHOWSET L 1-6.
  - *M1:[953, 955, 956, 960]*: QUIT, ACCESS/OTTER, ACCESS (multiple times in a row), HISTORY, SHOWSET
  - *M1:959*: QUIT, ACCESS/OTTER, HISTORY, SHOWSET, DATE, TIME (multiple times in a row).

Most of these commands happen repeatedly. For example, in Fig. 13, which represents the state machine of commands sent to ports 952 and 954 of M1, H1 sends

Table 6. Command summary of SEL devices

Command	Description
Access Level 0	
ACCESS	enters level 1; password OTTER
Access Level 1	
2ACCESS	enters level 2; password TAIL
HISTORY	DATE, TIME, etc., for the last 40 events
QUIT	returns control to Access Level 0
SHOWSET n	active group settings for Group n
SHOWSET C	calibration settings
SHOWSET G	global settings
SHOWSET L	active logic settings
SHOWSET P	active port settings
STATUS	self-test status
TIME	Shows or sets time
Access Level 2	
SHOWSET F	future re-calibration settings

ACCESS command 29 times, despite reaching access level 1. It is followed by QUIT and returns to access level 0. It again enters access to level 1 and requests the settings of these devices connected to 953 and 954.

References

1. Which TCP/UDP ports do i need to open to access my mgate gateway remotely? https://www.moxa.com/en/support/product-support/product-faq/which-tcp-udp-ports-to-open-to-access-mgate-gateway-remotely. Accessed 05 May 2022
2. Megatec product list (1998-2004). http://www.megatec.com.tw/Product_list.htm. Accessed 05 May 2022
3. Araujo, J., Lázaro, J., Astarloa, A., Zuloaga, A., García, A.: PRP and HSR version 1 (IEC 62439-3 ed.2), improvements and a prototype implementation. In: IECON 2013 - 39th Annual Conference of the IEEE Industrial Electronics Society, pp. 4410–4415 (2013). https://doi.org/10.1109/IECON.2013.6699845
4. Barbier, G., Conti, M., Tippenhauer, N.O., Turrin, F.: Assessing the use of insecure ICS protocols via IXP network traffic analysis. In: 2021 International Conference on Computer Communications and Networks (ICCCN), pp. 1–9 (2021). https://doi.org/10.1109/ICCCN52240.2021.9522219
5. Biswas, P.P., Tan, H.C., Zhu, Q., Li, Y., Mashima, D., Chen, B.: A synthesized dataset for cybersecurity study of IEC 61850 based substation. In: 2019 IEEE International Conference on Communications, Control, and Computing Technologies for Smart Grids (SmartGridComm), pp. 1–7 (2019). https://doi.org/10.1109/SmartGridComm.2019.8909783

6. Boakye-Boateng, K., Siahaan, I.S.R., Al Muktadir, A.H., Xu, D., Ghorbani, A.A.: Sniffing serial-based substation devices: a complement to security-centric data collection. In: 2021 IEEE PES Innovative Smart Grid Technologies Europe (ISGT Europe), pp. 1–6 (2021). https://doi.org/10.1109/ISGTEurope52324.2021.9640212
7. Buhagiar, T., Cayuela, J.P., Procopiou, A., Richards, S., Ramlachan, R.: Smart substation for the French power grid. In: 69th Annual Conference for Protective Relay Engineers (CPRE). IEEE (2016). https://doi.org/10.1109/CPRE.2016.7914915
8. Chang, J., Vincent, B., Reynen, M.: Protection and control system upgrade based on IEC-61850 and PRP. In: 2014 67th Annual Conference for Protective Relay Engineers, pp. 496–517 (2014)
9. CyberSecurity & Infrastructure Security Agency: Cyber-attack against Ukrainian critical infrastructure (2021). https://www.cisa.gov/news-events/ics-alerts/ir-alert-h-16-056-01/
10. Formby, D., Walid, A., Beyah, R.: A case study in power substation network dynamics. SIGMETRICS Perform. Eval. Rev. **45**(1), 66 (2017). https://doi.org/10.1145/3143314.3078525
11. Hoga, C., Wong, G.: Utilities and industries gain benefits as IEC 61850-implementation gains speed. In: 2005 IEEE Power Engineering Society Inaugural Conference and Exposition in Africa, pp. 176–179 (2005)
12. Huang, W.: Learn IEC 61850 configuration in 30 minutes. In: 2018 71st Annual Conference for Protective Relay Engineers (CPRE), pp. 1–5 (2018). https://doi.org/10.1109/CPRE.2018.8349803
13. Ingalalli, A., Silpa, K.S., Gore, R.: SCD based IEC 61850 traffic estimation for substation automation networks. In: 2017 22nd IEEE International Conference on Emerging Technologies and Factory Automation (ETFA), pp. 1–8 (2017). https://doi.org/10.1109/ETFA.2017.8247596
14. Juárez, J., Rodríguez-Morcillo, C., Rodríguez-Mondéjar, J.A.: Simulation of IEC 61850-based substations under omnet++. In: SIMUTOOLS 2012, ICST (Institute for Computer Sciences, Social-Informatics and Telecommunications Engineering), Brussels, BEL, pp. 319–326 (2012)
15. Kanabar, M.G., Sidhu, T.S.: Performance of IEC 61850-9-2 process bus and corrective measure for digital relaying. IEEE Trans. Power Deliv. **26**(2), 725–735 (2010)
16. Kaneda, K., Tamura, S., Fujiyama, N., Arata, Y., Ito, H.: Iec61850 based substation automation system. In: 2008 Joint International Conference on Power System Technology and IEEE Power India Conference, pp. 1–8 (2008). https://doi.org/10.1109/ICPST.2008.4745296
17. Konka, J.W., Arthur, C.M., Garcia, F.J., Atkinson, R.C.: Traffic generation of IEC 61850 sampled values. In: 2011 IEEE First International Workshop on Smart Grid Modeling and Simulation (SGMS), pp. 43–48. IEEE (2011)
18. Kwon, Y., Lee, S., King, R., Lim, J.I., Kim, H.K.: Behavior analysis and anomaly detection for a digital substation on cyber-physical system. Electronics **8**(3) (2019). https://doi.org/10.3390/electronics8030326
19. León, H., Montez, C., Stemmer, M., Vasques, F.: Simulation models for IEC 61850 communication in electrical substations using goose and SMV time-critical messages. In: 2016 IEEE World Conference on Factory Communication Systems (WFCS), pp. 1–8 (2016). https://doi.org/10.1109/WFCS.2016.7496500
20. León, H., Montez, C., Valle, O., Vasques, F.: Real-time analysis of time-critical messages in IEC 61850 electrical substation communication systems. Energies **12**(12) (2019). https://doi.org/10.3390/en12122272
21. Lozano, J., Koneru, K., Castellanos, J.H., Cardenas, A.A.: Timing analysis of goose in a real-world substation. In: 2022 IEEE International Conference on Communications, Control, and Computing Technologies for Smart Grids (SmartGridComm), pp. 160–165 (2022). https://doi.org/10.1109/SmartGridComm52983.2022.9961030

22. Mackiewicz, C.R.: Technical overview and benefits of the IEC 61850 standard for substation automation. https://api.semanticscholar.org/CorpusID:17966971
23. Mai, K., Qin, X., Ortiz, N., Molina, J., Cardenas, A.A.: Uncharted networks: a first measurement study of the bulk power system. In: Proceedings of the ACM Internet Measurement Conference, IMC 2020, pp. 201–213. Association for Computing Machinery, New York (2020). https://doi.org/10.1145/3419394.3423630
24. Nivethan, J., Papa, M., Hawrylak, P.: Modeling and simulation of electric power substation employing an IEC 61850 network. In: Proceedings of the 9th Annual Cyber and Information Security Research Conference, CISR 2014, pp. 89–92. Association for Computing Machinery, New York (2014). https://doi.org/10.1145/2602087.2602096
25. Ortiz, N., Rosso, M., Zambon, E., den Hartog, J., Cardenas, A.A.: From power to water: dissecting SCADA networks across different critical infrastructures. In: International Conference on Passive and Active Network Measurement, pp. 3–31. Springer (2024)
26. Ozansoy, C.R., Zayegh, A., Kalam, A.: Time synchronisation in a IEC 61850 based substation automation system. In: 2008 Australasian Universities Power Engineering Conference, pp. 1–7 (2008)
27. Schweitzer Engineering Laboratories: Changing the default passwords; table 4.3 sel-421 relay access levels; figure 4.5 access level structure - sel -421 user manual (2012). https://www.manualslib.com/manual/1645670/Sel-Sel-421.html. Accessed 13 May 2022
28. Shrestha, A., Silveira, M., Yellajosula, J., Mutha, S.K.: Understanding the impacts of time synchronization and network issues on protection in digital secondary systems. In: PAC World Global Conference 2021 (2021)
29. Siemens AG: Siemens SIPROTEC 5 Portfolio. https://new.siemens.com/global/en/products/energy/energy-automation-and-smart-grid/protection-relays-and-control/siprotec-5.html. Accessed 05 May 2022
30. GG Solutions: Reason RPV311. https://www.dsgenterprisesltd.com/product/ge-reason-rpv311-digital-recorder/
31. TW Solutions: At commands reference guide. https://www.sparkfun.com/datasheets/CellularModules/AT_Commands_Reference_Guide_r0.pdf. Accessed 05 May 2022
32. Stiller, N.: Mac address lookup (2019). https://www.macvendorlookup.com/
33. Wang, J., Wang, Z.: Research and implementation of virtual circuit test tool for smart substations. Procedia Comput. Sci. **183**, 197–204 (2021)
34. Wang, L., Huang, J., Zhang, C.: Design and development of SCD file management and control system for serving substation reconstruction and expansion projects. In: 2022 IEEE 10th Joint International Information Technology and Artificial Intelligence Conference (ITAIC), vol. 10, pp. 1782–1790 (2022). https://doi.org/10.1109/ITAIC54216.2022.9836958
35. Weibel, H.: Tutorial on parallel redundancy protocol (PRP) (2003). http://caxapa.ru/thumbs/767218/tutorial-on-prp.pdf
36. Wong, T.J., Das, N.: Modelling and analysis of IEC 61850 for end-to-end delay characteristics with various packet sizes in modern power substation systems. In: 5th Brunei International Conference on Engineering and Technology (BICET 2014), pp. 1–6 (2014). https://doi.org/10.1049/cp.2014.1073
37. Zhao, J., et al.: A network scheme for process bus in smart substations without using external synchronization. Int. J. Electr. Power Energy Syst. **64**, 579–587 (2015). https://doi.org/10.1016/j.ijepes.2014.07.066. https://www.sciencedirect.com/science/article/pii/S0142061514005018

RustBound: Function Boundary Detection over Rust Stripped Binaries

Ryan Evans, William Hawkins, and Boyang Wang[✉]

University of Cincinnati, Cincinnati, OH, USA
evans2ra@mail.uc.edu, hawkinwh@ucmail.uc.edu, boyang.wang@uc.edu

Abstract. Function boundary detection identifies start addresses and end addresses of functions in a binary. It is a critical step in binary analysis and is considered as a challenging task over stripped binaries. While existing studies have shown that it is feasible to efficiently and accurately perform function boundary detection over C stripped binaries, it remains unknown whether these methods will perform well over Rust stripped binaries. In this paper, we experimentally evaluate and compare four methods/tools, including two industry reverse engineering tools (Ghirda and IDA Pro) and two neural-network-based methods, in the context of function boundary detection over Rust binaries. We establish a large-scale dataset consisting of 2,471 Rust binaries (with over 8.69 million functions) across five optimization levels and develop two tools to perform analyses automatically. We derive two major findings based on our experimental results. First, one of the two neural-network-based methods, named XDA, can achieve promising results (e.g., 94.8% precision and 85.5% recall over binaries compiled with O0) and outperform other methods/tools in detecting function boundaries over Rust binaries, except over binaries from Oz optimization. Second, although Ghidra and IDA Pro can accurately detect function starts, they are not effective on precisely distinguishing function ends over Rust binaries.

Keywords: Function detection · Rust · Stripped binaries

1 Introduction

Function boundary detection identifies addresses of function starts and function ends over binaries. It is a fundamental component in binary analysis and is critical for downstream tasks, such as function similarity analyses, binary rewriting, and malware detection [8,15]. However, function boundary detection over *stripped binaries* is considered as a challenging task as function information are no longer available in stripped binaries.

Several studies [4,5,7–9,16,18,20] have proposed to utilize neural networks to address function boundary detection over stripped binaries. These methods can even outperform state-of-the-art reverse engineering tools, including Ghidra

[1] and IDA Pro [2]. Despite promising results over binaries compiled from C programs in the current literature, *it remains unknown whether neural-network-based methods would perform well over Rust binaries—binaries compiled from Rust programs.*

Rust is considered as a safer system-level programming language than C by enforcing memory safety and concurrency safety [3,12]. Rust has become more popular in the past several years, especially for embedded systems [6]. Despite its safety enhancement, some recent research also demonstrate that critical security vulnerabilities can still be found in Rust programs [11,13,14]. Therefore, we believe that understanding the capability of binary analysis, including function boundary detection, over Rust binaries is critical and necessary.

In this paper, we leverage four existing methods/tools in the context of function boundary detection and experimentally compare their performance over large-scale stripped Rust binaries. Specifically, we investigate two state-of-the-art reverse engineering tools, including Ghidra and IDA Pro, and two neural-network-based methods, denoted as BiRNN [18] and XDA [16]. Both BiRNN and XDA are originally designed for function detection over C binaries. BiRNN is built upon bi-directional Recursive Neural Networks while XDA is based on masked Language Modeling. Our contributions and observations are summarized as below:

- We establish a large-scale dataset, named RUBIN, which consists of 2,471 Rust binaries (x86-64 in ELF – Executable and Linkable Format) with over 8.69 million functions across five optimizations, including O0, O1, O2, O3, and Oz. The entire dataset (including stripped and non-stripped binaries) is over 66 GBs, where non-stripped binaries are kept for reproducibility and future expansion.
- We develop two open-source tools, named `ripkit` and `cargo_picky`, that can automatically obtain Rust stripped binaries and obtain results of function boundary detection over the four methods we examine. These tools can be utilized to reproduce our results and expand our findings over new datasets/methods over Rust binaries. We believe that both our dataset and the tools are valuable contributions to the research community.
- Our experimental results suggest that (1) XDA can achieve promising results (e.g., 94.8% precision and 85.5% recall over binaries compiled with O0) in function boundary detection and outperform the other 3 methods/tools, except over stripped binaries from Oz optimization; (2) both Ghidra and IDA Pro perform very well on detecting function starts (e.g., $\geq 92.7\%$ precision and $\geq 97.7\%$ recall) but are less promising on identifying function ends (e.g., as low as 55.8% precision and 56.4% recall).

Reproducibility. Our source code and dataset are publicly available at https://github.com/UCdasec/RustBound.

2 Background

Rust. Rust is a relatively new programming language which has a similar syntax and performance as C and C++. However, Rust is safer than C and C++ by offering memory safety and concurrency safety. Specifically, Rust catches vulnerabilities, such as memory corruption, race conditions, and data races, at compile time by leveraging the concept of *ownership* and the *borrower checker*. To compile a Rust program into a binary (either a library or an executable), Rust utilizes a specific compiler named rustc. Due to its enhanced safety, it has become more popular among developers and has been recently added to the Linux kernel and Microsoft Windows.

A *crate* is the smallest amount of code that the Rust compiler considers at a time[1]. A crate can be a binary crate or a library crate. A binary crate must have a main function and can be compiled to an executable. A library crate does not consist of a main function and cannot be compiled to an executable. All the crates we evaluated in this study are binary crates.

Stripped v.s. Non-Stripped. When producing a binary from a program, a compiler has an option (-s) to remove debugging and symbol information that are not essential for execution. This process, named *stripping*, is typically done to greatly reduce the binary file size as well as making the binary much more difficult to disassemble. If stripping is applied, a binary generated by a compiler is referred to as a *stripped binary*. Otherwise, it is a *non-stripped binary*. In addition to applying at compiler time, stripping can also be performed independently over a non-stripped binary after it has been compiled with the strip command. Besides performance improvement, another major benefit of applying stripping is to make it much more difficult for reverse engineering, either from a benign perspective for better protecting intellectual property or from a malicious perspective for hiding malicious code.

Function Boundary Detection. Given a binary (in essence, a sequence of bytes), function boundary detection is a task for outputting a label to every byte in this sequence, where a label is *function start*, *function end*, or *neither*. More formally speaking, given a sequence of bytes $B = (b_1, ..., b_n)$, function boundary detection F outputs a sequence of labels $L = (l_1, ..., l_n)$ as

$$F(B) \rightarrow L = (l_1, ..., l_n), l_i \in \{\text{S}, \text{E}, \text{N}\} \text{ for } 1 \leq i \leq n \tag{1}$$

where l_i is the label of byte b_i and l_i is either S (function start), E (function end), or N (neither). Function boundary detection often happens concurrently with disassembly or directly after. It is a fundamental component in binary analysis and is critical for downstream tasks, such as function similarity analyses, binary rewriting, and malware detection [8].

Function boundary detection is a trivial task over a non-stripped binary as function names, addresses of function starts, and function lengths are available

---

[1] https://doc.rust-lang.org/book/title-page.html.

in the headers of a binary. A debugger, such as gdb, can interpret these information easily and locate the addresses of function starts and ends. However, it is considered much more challenging to perform over stripped binaries, which do not consist of the information of function names, addresses of function starts, and function lengths.

Examples. We provide several examples of function starts in Rust and show that the (potential) signatures, i.e., the beginning bytes, of functions vary and are not trivial to distinguish. First, we present three examples of function starts where the binaries are compiled with O0 optimization. In Listing 1.1, a function starts with a sub instruction followed with a lea instruction. In Listing 1.2, a function starts with an or instruction followed with a mov instruction. In Listing 1.3, a function starts with a mov instruction.

```
0x71a0 <rpn_reckoner_function>:
sub $0x18, %rsp
lea -0x1(%rsi), %rax
bsr %rax, %rcx
...
```

Listing 1.1. Function Start Example 1 (Assembly O0)

```
6a90 <exa_function>:
or %esi, %edi
mov %edi, -0x4(%rsp)
mov -0x4(%rsp), %eax
...
```

Listing 1.2. Function Start Example 2 (Assembly O0)

```
0x169800 <exa_function>:
mov 0x(%rdi), %ax
ret
cs nopw
...
```

Listing 1.3. Function Start Example 3 (Assembly O0)

It is worth to mention that the signatures of function starts (and ends) are constantly complex and difficult to observe across different optimizations. For instance, the following three function starts in binaries compiled with Oz optimization show various patterns of function starts.

```
0x52ff2 <exa_function>:
push %r14
push %rbx
sub $0x38, %rsp
...
```

Listing 1.4. Function Start Example 4 (Assembly Oz)

```
0x52ff2 <exa_function>:
push %rcx
cmp $0x2, %edi
je 53018
...
```

Listing 1.5. Function Start Example 5 (Assembly Oz)

```
0xe40cb <mgart_function>:
shr %rsi
cmp 0x8(%rdi), %rsi
jae e40d7
...
```

Listing 1.6. Function Start Example 6 (Assembly Oz)

Functions with Padding Bytes. It is typical for a compiler to add padding bytes (0x00s or NOPs) at the end of a function. In this study, *our definition of a function end indicates the address of the last byte of a function excluding padding bytes.* As shown in the following example, a function (`foo`) with function length 0x2a starts at 0x8970 and ends at 0x8999 (i.e., 0x8970 + 0x2a - 1). There are several padding bytes starting from 0x899a but before the start of the next function (main) at 0x89a0.

```
0x8970 <foo>:
...
0x8995: 48 83 c4 38              add $0x38, %rsp
0x8999: c3                       ret
0x899a: 66 0f 1f 44 00 00        nopw

0x89a0 <main>:
...
```

Listing 1.7. Example of A Function with Padding Bytes (Assembly O0)

Evaluation Metric. By following the evaluation metric from existing studies [4, 5, 7–9, 16, 18, 20], we examine the performance of a method from three aspects (1) function starts, (2) function ends, and (3) function boundaries, using precision, recall, and F1 score. Precision, recall, and F1 score are defined as below.

$$Pecision = \frac{TP}{TP + FP}$$

$$Recall = \frac{TP}{TP + FN}$$

$$F1 = \frac{2 \cdot P \cdot R}{P + R}$$

where TP is true positive, FP is false positive, and FN is false negative.

For function starts, it measures the capability of a method on detecting start addresses of functions in a binary. Let set $S = \{s_1, ..., s_m\}$ be the start addresses of all the m functions in binary B, where s_i is the start address of function f_i in B. Given a set of start addresses $S' = \{s'_1, ..., s'_k\}$ reported by a method over binary B, the true positive, false positive, and false negative of function starts are defined as below:

- TP: the number of start addresses that are in both set S and S', where $TP = |S \cap S'|$.
- FP: the number of start addresses that are in set S' but not in set S, where $FP = |S' - S|$.
- FN: the number of start addresses that are in set S but not in set S', where $FP = |S - S'|$.

TP, FP, and FN of function ends are defined similarly by using the end addresses of functions.

A function boundary is reported correctly if both the function start and function end are correct. In other words, the performance of function boundaries can be aggregated based on the performance of function starts and ends. Specifically, we assume that there are m functions in a binary B, where the function starts and ends can be described with a set of start-end address pairs, $D = \{(s_1, e_1), ..., (s_n, e_m)\}$. Let s_i and e_i be the start address and end address of function f_i respectively. Given a set of start-end address pairs $D' = \{(s'_1, e'_1), ..., (s'_n, e'_k)\}$ reported by a method over binary B, we define the true positive, false positive, and false negative of function boundaries over binary B as below

- TP: the number of start-end address pairs that are both in set D and D', where $TP = |D \cap D'|$.
- FP: the number of start-end address pairs that are in set D' but not in set D, where $FP = |D' - D|$
- FN: the number of start-end address pairs that are in set D but not in set D', where $FN = |D - D'|$.

3 Function Boundary Detection Methods

Existing Reverse Engineering Tools. Popular reverse engineering tools, such as Ghidra and IDA Pro, have historically struggled with disassembling stripped binary files. Traditionally, both tools rely on sophisticated dynamic analysis techniques, heuristics, and complex algorithms to first reconstruct the source code of a given binary file in order to perform program boundary detection. When debugging information are unavailable, it introduces many challenges to reconstruct code and detect function boundaries correctly.

Recently, both Ghidra and IDA Pro have began leveraging signature-based approaches to complement dynamic analysis in function boundary detection for functions in standard libraries. For instance, IDA Pro leverages FLIRT algorithm[2] and Ghidra utilizes Function ID in their signature-based approaches[3] respectively. By searching a large-scale signature database and comparing hash values of bytes in a given binary, these tools can detect function entries and exits more effectively.

Function Boundary Detection using Neural Networks. There are several studies have investigated neural networks in function boundary detection over stripped binaries [4,5,7–9,16,18,20]. However, all the findings are based on binaries from C programs. In this study, we specifically discuss two methods and examine their performance over Rust stripped binaries. One method, referred to as BiRNN [18], is the first study utilizing neural networks for function boundary detection. The other method, referred to as XDA [16], is one of the state-of-the-art methods in disassembly and function boundary detection using neural networks.

Details of BiRNN. Shin et al. [18] first leveraged neural networks to address function boundary detection. Specifically, binaries are pre-processed into 1000-byte sequences, where each byte is encoded into a $\mathbb{R}^{256}$ binary vector with one-hot encoding. These byte sequences are passed to a Bi-directional Recursive Neural Network (BiRNN) to train in order to locate function starts or function ends. Once a BiRNN is trained, it predicts labels over bytes in 1000-byte sequences. It is also worth mentioning that one BiRNN can only report function starts or function ends but not both. When reporting results from both function starts and ends, two BiRNNs will need to be trained separately.

Details of XDA. XDA leverages masked Language Modeling – to address function boundary detection [16]. Specifically, XDA applies a specific implementation of BERT (Bidirectional Encoder Representations from Transformers), referred to as RoBERTa. Bytes in binaries are first pre-processed into 512-byte sequences and each byte is encoded into a $\mathbb{R}^{256}$ binary vector with one-hot encoding.

The neural network model in XDA is trained in two phases, including (1) pre-training phase and (2) fine-tuning phase. In the pre-training phase, the model receives byte sequences as inputs where some bytes are marked as missing (i.e., blanked out intentionally). The model is pre-trained to predict the value of missing bytes. The goal of this pre-training phase is to learn semantics in binaries. In the fine-tuning phase, the model is fine-tuned to transfer the knowledge learned in the pre-training phase to a specific task—function boundary detection. Specifically, each byte is assigned three probabilities, including the probabilities of function start, function end, and neither. The one with the highest probability among the three is the predicted label for the byte. The fine-tuning phase updates the parameters of the model by minimizing discrepancies between predicated labels and ground truth labels. Once the model is fine-tuned, it is utilized to predicate labels over bytes in the testing phase.

---

[2] https://hex-rays.com/products/ida/tech/flirt/in_depth/.
[3] https://github.com/NationalSecurityAgency/ghidra/blob/master/Ghidra/Features/FunctionID/src/main/doc/fid.xml.

4 Our Dataset and Tools

Overview of Our Rust Binary Dataset. To investigate function boundary detection over stripped binaries, especially with neural networks, we need to first build a large-scale dataset of Rust binaries as no public Rust binary datasets are available in the current literature. In this paper, a Rust binary indicates that a binary was compiled from a program written in Rust.

Our dataset, referred to as **RUBIN dataset**, includes a total of 2,471 Rust binaries generated with multiple optimizations, including O0, O1, O2, O3, and Oz, using compiler `rustc` (version 1.70) provided in `Cargo`—Rust's build system and package manager. Each stripped binary is a x86-64 ELF binary and its associated non-stripped binary is also included in our dataset for generating/validating ground truth labels. The total number of functions in `.text` sections of all the stripped binaries in our dataset is 8,764,123. The size of our entire dataset is over 66 GBs (including all the stripped binaries and non-stripped binaries). We split the entire dataset into two subsets, one for training (if needed) and one for testing. More specific information about our dataset is presented in Table 1.

In addition, we develop two tools, named `ripkit` and `cargo_picky`, to generate our dataset and results automatically in a large scale. These two tools can also be leveraged independently to expand our dataset or create new large-scale datasets for function boundary detection over Rust stripped binaries. Specifically, `cargo_picky` can automatically clone and compile crates with `cargo` using various optimization levels in a large scale. `ripkit` serves as a pre-processing tool to transform bytes in binaries into various formats (especially for neural network approaches). In addition, `ripkit` also serves as a database for maintaining the original binaries and their pre-processed data, profiles byte sequence patterns in large binary datasets, and summarizes experimental results automatically.

How Binaries Are Generated in Our Dataset. The binaries in our dataset are generated by following the steps below.

1. We randomly select and retrieve 300 Rust crates from `crates.io`—the largest Rust crate registry. There are more than 133,000 crates available on `crates.io` as of August 2024. A comprehensive list of all the 300 crates we examine can be found at our GitHub repository along with our dataset.
2. Given compiler `rustc` and compile target `x86-64-unknown-linux-gnu`, we select one optimization level (either O0, O1, O2, O3, or Oz), compile all the 300 crates, and obtain associated non-stripped binaries. Note that there is a small number of crates that cannot be compiled successfully given each optimization level (except O0). In addition, our test dataset excludes binaries that have a `.text` section with less than 1,000 bytes[4]. We repeat this process for every optimization level.
3. Next, we apply `strip` command over the non-stripped binaries we derived in the previous step to obtain stripped binaries.

---

[4] BiRNN does not support a `.text` section with less than 1,000 bytes.

Table 1. Overview of Binaries in RUBIN Dataset.

Optimization	Training Set			
	No. of Binaries	No. of Functions	Size (GB) Non-Stripped	Size (MB) Stripped
O0	100	1,031,919	3.6	369
O1	100	425,596	4.7	235
O2	100	369,387	4.1	205
O3	100	436,519	5.1	232
OZ	100	433,901	2.5	143
Total	500	2,697,322	20.0	1,184
Optimization	Test Set			
	No. of Binaries	No. of Functions	Size (GB) Non-Stripped	Size (MB) Stripped
O0	200	2,574,316	8.8	959
O1	195	891,936	10.0	474
O2	190	737,865	8.7	437
O3	193	644,359	8.2	410
OZ	193	1,218,525	7.3	462
Total	1,971	6,066,801	43.0	2,742

The stripped binaries are utilized for the evaluation of function boundary detection and their non-stripped binaries are kept to generate ground truth labels.

How Ground Truth Labels Are Generated in Our Dataset. As in previous studies [8,16], we leverage two existing tools, including `lief`[5] and `pyelftools`[6], to create ground truth labels for bytes in .text sections. Specifically, given a non-stripped binary, we first leverage `lief` to extract all the bytes in .text section. Next, we utilize `pyelftools` to extract the addresses of all the function starts and function sizes from the headers of this binary. In addition, we derive the address of each function end by adding its function size to the address of its function start. Finally, we label each byte in .text section as function start, function end, or neither based on the addresses. As all the bytes in .text section do not change after applying stripping, the label for each byte remains the same in the corresponding stripped binary.

In addition to `pyelftools`, other existing tools, such as `objdump`, can also be leveraged to identify ground truth labels for function boundaries over bytes in a binary [8,16]. *It is worth mentioning that different tools could lead to minor discrepancies in ground truth labels.* For instance, given the 200 binaries compiled with O0 in our training data, `objdump` reports 2,583,853 functions while `pyelftools` reports 2,574,316 functions (i.e., 0.37% less). Given the large number of functions in binaries, it is infeasible for us to manually verify which tool is more accurate than others for generating ground truth labels over our dataset. On the other hand, since the discrepancy is minor and it has a

---

[5] https://lief-project.github.io/.
[6] https://github.com/eliben/pyelftools.

negligible impact to the overall evaluation and comparison among methods, we believe choosing either tool to produce labels is sound and reasonable.

5 Evaluation

In this study, we evaluate and compare the performance of four methods/tools for function boundary detection over stripped binaries compiled from Rust program using our RUBIN dataset. Specifically, we reports results from two state-of-the-art industry tools, including Ghidra and IDA Pro, and two neural-network-based methods, including BiRNN and XDA. For a fair comparison, we compare results of the four methods in terms of precision, recall, F-1 score, and running time over our test data in every optimization.

5.1 Experiment Settings

All the analyses and experiments are performed on a Linux desktop running Ubuntu 22.04.3 with Intel i9-14900K CPU, 128 GB Memory, and NVIDIA 4080 GPU.

Experiment Settings in Ghidra and IDA Pro. We leverage command-line tools and APIs to automatically decompile a stripped binary and report the addresses of function starts and ends (or function lengths). Specifically, we leverage `idautils.Function()` and `ida_funcs.calc_func_size()` to extract function starts and lengths in IDA Pro. Similarly, we utilize `FunctionManager`, `getBody()`, and `getNumAddress()` in Ghidra's API to derive function starts and lengths. Since both Ghidra v11.0 and IDA Pro v8.3 do not support compiler `rustc` officially, we do not specific the compiler information but use the default setting during the decompilation process.

Experiment Settings for BiRNN. For BiRNN, we implement it based on the description from [18]. We follow the same pre-processing steps as in [18]. Specifically, given bytes from the .text section of a stripped binary, we divide the bytes into byte sequences with no overlaps, where each byte sequence has 1,000 consecutive bytes. If a byte sequence does not have 1,000 bytes (e.g., towards the end of the .text section), we discard it. We randomly select a total number of 1,000 byte sequences from binaries in the training data of each optimization to train a BiRNN. Each byte in a byte sequence is then encoded into a R^{256} binary vector with one-hot-encoding before passing it as an input to a BiRNN.

The BiRNN model consists of 1 bi-directional RNN layer with 16 RNN hidden units, a single linear layer, and an output layer with `sigmoid` as the activation function. We choose a learning rate of 0.0005 and Binary Cross-Entropy (BCE) as the loss function. The model includes a total of 8,796 parameters.

When reporting the results of BiRNN for each optimization, we obtain all the 1000-byte sequences from the stripped binaries reserved for testing and pass these byte sequences to a BiRNN to obtain the output on each byte. Similar as in

[18], sequences with less than 1,000 bytes cannot report outputs and are excluded from results. It is also worth to mention that one BiRNN can only report function starts or function ends but not both. *When reporting results from both function starts and ends, two BiRNNs will need to be trained separately.*

Experiment Settings for XDA. We follow the same pre-processing steps as in [8] and leverage their source code[7]. Specifically, given one optimization, we first randomly select 50 binaries as pre-training data and another 50 binaries as fine-tuning data. These binaries are from RUBIN training set. Second, we extract and form one super byte sequence by concatenating all the bytes from .text sections of the 50 binaries in pre-training data. Next, we divide the super sequence into 512-byte sequences and pass these sequences to the model for pre-training. Masking is randomly applied to 20% of the bytes during the pre-training by the original implementation of XDA. In the fine-tuning phase, a super sequence is formed from the 50 binaries in the fine-tuning data and is divided into 512-byte sequences. In addition, labels are attached to all the bytes. All the 512-byte sequences with labels are passed to the model for fine-tuning. We pre-train each model for 300 epochs and fine-tune each model for 20 epochs.

XDA leverages a transformer-based model, utilizing multi-headed self attention to increase the models attention to long range dependencies. In total, XDA includes 12 self-attention layers, each with 12 self-attention heads. These self-attention layers use the GeLu activation function. while the fine-tuning layers use the tanh function. The encoding layer outputs an embedded with 786 dimensions. The final decoder layer has 2 feed forward networks stack one top of it, one of which is used to predict mask bytes, the other is used to fine-tune to predict the function boundary labels.

When reporting the results of XDA for each optimization, we obtain all the 512-byte sequences from the stripped binaries reserved for testing and pass these byte sequences to XDA to obtain the output on each byte. Similar as in [8], sequences with less than 512 bytes cannot report outputs and are excluded from results. *A single XDA model can reports results in both function starts and ends.*

Reporting Results on Function Bounds. When we report the results on function bounds from Ghidra or IDA, it is straightforward as each tool reports a start address and the length for every function in a binary. This information can be easily transform to the end address of a function as well as a start-end address pair. In addition, there are no overlaps between start-end address pairs reported from Ghidar or IDA. However, for neural-network-based methods, including BiRNN and XDA, there could be cases that a method reports multiple function starts before it identifies the next function end or a function start happens at a lower address than the function end of a previous function. Therefore, for neural-network-based methods, we perform additional post processing based on the results of function starts and function ends to obtain start-end address pairs such that the start address of a function is the first start address that

---

[7] https://github.com/CUMLSec/XDA

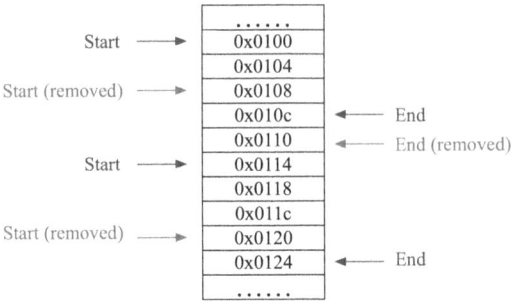

Fig. 1. Example of deriving start-end address pairs based on reported function starts and function ends (for neural-network-based methods only). Given four function starts and three function ends, two start-end address pairs, (0x0100, 0x010c) and (0x0114, 0x0124), are assembled in this example.

Table 2. The number of start 2-grams (first two bytes for function starts) and the number of end 2-grams (last two bytes for function ends) in RUBIN Dataset (training data only).

	No. of Binaries	No. of Functions	No. of Start 2-Grams	No. of Start 2-Grams (Elsewhere)	No. of End 2-Grams	No. of End 2-Grams (Elsewhere)
O0	100	1,031,919	212	212	27	26
O1	100	425,596	406	406	37	36
O2	100	369,387	394	394	26	25
O3	100	436,519	473	473	36	35
OZ	100	433,901	790	790	143	142

does not overlap with a previous function. Additional start addresses before the next end address are also removed. An example of post-processing is presented in Fig. 1.

5.2 Experiments

Experiment 1: Understanding the Variety of Function Start and Ends. We first run analyses over our dataset to show the variety of bytes associated with function starts and ends and why function boundary detection is non-trivial. Specifically, we show the number of *start 2-grams* (the first two bytes of a function) and the number of *start 2-grams elsewhere* (the first two bytes of a function but are also found at non-function starts) for every optimization level in Table 2. For instance, given 425,596 functions across 100 binaries from O1 optimization, there are 406 unique start 2-grams. However, all of these 2-grams can also be found at other addresses which are not function starts. This indicates that leveraging these 2-grams as trivial signatures for detecting function starts

Table 3. Comparison of the four methods/tools (cells highlighted with green are the highest within each optimization).

Train (if needed) & Test	Methods	Function Starts			Function Ends			Function Bounds		
		PR	RE	F1	PR	RE	F1	PR	RE	F1
O0	Ghidra	0.996	0.996	0.996	0.771	0.771	0.771	0.771	0.771	0.771
	IDA Pro	0.992	0.999	0.995	0.779	0.779	0.779	0.779	0.779	0.779
	BiRNN	0.999	0.876	0.934	0.999	0.949	0.973	0.958	0.804	0.875
	XDA	0.999	0.999	0.999	0.999	0.901	0.948	0.948	0.855	0.899
O1	Ghidra	0.998	0.997	0.998	0.631	0.632	0.632	0.631	0.631	0.631
	IDA Pro	0.975	0.986	0.981	0.564	0.564	0.564	0.564	0.564	0.564
	BiRNN	0.997	0.784	0.878	0.997	0.823	0.902	0.879	0.635	0.737
	XDA	0.999	0.999	0.999	0.999	0.883	0.938	0.931	0.823	0.874
O2	Ghidra	0.940	0.941	0.940	0.574	0.574	0.574	0.574	0.574	0.574
	IDA Pro	0.972	0.984	0.978	0.575	0.576	0.575	0.570	0.571	0.571
	BiRNN	0.991	0.841	0.910	0.987	0.881	0.931	0.906	0.709	0.795
	XDA	0.999	0.999	0.999	0.998	0.932	0.964	0.939	0.875	0.906
O3	Ghidra	0.927	0.932	0.929	0.558	0.561	0.559	0.557	0.561	0.559
	IDA Pro	0.969	0.985	0.977	0.570	0.571	0.571	0.570	0.571	0.571
	BiRNN	0.995	0.827	0.903	0.992	0.821	0.899	0.876	0.660	0.752
	XDA	0.998	0.999	0.999	0.998	0.927	0.961	0.939	0.871	0.904
Oz	Ghidra	0.998	0.998	0.998	0.739	0.739	0.739	0.739	0.739	0.739
	IDA Pro	0.979	0.977	0.978	0.653	0.653	0.653	0.654	0.654	0.654
	BiRNN	0.989	0.761	0.860	0.974	0.832	0.898	0.637	0.401	0.492
	XDA	0.998	0.999	0.999	0.999	0.184	0.311	0.952	0.176	0.296

is not sufficient. More comprehensive signatures should be considered to tackle the problem. We have similar observations from function ends as well when we compare the number of *end 2-grams* (the last two bytes of a function) and the number of *end 2-grams elsewhere* (the last two bytes of a function but are also found at non-function ends) in each optimization.

Experiment 2: Comparison among Different Methods. We report and compare the precision, recall, and F1 score of each method, including Ghirda, IDA Pro, BiRNN, and XDA, over stripped binaries from each optimization level. The detailed results are summarized in Table 3. Overall, we have two major observations.

- ***Observation 1.1:*** *XDA* performs the best in detecting function starts across all the five optimizations with over 99.8% precision and 99.8% recall. In addition, it derives the best results in identifying function ends in O1, O2, and O3. On the other hand, BiRNN performs the best in O0 and Ghidra outperforms other methods in Oz in terms of detecting function ends.

Table 4. True Positive, False Positive, and False Negative of Function Bounds (Ghidra and IDA Pro from Table 3).

	Methods	TP	FP	FN	PR	RE
O0	Ghidra	1,983,879	591,009	590,437	0.771	0.771
	IDA Pro	2,004,180	568,008	570,136	0.779	0.779
O1	Ghidra	563,513	328,959	328,618	0.631	0.632
	IDA Pro	503,400	389,207	388,731	0.564	0.564
O2	Ghidra	423,886	314,675	314,169	0.574	0.564
	IDA Pro	367,857	277,053	276,695	0.570	0.571
O3	Ghidra	361,671	287,277	282,881	0.557	0.561
	IDA Pro	367,857	277,053	276,695	0.570	0.571
Oz	Ghidra	847,384	299,257	298,886	0.739	0.739
	IDA Pro	748,798	398,408	397,472	0.654	0.654

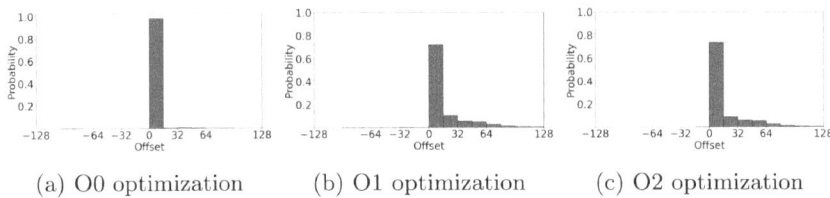

(a) O0 optimization   (b) O1 optimization   (c) O2 optimization

Fig. 2. The PDFs of address offsets (distance) of IDA Pro on function ends

- *Observation 1.2:* Ghidra and IDA Pro offers promising performance with high precision and recall in function starts but performs much worse in detecting function ends than XDA (except for binaries with Oz optimization). For instance, IDA Pro achieves 99.2% precision and 99.9% recall in detecting function starts but only 77.9% precision and 77.9% recall in identifying function ends in O1.

In addition, we also present detailed results from Ghidra and IDA Pro to provide more insights. First, we present true positives, false positives, and false negatives of these two methods on function bounds in Table 4 to interpret why the precision and recall of each tool is almost the same in Table 3. This is because the number of functions reported by each tool is very close to the ground truth, which leads to only a minor difference between false positives and false negatives.

Second, we also show to what degree IDA Pro (or Ghidra) fails to detect function ends. Specifically, we report the PDF (Probability Dense Function) of address offsets, where each address offset is the offset between a ground-truth function end and an inferred function end from IDA Pro (or Ghidra). We define an address offset (for function ends) as below. Given a set of (ground-truth) end addresses $E = \{e_1, ..., e_m\}$ and a set of (inferred) end addresses $E' = \{e'_1, ..., e'_k\}$, we obtain a set of address offsets $O = \{o_1, ..., o_k\}$, where

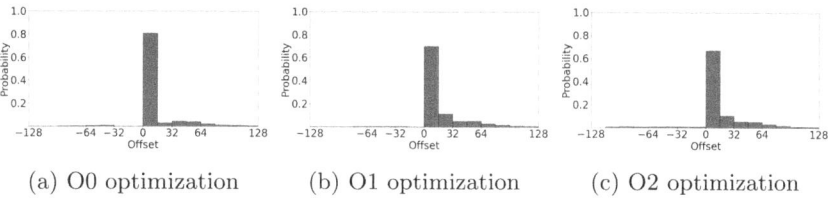

(a) O0 optimization (b) O1 optimization (c) O2 optimization

Fig. 3. The PDFs of address offsets (distance) of Ghidra on function ends

Table 5. Comparison of the Four Methods/Tools in Training Time and Test Time.

	Ghirda	IDA Pro	BiRNN	XDA
Training Time (hours)	NA	NA	0.03	144
Test Time (KBs/sec)	2.6	4.2	14.13	9.99

$$o_i = e'_i - e_j, \quad s.t. \quad j = \arg\min_{k \in [1,m]} |e'_i - e_k|$$

for $1 \leq i \leq m$. As we can see from Fig. 2, although the precision and recall of IDA Pro on function ends are less promising, the address offsets are small, which suggests that the inferred addresses are not far from the ground truth in general. We also have similar observations from Ghidra on function ends as shown in Fig. 3.

While XDA performs the best among the four methods/tools, we would like to acknowledge that XDA requires significant amount of training time (more specifically, pre-training time and fine-tuning time). For instance, it takes about 144 h to complete the pre-training phase (96 h) and fine-tuning phase (48 h) per XDA model with our GPU machine. In terms of test time, the four methods are all efficient and report results in large-scales within reasonable amounts of time as shown in Table 5.

Experiment 3: Performance of XDA (Cross-Optimization Scenario). One common limitation of neural-network-based methods is that the performance of a model can drop significantly when there are domain shifts between training data and test data. In binary analysis, a common factor that could lead to domain shifts is compiler optimization [17,19]. To examine whether XDA is robust again domain shifts due to compiler optimization, we measure the performance of XDA in cross-optimization scenarios. Specifically, we leverage a trained model using binaries from one optimization (e.g., O0) in the previous experiment but test the model with binaries from a different optimization (e.g., O1).

As shown in Table 6, when a trained neural network is tested over binaries with lower optimization levels, such as O0 or O1, XDA still performs well (with slightly performance drops) when training binaries and test binaries are compiled with different optimizations. On the other hand, when a trained neural network is tested over binaries with higher optimization levels, such as O2, O3 or Oz, but

Table 6. Performance of XDA in cross-optimization scenarios (cells highlighted with green are results from same-optimization for easy comparison; cells highlighted with red are results with more than 10% drops than same-optimization scenarios.)

Train	Test	Function Starts			Function Ends			Function Bounds		
		PR	RE	F1	PR	RE	F1	PR	RE	F1
O0	O0	0.999	0.999	0.999	0.999	0.901	0.948	0.948	0.855	0.899
O1	O0	0.997	0.985	0.991	0.999	0.897	0.946	0.944	0.843	0.891
O2	O0	0.996	0.985	0.990	0.992	0.898	0.942	0.938	0.841	0.887
O3	O0	0.994	0.989	0.991	0.998	0.901	0.947	0.945	0.850	0.895
Oz	O0	0.989	0.986	0.988	0.999	0.896	0.945	0.943	0.842	0.890
O0	O1	0.989	0.989	0.989	0.998	0.883	0.937	0.929	0.820	0.871
O1	O1	0.999	0.999	0.999	0.999	0.883	0.938	0.931	0.823	0.874
O2	O1	0.999	0.995	0.997	0.999	0.838	0.938	0.999	0.883	0.938
O3	O1	0.999	0.995	0.997	0.999	0.883	0.938	0.931	0.822	0.874
Oz	O1	0.996	0.999	0.997	0.999	0.883	0.938	0.932	0.823	0.874
O0	O2	0.977	0.995	0.986	0.976	0.931	0.953	0.909	0.859	0.883
O1	O2	0.970	0.234	0.378	0.998	0.216	0.355	0.941	0.204	0.335
O2	O2	0.999	0.999	0.999	0.998	0.932	0.964	0.939	0.875	0.906
O3	O2	0.998	0.999	0.999	0.999	0.932	0.964	0.999	0.932	0.905
Oz	O2	0.993	0.999	0.996	0.994	0.932	0.962	0.933	0.872	0.902
O0	O3	0.977	0.995	0.986	0.974	0.926	0.950	0.909	0.856	0.882
O1	O3	0.304	0.004	0.008	0.604	0.001	0.001	0.221	0.000	0.000
O2	O3	0.998	0.999	0.998	0.999	0.927	0.961	0.939	0.871	0.903
O3	O3	0.998	0.999	0.999	0.998	0.927	0.961	0.939	0.871	0.904
Oz	O3	0.993	0.999	0.996	0.994	0.927	0.959	0.933	0.868	0.900
O0	Oz	0.988	0.808	0.889	0.993	0.185	0.312	0.940	0.175	0.294
O1	Oz	0.766	0.014	0.027	0.556	0.000	0.001	0.389	0.000	0.000
O2	Oz	0.994	0.832	0.906	0.999	0.186	0.314	0.943	0.176	0.296
O3	Oz	0.993	0.885	0.936	0.998	0.213	0.352	0.840	0.214	0.295
Oz	Oz	0.998	0.999	0.999	0.999	0.184	0.311	0.952	0.176	0.296

the neural network is trained based on O0 or O1, noticeable performance drops (e.g., more than 10% drops) can be observed.

Experiment 4: Performance of XDA (Multi-optimization Training)
To overcome the limitation above, one typical approach is to train a neural network with data from multiple domains, more specifically, bytes from multiple optimizations, such that the neural network is able to generalize, at least for each optimization that is considered during the training. Specifically, we pretrain a unified model of XDA by utilizing bytes from 50 binaries from all the

Table 7. Performance of XDA (Multi-Optimization Training with O0, O1, O2, O3 and Oz).

Test	Function Starts			Function Ends			Function Bounds		
	PR	RE	F1	PR	RE	F1	PR	RE	F1
O0	0.998	0.999	0.999	0.999	0.901	0.948	0.948	0.854	0.899
O1	0.996	0.998	0.997	0.999	0.883	0.938	0.932	0.822	0.874
O2	0.997	0.998	0.997	0.998	0.932	0.964	0.937	0.874	0.905
O3	0.997	0.999	0.998	0.998	0.927	0.961	0.938	0.870	0.902
Oz	0.994	0.997	0.996	0.999	0.184	0.311	0.952	0.176	0.296

5 optimizations (including O0, O1, O2, O3, O4, and O5) with 50 binaries per optimization. Next, we fine-tune this model with bytes from 100 total binaries from all the 5 optimizations (20 binaries per optimization). One the model is fined tuned, we report the performance of the model on the test data from each optimization.

We observe that training a single neural network with binaries from multiple optimizations can effectively overcome domain shifts and offer high precision, recall and F1 score over binaries from all optimization except for Oz optimization. The performance of this unified model is almost the same as training each optimization and test each optimization separately shown in Table 3. However, training one unified neural network obviously can reduce computational overhead rather than training five neural networks separately (Table 7).

6 Related Work

Function Boundary Detection over C Binaries. Bao et al. [7] designed a function boundary detection boundary method, named ByteWeight. It learns signatures of function starts based on weighted prefix tree and identify function starts by matching binary segments with signatures. Andriesse et al. [5] propose Nucleus, which can detection functions based on control flow graphs and is compiler-agnostic. Alves-Foss and Song [4] developed a function boundary detection method and integrated it into Jima, a tool suite for binary vulnerability analysis and repair. This method leverages explicit calls and jumps as indicators of function starts. It does not require extensive training time compared to neural-network-based approaches. Bundt et al. [8] investigated black-box attacks on neural-network-based binary function detection. Specifically, the authors consider inadvertent attacks caused by compiler options and adversarial attacks by instruction rewriting (e.g., replacing NOPs with jumps or mov instructions). Yu et al. [20] proposed a method, named DeepDi, which leverages graph neural networks to capture instruction relations. DeepDi is primarily used for disassembly. In addition, it also offers capabilities to identify function starts using heuristics. *All these existing studies report experimental results over C stripped binaries.*

Vulnerabilities in Rust. Li et al. [13] designed a method to detect bugs in Rust programs using static analysis. Liu et al. [14] proposed XRust, a method that changes unsafe Rust code into safe Rust code by leveraging a novel heap allocator. Felix et al. [11] developed a proof-of-concept to show that it is feasible to force a bounds-checked Rust array variable access to read any byte in the memory.

7 Discussions and Future Work

While we examine the performance of four methods/tools over different optimization levels, we did not examine the impacts of the version of the compiler (`rustc`) or compiler flags. Recent work in [8] shows that binaries produced by various compiler flags can lead to domain shifts and could affect the performance of boundary detection over C binaries. These impacts over Rust binaries remain unknown and will be interesting to explore in future research. Whether it is feasible for an attacker to modify Rust stripped binaries and force a neural-network-based method to predict incorrectly on function starts and function ends remains open. Techniques, such as binary rewriting [10], can be examined to address this problem. One of the key challenges would be how to modify Rust binaries automatically in a large scale without affecting functionalities of binaries. It would also be interesting to explore the performance of function boundary detection over Rust binaries with other standard instruction set architectures, such as RISC-V, in future work.

GhidRust[8] is a relatively new open-source tool, which is a Ghidra extension specifically for analyzing Rust binaries. It can decide whether a binary is a Rust binary and report functions in a binary based on signatures using Ghidra's Function ID. Specifically, it creates a Function ID database for Rust's `libstd` on x86-64 for Rust version 1.58.0 and perform function detection (including function starts, function size, and function names)[9]. Unfortunately, no comprehensive analyses are reported regarding the accuracy of GhidRust's function detection. While the development of this tool has been paused, it will be still interesting to examine the detection performance of this tool over our dataset RUBIN in our future work.

8 Conclusion

We investigate the problem of function boundary detection over Rust stripped binaries. Our experimental results show that a neural-network-based method can achieve very high precision, recall and F1 score and outperform start-of-the-art industry reverse engineering tools in function boundary detection. The neural-network-based method can also render outstanding performance across different optimization levels when it is trained with binaries from multiple optimization

---

[8] https://github.com/DMaroo/GhidRust.
[9] https://github.com/DMaroo/GhidRust/blob/master/media/report.pdf.

levels. Moreover, we develop two automatic tools and one large-scale dataset that can be utilized by the research community to expand and extend research findings on function boundary detection over Rust binaries.

Acknowledgement. The authors thank the anonymous shepherd and reviewers for their comments and suggestions. This work was partially supported by National Science Foundation (CNS-2150086) and UC (University of Cincinnati) Undergraduate Research Fellowship.

References

1. https://ghidra-sre.org/
2. Ida pro. https://hex-rays.com/ida-pro/
3. Rust. https://www.rust-lang.org/
4. Alves-Foss, J., Song, J.: Function boundary detection in stripped binaries. In: Proceedings of ACSAC 2019 (2019)
5. Andriesse, D., Slowinska, A., Bos, H.: Compiler-agnostic function detection in binaries. In: Proceedings of Euro S&P2017 (2017)
6. Ayers, H., et al.: Tighten rust's belt: shrinking embedded rust binaries. In: Proceeedings of the 23rd ACM SIGPLAN/SIGBED International Conference on Languages, Compilers, and Tools for Embedded Systems (LCTES 2022) (2022)
7. Bao, T., Burket, J., Woo, M., Turner, R., Brumley, D.: ByteWeight: learning to recongnize functions in binary code. In: Proceedings of of USENIX Security 2014 (2014)
8. Bundt, J., Davinroy, M., Agadakos, I., Oprea, A., Robertson, W.: Black-box attacks against neural binary function detection. In: Proceedings of RAID 2023 (2023)
9. Chua, Z.L., Shen, S., Saxena, P., Liang, Z.: Neural nets can learn function type signatures from binaries. In: Proceedings of USENIX Security 2017 (2017)
10. Duck, G.J., Gao, X., Roychoudhury, A.: Binary rewriting without control flow recovery. In: Proc. of ACM SIGPLAN International Conference on Programming Language Design and Implementation (PLDI 2020) (2020)
11. Felix, C., Benti, D., Austin, T.: Spectre v1 proof-of-concept attack in the rust language. https://github.com/toddmaustin/spectre-rust
12. House, T.W.: Black to the building blocks: a path toward secure and measureable software. https://www.whitehouse.gov/wp-content/uploads/2024/02/Final-ONCD-Technical-Report.pdf
13. Li, Z., Wang, J., Sun, M., Lui, J.C.: MirChecker: detecting bugs in rust programs via static analysis. In: Proceedings of ACM CCS 2021 (2021)
14. Liu, P., Zhao, G., Huang, J.: Securing unsafe rust programs with XRust. In: Proceedings of IEEE/ACM ICSE 2020 (2020)
15. Pang, C., Yu, R., Chen, Y., Koskinen, E., Portokalidis, G., Mao, B., Xun, J.: SoK: all you ever wanted to know about x86/x64 binary disassembly but were afraid to ask. In: Proceedings of IEEE S&P 2021 (2021)
16. Pei, K., Guan, J., Williams-King, D., Yang, J., Jana, S.: XDA: accurate, robust disassembly with transfer learning. In: Proceedings of NDSS 2021 (2021)
17. Ren, X., Ho, M., Ming, J., Lei, Y., Li, L.: Unleashing the hidden power of compiler optimization on binary code difference: an empirical study. In: Proceedings of PLDI 2021 (2021)

18. Shin, E., Song, D., Moazzezi, R.: Recognizing functions in dinaries with neural networks. In: Proceedings of USENIX Security 2015 (2015)
19. Wang, C., et al.: Portability of deep-learning side-channel attacks against software discrepancies. In: Proceedings ACM WiSec 2023 (2023)
20. Yu, S., Qu, Y., Hu, X., Yin, H.: DeepDi: learning a relational graph convolutional network model on instructions for fast and accurate disassembly. In: Proceedings of USENIX Security 2022 (2022)

Adversarial Attacks in Autonomous Systems

Transient Adversarial 3D Projection Attacks on Object Detection in Autonomous Driving

Ce Zhou, Qiben Yan[✉], and Sijia Liu

Michigan State University, East Lansing, MI 48823, USA
{zhouce,qyan,liusiji5}@msu.edu

Abstract. Object detection is a crucial task in autonomous driving. While existing research has proposed various attacks on object detection, such as those using adversarial patches or stickers, the exploration of projection attacks on 3D surfaces remains largely unexplored. Compared to adversarial patches or stickers, which have fixed adversarial patterns, projection attacks allow for transient modifications to these patterns, enabling a more flexible attack. In this paper, we introduce an adversarial 3D projection attack specifically targeting object detection in autonomous driving scenarios. We frame the attack formulation as an optimization problem, utilizing a combination of color mapping and geometric transformation models. Our results demonstrate the effectiveness of the proposed attack in deceiving YOLOv3 and Mask R-CNN in physical settings. Evaluations conducted in an indoor environment show an attack success rate of up to 100% under low ambient light conditions, highlighting the potential damage of our attack in real-world driving scenarios.

Keywords: Autonmous driving · 3D projection attack · Object detection · Adverarial patch

1 Introduction

Object detection is a crucial task of autonomous driving systems, playing a pivotal role in ensuring the safety of human life. Accurate and reliable object detection enables autonomous vehicles (AVs) to perceive and respond to their environment, and to recognize and track objects such as pedestrians, vehicles, and obstacles. This capability is essential for making informed decisions and taking appropriate actions to navigate through complex and dynamic traffic scenarios.

However, it is a well-recognized challenge that Deep Neural Networks (DNNs) based object detectors are vulnerable to adversarial perturbations. Previous research has designed various adversarial examples (AEs) within the digital domain to attack object detection models. To ensure physical realizability, existing studies create adversarial patches or stickers [6,22,23] as AEs that take real-world environmental conditions into account, including diverse viewing distances

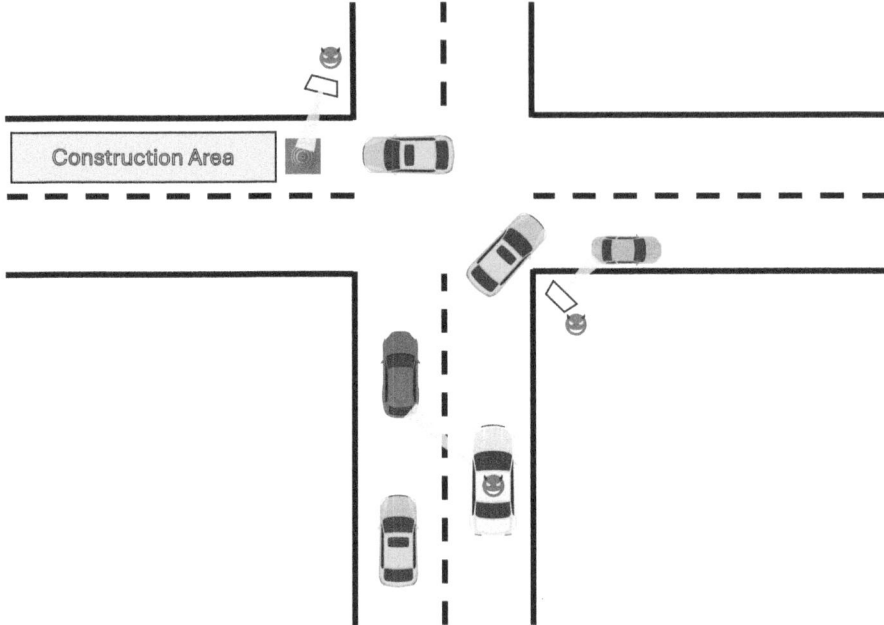

Fig. 1. The attack scenarios of 3D projection attacks: ❶ The blue car and the white car are driving relatively static. ❷ The green car parks at the roadside. ❸ The traffic cone is placed at the entrance of the road construction area. The attackers position themselves either on the roadside or in another car. From these vantage points, the attacker can project an adversarial patch (projection light beams in yellow) onto the target vehicles or traffic cones, to render it undetectable by the victim AV (vehicles in yellow). This strategy could potentially lead to a collision between the victim AV and the target vehicles, or lead the victim vehicle into a dangerous road construction area. (Color figure online)

or angles. However, patch-based attacks are indiscriminate toward all AVs, lacking flexibility in the physical world. This type of attack usually cannot bypass the defense mechanisms [4], as these adversarial attacks would largely perturb the model input. To overcome these limitations, the researchers proposed projection attacks to project the short-lived perturbations onto the target object to render it undetectable by the object detector [9,16]. These adversarial perturbations are crafted for use on 2D surfaces or relatively flat small 3D surfaces. However, most objects have curved or uneven large 3D surfaces, making the existing perturbations ineffective as their shapes can become distorted when projected onto a 3D surface.

In this paper, we propose an adversarial 3D projection attack against object detection in autonomous driving scenarios. The proposed attack can work on a curved or uneven 3D surface with varying viewing angles. To the best of our knowledge, this is the first physically-realizable 3D projection attack towards AVs. We illustrate three attack scenarios in Fig. 1. The attack goal is to make the target vehicle undetectable by the victim AV, which could potentially lead to a collision between the victim AV and the target objects. In the first attack scenario, the attacker is driving the white car at the same driving speed as the blue car. The attacker projects the adversarial patch onto the surface of the blue car, causing it to disappear from the victim AV's object detector. In the second attack scenario, the attacker projects an adversarial patch onto a green car parked on the roadside. The victim AV fails to detect the green car, potentially leading to a crash. The third attack scenario is to project an adversarial patch onto the traffic cone to render it undetected by the victim AV's object detector. This could result in the victim AV driving into a dangerous road construction area.

Several research challenges arise in implementing the proposed attack. When projecting the image onto the object's surface, the resulting images are influenced by various factors, including surface color, material composition, ambient light, projection distortion, viewing angle, etc. First, we need to solve the research question: *"How to map the projected image to the resulting image on the object surface to preserve the intended color schemes?"*. Second, the 3D surface is always curved or uneven, which leads to the distortion of the projected adversarial patch. Thus, the research question is: *"How to transform the projected patch onto the 3D surface to ensure its effectiveness?"*. Third, when projecting the attack pattern onto the target vehicle, the attacker and the target vehicle may both move. So, another research question is: *"How to ensure the attack's robustness in a dynamic environment?"*.

To address these challenges, we employ a color mapping model to map the projection color onto the 3D attack surface. Additionally, we introduce a geometric transformation model that employs the Thin Plate Spine (TPS) algorithm to perform the geometric transformation from the projected patch to the 3D attack surface. To enhance attack robustness, we enhance the training dataset by applying different transformations, i.e., Expectation Over Transformation (EOT), to input images captured with various environmental factors, such as view angles, distances, etc.

We conduct experimental evaluations in an indoor environment with a model vehicle and present a video demo[1] to demonstrate the 3D projection attack. We list our contributions as follows:

- We are the first to propose a transient and physically-realizable projection attack by projecting an adversarial patch on a 3D object surface in autonomous driving scenarios.
- We formulate the adversarial patch generation as an optimization problem by combining a color projection model and a geometric transformation model with the target object detection network. We enhance the training data by considering various environmental conditions and different viewing angles of the adversarial patch, which ensures the effectiveness and robustness of the attack.

---

[1] https://youtu.be/8RbDpAAmsjs.

- We evaluate our attack on two object detection models using a 1/10 scale remote control (RC) car in an indoor environment. We run the experiments under varying attack distances, angles, and ambient light conditions. Evaluations show an attack success rate of up to 100% under low ambient light conditions, demonstrating the potentially damaging consequences of the proposed attack in real-world driving scenarios.

2 Background

We first introduce the background knowledge of projector technology and object detection. Then, we present existing literature related to physically-realizable AEs.

2.1 LCD Projector Technology

Liquid Crystal Display (LCD) projectors, widely employed in business presentations, classrooms, and home entertainment, utilize liquid crystal technology to project images. The fundamental concept involves splitting white light emitted by a lamp into its red, green, and blue components through dichroic mirrors. These mirrors reflect specific wavelengths while allowing others to pass through. Consequently, the white light transforms into the primary colors of red, green, and blue, forming the basis for deriving all other colors [28].

Lumens of the projector measure the brightness of a projector, indicating the total amount of light it emits. A higher lumen rating in LCD projectors enhances image visibility, especially in well-lit surroundings. Indoor projectors typically fall within the 2,000 to 3,000 lumens range for emitted light, whereas some outdoor projectors can achieve super-high lumens.

The throwing ratio of an LCD projector is defined as the distance between the projector and the screen relative to the width of the projected image. Lux is the unit for one lumen per square meter. Therefore, for a determined projecting distance, the throwing ratio influences the size and brightness of the projected image.

When we project a 2D image onto a 3D surface using a projector, the appearance of the image can be affected by the shape and texture of the surface. On smooth surfaces, the image is likely to be more uniform. However, any irregularities in the surface might cause slight distortions or uneven brightness. Textured surfaces can cause the projected image to look uneven, as the texture may interfere with the clarity of the image. If the surface is flat, the projected image will appear as it does on a flat screen, whereas, on curved 3D surfaces, the image may appear distorted, especially towards the edges. The amount of distortion depends on the degree and nature of the curvature. In a real-world attack scenario, both vehicles and road objects, such as traffic cones, have 3D curved surfaces, making existing adversarial attacks ineffective. This paper aims to develop a new projection attack that ensures attack effectiveness on these 3D surfaces.

2.2 Object Detection

Object detection, a computer technology within the domain of computer vision and image processing, is centered on identifying instances of semantic objects belonging to specific classes, such as pedestrians and cars, in images and videos [27]. In this paper, we focus on Mask-RCNN [21] and Yolov3 [20], as they are representative object detectors widely used in computer vision research.

Faster-RCNN is an evolution of the initial R-CNN object detector network [8]. Utilizing a two-stage detection method, the first stage is to generate region proposals, and the second stage is to predict labels for the proposals. Additionally, Mask-RCNN [11] expands Faster-RCNN to incorporate instance segmentation, providing precise pixel-level masks alongside accurate object detection, making it ideal for applications requiring detailed object boundaries. Its flexible architecture and strong performance on benchmark datasets make it a versatile choice.

Yolo object detectors are designed to detect objects in real-time, making them highly suitable for various applications [13]. Yolov3 is a one-stage detector utilizing a single convolutional neural network (CNN), which incorporates a backbone network to compute feature maps for each square-grid cell in the input image. It employs three different grid sizes to enhance the accuracy of detecting smaller objects. Its unified architecture simplifies implementation, and its ability to perform multi-scale detection ensures robustness across various object sizes. Both Yolov3 and Mask-RCNN employ non-maximum suppression in post-processing to eliminate highly overlapped redundant boxes.

2.3 Physical Adversarial Examples

Athalye et al. [2] introduce the EOT framework, focusing on crafting adversarial perturbations resilient to random linear transformations. Although this approach uses a 3D printer to print out the 3D object with the AE, which can successfully deceive the image classifiers, it is not flexible in the context of dynamic driving scenarios.

Other studies [6,23,30] design robust physical perturbations, like stickers or patches, for traffic signs. These perturbations are proven resilient to changes when reproduced in the physical world, such as alterations in distance and viewing angle. Sharif et al. [22] showcase the feasibility of physical AEs for face recognition using colored eye-glass frames and enhance the perturbation realizability in the presence of input noise. Xu et al. [7,29] design a patch on a non-rigid object (i.e., T-shirt) to make the person undetected in the object detector. However, such patch-based attacks also lack the flexibility in a complex autonomous driving scenario.

Patches have drawbacks that can be mitigated by the projectors, particularly due to the short-lived and dynamic nature of projections. This allows attackers to control and adjust the projection easily, targeting specific vehicles without leaving traces of the malicious attack. Nassi et al. [19] project 2D images directly into the real world to deceive object detectors. Man et al. [18] project optimized adversarial perturbations directly to the camera using the flare lens effect to deceive image classification. Similarly, Zhou et al. [31] exploit vulnerabilities in stereo depth estimation algorithms and use two projectors to deceive depth estimation results for AVs. In [9,16], the researchers project the perturbation on the traffic sign to make it undetectable by the object detector. In these cases, adversarial perturbations are crafted for use on 2D surfaces, despite most objects having 3D surfaces, rendering flat patches ineffective on curved 3D object surfaces.

3 Threat Model

Our proposed attack targets an autonomous driving scenario, in which the decision-making relies exclusively on camera sensors for the AV, such as Tesla vehicles with Full Self-Driving features and Active Safety Features [26]. The AV utilizes these cameras to identify and monitor 3D objects within the driving scene, including other moving vehicles.

Attack Goal. The adversaries aim to cause a traffic accident by projecting a transient adversarial patch onto a 3D object, making it undetectable by a DNN-based object detector model that processes the camera feeds within the AV. For instance, the lack of detection may cause the surrounding vehicles to collide with the victim AV without engaging the brakes.

Attacker's Capabilities. Our attack is a *white-box, hiding attack*. We consider a scenario where an attacker seeks to promptly generate and project an attack patch onto an existing real-world 3D object to *hide* it from an object detection system employing a DNN model. During this timeframe, every captured frame of the scene is anticipated to yield a deceptive object detection outcome. We assume the attacker possesses comprehensive knowledge of the target DNN, including the model parameters and architecture, while remaining unaware of the technical specifications of the cameras. Additionally, the attacker is assumed to have physical proximity to the target 3D objects, such as other vehicles. This proximity enables the attacker to capture images of the target 3D objects for adversarial patch training and facilitates the projection of the attack patch into the real world. We also assume that the attacker has the capability to purchase or locate an identical object as a target for the projection attack, thereby gathering all the required data for the attack preparation (i.e., adversarial patch generation).

4 Attack Design

We first present an overview of the attack pipeline, and then we illustrate the design details regarding the color mapping model, geometric transformation model, adversarial patch generation, and training data enhancement.

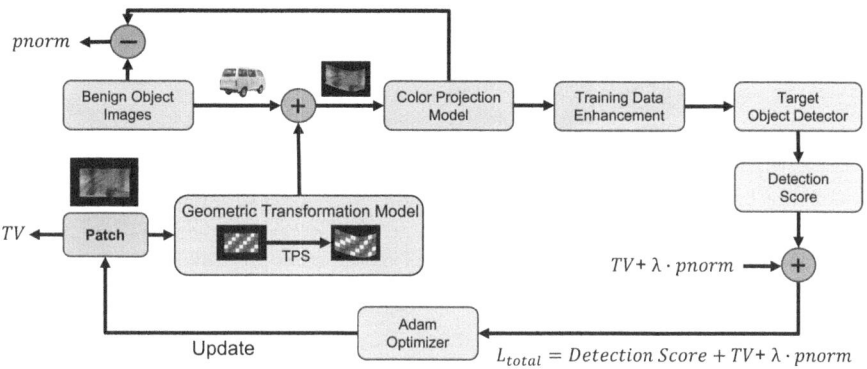

Fig. 2. Overview of the adversarial patch generation pipeline.

4.1 Attack Overview

Figure 2 shows the adversarial patch generation pipeline. The goal is to optimize the adversarial patch using a geometric transformation model and a color mapping model to minimize the target object detection score on a specific DNN-based object detector. The adversarial patch undergoes geometric transformation, and then the color mapping model is applied to simulate the patch's appearance on the target object. The geometric transformation model simulates the distortion on the 3D object surface, applying the transformation on the patch based on the 3D structure of the target object. The color mapping model simulates the projected color on the target surface. The transformed patch and the benign target object are combined through the color projection model to generate the patched final object. We then perform training data enhancement to place the patched target object in different backgrounds to enhance robustness in real-world driving scenarios. Finally, the optimization process is conducted iteratively across the enhanced training dataset to generate the adversarial patch (for projection) by minimizing the loss function.

4.2 Color Mapping Model

To create physically-realizable perturbations, prior studies [6,22,23] typically leverage the non-printability score (NPS) to characterize printable colors by printers. In our attack scenarios, projecting different colors onto the target 3D object surface yields results influenced by multiple factors, such as projection strength, ambient light, projection surface, etc. Some previous investigations on the projection-based physical attacks [9,16] propose color mapping models to simulate the color overlays. The attainable range of colors is significantly reduced in comparison to the spectrum accessible to printed patches, primarily because of the color and materials of the target surface.

(a) Color board patch. (b) Viewing angle 1. (c) Viewing angle 2.

Fig. 3. Examples of two viewing angles (b)(c) of the same (a) color board patch projection on the same vehicle.

Our objective in developing the color mapping model is to establish a mapping that associates colors from the projected image onto the target 3D surface. Therefore, we formulate our color mapping model $\mathcal{P}_c$ as follows:

$$\mathcal{P}_c(\theta_p, S, P) = O, \qquad (1)$$

where θ_p is the model parameters, S is the 3D target surface, P is the projected image, and O is the color overlays captured by the camera on the target 3D surface.

Subsequently, we collect projected data to train the color mapping model, with the goal of approximating the resulting output color for specific projected images and target 3D surfaces. The fundamental approach involves projecting various colors with RGB values ranging from 0 to 255 onto the target object surface, capturing images each time to document the outcomes. The distinct color mappings are denoted as $(S, P_i) \Rightarrow O_i$, where i represents each color index.

To train the color mapping model, we utilize triples (S, P_i, O_i) to fit a neural network consisting of two hidden layers with ReLU activation. Then, Eq. (1) is transformed into an optimization problem with the following loss function:

$$Loss_{\mathcal{P}_c} = \underset{\theta_p}{\mathrm{argmin}} \sum_{\forall i} ||\mathcal{P}_c(S, P_i) - O_i||_1, \qquad (2)$$

where $||\cdot||_1$ denotes the L1 norm. We optimize the network parameters θ_p using gradient descent and the Adam optimizer to derive the color mapping model.

4.3 Geometric Transformation Model

The EOT process [2] involves using images with various transformations, encompassing scaling, translation, rotation, brightness, additive Gaussian noise, etc. This collection aids in creating a single perturbation effective in deceiving the target neural network across diverse views. However, this approach may not adequately capture the projected image warping on a 3D object surface, particularly when viewed from different angles. For instance, in Fig. 3, the projected patch, i.e., the color board, on the 3D surface is unevenly distorted on the car's surface.

Thus, it is imperative to consider this viewing angle aspect when modeling the transformation.

Different viewing angles or relative movements between the projected 3D surface and the projector can result in varying distortions in each video frame. To address this, we incorporate TPS [3] to model the projection distortion on the 3D surface. TPS has been widely employed as a non-rigid transformation model in prior studies [7,12,29]. TPS transformation is a mathematical model used for non-linear spatial transformations, often applied in image warping and deformation. It is particularly useful in morphing one image into another while preserving local structures. TPS transformation is defined as a combination of affine, rigid, and non-rigid components. Our exploration will reveal that the non-rigid warping aspect of TPS proves to be an effective means of modeling projection distortion on 3D surfaces in learning adversarial patterns.

TPS involves learning a parametric deformation mapping that describes the displacement of each pixel from an original image x to a target image z using a set of control points with predefined positions. Suppose that the control points in the source image are (x_i, y_i) and their corresponding positions in the target image are (z_i, w_i). The TPS transformation $\mathcal{P}_g$ is defined as:

$$\mathcal{P}_g(x,y) = f(x,y) + \sum_{i=1}^{N} w_i \cdot \phi(r_i), \tag{3}$$

where $f(x,y)$ is an affine function, and $\phi(r_i)$ is a radial basis function defined as $\phi(r) = r^2 \cdot \log(r)$. The Euclidean distance r_i between the point (x,y) and the control point (x_i, y_i) is given by:

$$r_i = \sqrt{(x-x_i)^2 + (y-y_i)^2}.$$

With the determined transformation parameters, it also allows the reverse transformation from the target image z back to the original image x. The reverse transformation is given by:

$$\mathcal{P}_g^{-1}(z) = \sum_{i=1}^{N} w_i \cdot \phi(r_i).$$

Please note that Eq. (3) does not explicitly define z_i. Instead, it uses the notation $w_i \cdot \phi(r_i)$ to represent the non-rigid component of the transformation. z_i is implicitly part of the control points (z_i, w_i) in the target image but does not appear directly in the equation. In other words, $\sum_{i=1}^{N} w_i \cdot \phi(r_i)$ already incorporates the influence of these control points in the transformation.

Implementing TPS to design a 3D adversarial projection patch is challenging due to the difficulty in determining the control points on both the projected and captured target surface images. To tackle this challenge, we project a checkerboard onto the target object and identify the intersection points between adjacent grid regions as the control points by manually marking them. We then scale the projected image and mark the control points using the same procedure.

4.4 Adversarial Patch Generation

Our approach for generating the adversarial projection pattern involves combining the color projection model and the geometric transformation model with the target network. We employ gradient descent along both dimensions to optimize the projected image. Our objective function is defined as:

$$\underset{\delta}{\mathrm{argmin}} J(f_\theta(b + \mathcal{P}(x,\delta))), \qquad \text{s.t. } 0 \leq \delta \leq 1, \qquad (4)$$

where δ is the projected image, x is the target object image, J is the detection loss, f_θ is the target object detector, and b is the input image background. $\mathcal{P}$ is the projection model which is defined as:

$$\mathcal{P}(x,\delta) = (1-M) \cdot x + \mathcal{P}_c(M \cdot x, \delta_g), \qquad (5)$$

where M is the mask of the patch on the target object image, and δ_g is the patch on the target object image. Here, $M = \mathcal{P}_g(S_\delta)$, and $\delta_g = \mathcal{P}_g(\delta)$, respectively. S_δ denotes the shape of the projected image. We then rewrite Eq. (5) as:

$$\mathcal{P}(x,\delta) = (1-\mathcal{P}_g(S_\delta)) \cdot x + \mathcal{P}_c(\mathcal{P}_g(S_\delta) \cdot x, \mathcal{P}_g(S_\delta)). \qquad (6)$$

We predefine the projected patch shape. The projected patch goes through the geometric transformation model and color mapping model before it is applied to the input image. We mask the attack area on the object to ensure the perturbations only apply to the region of interest (ROI) on the 3D object surface.

To improve the practical feasibility of the projection, we impose a constant grid granularity of $n \times n$ cells on the projection, similar to the procedure in [16]. This guarantees that each cell consists of pixels with identical colors, promoting uniform projections across various viewing distances of the target object. Additionally, we integrate the total variation in the loss function. This inclusion is intended to alleviate the effects of camera smoothing and/or blurring on the overall projection [17].

To simplify the flowing of gradients when backpropagating, we substitute δ with a new variable w such that:

$$u = \frac{\tanh \delta}{2} + 0.5. \qquad (7)$$

Since δ is bounded in [0,1], therefore, u is bounded in [−1,1], which leads to faster convergence in the optimization [16].

Now, we can update our loss function Eq. (4) to constrain the amount of perturbations as:

$$\underset{\delta}{\mathrm{argmin}} J(f_\theta(b + \mathcal{P}(x,u))) + \lambda ||\mathcal{P}(x,u) - x||_p + TV(u)$$

$$\text{s.t. } -1 \leq u \leq 1, \qquad (8)$$

where λ is a parameter utilized to regulate the significance of the p-norm $||\cdot||_p$ and TV represents the total variation. The loss J is based on bounding boxes

$b \in B$ outputted from the target object detection network. Each bounding box has a confidence score indicating the probability of containing an object of class j, denoted as $p_j^{(b)}$. The loss J is then defined as the sum of the confidence scores for the target object, expressed as $\sum_{b \in B} p_j^{(b)}$.

4.5 Training Data Enhancement

Generating AEs that effectively function in autonomous driving scenarios necessitates the consideration of various environmental conditions, such as different viewing angles, distances, rotation, and brightness. To craft a robust adversarial patch, the optimization process needs to incorporate different input transformations. We adopt EOT to generate a set of training images synthetically. Our final loss is formulated as follows:

$$\operatorname{argmin}_{\substack{x_q \sim X \\ \delta}} \mathbb{E}_{b_i \sim B, m_j \sim M} J(f_\theta(b + m_j \cdot \mathcal{P}(x_q, u))) + \lambda ||\mathcal{P}(x_q, u) - x_q||_p + TV(u),$$

$$\text{s.t. } -1 \leq u \leq 1. \tag{9}$$

where $\mathcal{P}(x_q, \delta) = (1 - \mathcal{P}_g^q(S_\delta)) \cdot x_q + \mathcal{P}_c(\mathcal{P}_g^q(S_\delta) \cdot x_q, \mathcal{P}_g^q(S_\delta))$. X represents a collection of input images with various viewing angles, B is a set of background images, and M denotes a number of linear transformations to the target object with the patch. $\mathcal{P}_g^q$ denotes the geometric transformation model corresponding to the input object image x_q.

4.6 Overall Attack Process

We summarize the detailed attack process of our attack step by step as follows:

- Step 1: The attackers determine the patch shape and obtain the parameters for the geometric transformation model. First, the attackers design a patch shape, such as a rectangle, to be used as input to the projector. They use a color board with the same shape as the patch to project it onto the surface of the target 3D object. They collect images of the object both with and without the projection. Finally, they select the control points for the TPS and derive the parameters for the geometric transformation model based on these control points.
- Step 2: The attackers collect data for the color projection model and train the model. They project different colors onto the target object and capture images of the object with these projections. This data is then fed into the color projection model for training.
- Step 3: The attackers generate the patch using the enhanced training dataset. The transformed adversarial patch and collected benign images are processed through the color mapping model and applied to the target objects. They then enhance the training dataset by incorporating EOT, placing the target objects with the adversarial patches in different backgrounds.

- Step 4: The attackers run the optimization process and update the patch. They conduct the optimization process across the enhanced training dataset using the Adam optimizer. The goal is to minimize the loss function, which includes the summation of the detection score, total variation, and p-norm of the patch. Finally, they update the patch accordingly based on the optimization results.

5 Evaluation

To evaluate the proposed attack, we conduct experiments on 3D projection attacks in the physical world. We investigate our attack under different environment settings to verify its feasibility and effectiveness.

5.1 Experimental Setup

Attack Devices. Figure 4 illustrates the experimental setup. The target object is a 1/10 scale RC car [1]. For the projection, we use a PowerLite 1771W WXGA 3LCD Projector [5], an indoor projector priced around $740, which offers a maximum brightness of 3,000 lumens and a maximum resolution of $1,280 \times 800$ pixels. The projector has a throw ratio range of 1.04–1.26. The initial experiments were conducted in a lecture room, with the projector positioned 1.5 to 2.5 m away from the target object. Given the 1/10 scale of the RC car, this setup simulates an attack distance of 15 to 25 m in a real-world driving scenario.

To capture images and videos of the target object, we use the default camera on an iPhone 12 Pro Max. The iPhone is mounted on a phone slider [24] to capture videos from viewing angles ranging from $-20°$ to $20°$. These videos are recorded at various distances and angles while the projection is active, under ambient light conditions of 100 lux, 200 lux, and 500 lux.

Additionally, the data for the color mapping model are collected using the built-in webcam on an Alienware m15 R3 laptop [25]. The data for the geometric transformation model is collected by iPhone. The evaluation of the attack is conducted on the images and videos captured with the iPhone.

Target Attack Models. In our experiments, we evaluate our attack on YOLOv3 and Mask R-CNN object detectors. For YOLOv3, we utilize the Darknet-53 backbone [20]. For Mask R-CNN, we use ResNet-101 as a backbone along with a feature pyramid network for the region proposals [14]. Both object detectors produce a set of bounding boxes with associated confidence scores for each output class. We establish the detection threshold at 0.6, meaning that an object is detected as a "car" if the confidence score is higher than 0.6.

Evaluation Metrics. We feed each frame from the videos into the target attack models and count the instances where a "car" is detected at the output. We use

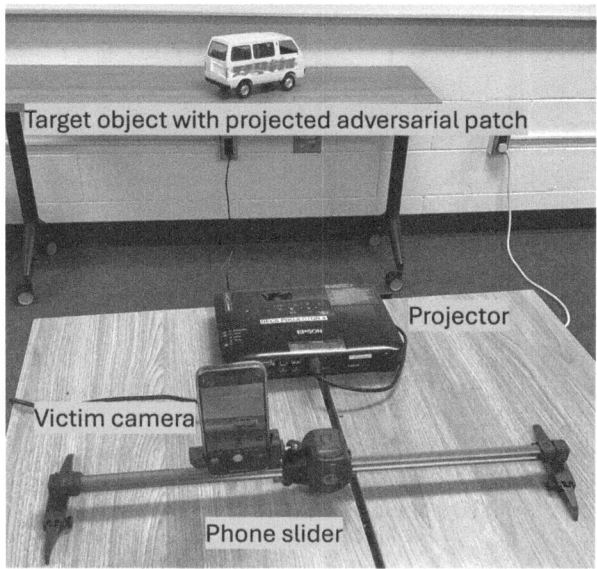

Fig. 4. Attack setup. We use the projector to project the simulated adversarial patch on the target vehicle. The phone slider is used to collect videos when the victim's camera is moving.

object misdetection rate (OMDR) as the main evaluation metric in the experiments. It is expressed as:

$$\text{OMDR} = \frac{\text{The number of frames that do not output a class of car}}{\text{Total number of frames}}$$

Experimental Procedure. We follow these essential steps to conduct our experiments:

- Designing the Attack Pattern: we determine an attack pattern shape (a rectangular shape) as the input to the projector. We design a color board based on this pattern shape and project it onto the target object. Using the iPhone, we capture images of the target object both with and without the projection to obtain the benign object images and the images needed to identify the target control points, respectively. With these collected data, we run the geometric transformation model to obtain the model parameters.
- Running the Color Mapping Procedure: we run the color mapping procedure to collect different color projection images using the webcam. The collected data is then used to train and construct a color mapping model.
- Generating the Adversarial Pattern: we use the color mapping model and the geometric transformation model to generate the adversarial pattern and obtain the optimized attack pattern to input into the projector.

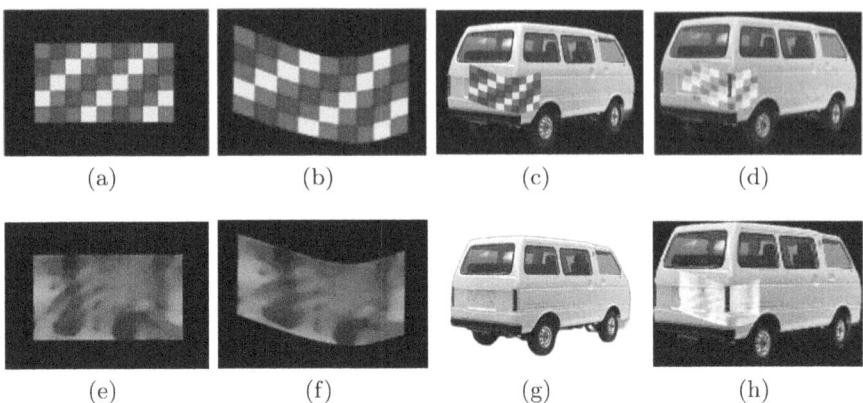

Fig. 5. A visualization example of geometric transformation using TPS. (a) The color board used to collect TPS source control points; (b) The simulated color board after TPS transformation; (c) The simulated color board projected on the vehicle; (d) The projection of the color board on the real vehicle to collect TPS target control points; (e) The trained adversarial patch; (e) The simulated adversarial patch after geometric transformation; (g) The benign vehicle image; (h) The simulated patch projected on the vehicle.

– Launching Attacks and Recording: we project the trained patch onto the target vehicle and record a set of videos at different distances and angles under various ambient light conditions.

5.2 Visualization of the Geometric Transformation

Figure 5 illustrates an example of geometric transformation using TPS. Figures 5(a)(d) are used to get source and target control points for TPS, respectively. We can then build up the geometric transformation model and obtain its corresponding parameters. Figures 5(b)(c) are the simulated projected patch after geometric transformation, and the simulated projected patch on the vehicle. It can be seen that the simulated projected patch on the vehicle (Fig. 5(c)) and the data collected in real life (Fig. 5(d)) match very well. We utilize the parameters in the geometric transformation model and the color mapping model to convert the adversarial patch in the projector (Fig. 5(e)) to the adversarial patch on the vehicle (Fig. 5(f)). We add the transformed adversarial patch (Fig. 5(f)) onto the benign object image (Fig. 5(g)) to generate the final projected patch on the vehicle, which becomes the input to the object detector (Fig. 5(h)).

We also show visualization examples of attacks on Yolov3 and Mask R-CNN in Fig. 6. The left column on both Fig. 6(a) and Fig. 6(b) is the well-trained projected attack patch, the simulated patch after geometric transformation, and the simulated patch on the vehicle from top to bottom. The right columns are the captured vehicle with the projected patch in the physical world. It can be

(a) Attack on Yolov3   (b) Attack on Mask R-CNN

Fig. 6. Visualization examples of attacks on Yolov3 and Mask R-CNN. The left column on both (a) and (b) is the well-trained projected attack patch, the simulated patch after geometric transformation, and the simulated patch on the vehicle from top to bottom. On the right side of each figure is the captured vehicle with the projected patch.

seen that the simulated patch on the vehicle and the physical world match well regarding the the patch shape and color.

5.3 Attack Performance Under Different Settings

We evaluate the 3D projection attack under three ambient light conditions with varying distances and angles for Yolov3 as shown in Fig. 7. The first row shows the baseline results of Yolov3. When there is no attack present, OMDR is usually very close to 0. Only when the ambient light is 500 lux and the viewing angle is between $-20°$ to $7.5°$ at the attack distance of 2 m, OMDR is relatively higher. The results might be attributed to the fact that the back view of the vehicle becomes harder to detect in Yolov3 under strong ambient light.

For the scenarios with the attacks (Figs. 7(d)(e)(f)), it shows that OMDR significantly increases. With lower ambient light, the attack is usually more successful. This is because, as ambient light increases, the range of achievable colors diminishes due to the reduced impact of the projector-emitted light on the vehicle's appearance. Besides, when the attack distance is between 1.5 to 2 m and the attack angle is between $-7.5°$ to $7.5°$, the average OMDR is around 96%. Notably, when the ambient light is 100lux, we can achieve 100% OMDR in this attack setting. A video demo of this attack can be viewed through the link: https://youtu.be/8RbDpAAmsjs.

Moreover, with the attack angle between $-20°$ to $7.5°$, it is usually easy to make the vehicle disappear in the detector. It is because that when the attack angle is between $7.5°$ to $20°$, part of the attack patch on the backside of the vehicle has less portion of the view as shown in Fig. 3(c). On the other hand, the attack patch has a better view in the rest of the viewing angles in our attack setting (Fig. 3(b)).

We also examine the performance of our attack on different object detectors under varying ambient light conditions at an attack distance of 1.5 m. The average OMDR results for YOLOv3 and Mask R-CNN are summarized in Table 1. Our findings indicate that as ambient light increases, the success rate of the attack generally decreases for both models. Furthermore, Mask R-CNN consistently demonstrates greater resilience compared to YOLOv3, likely due to its

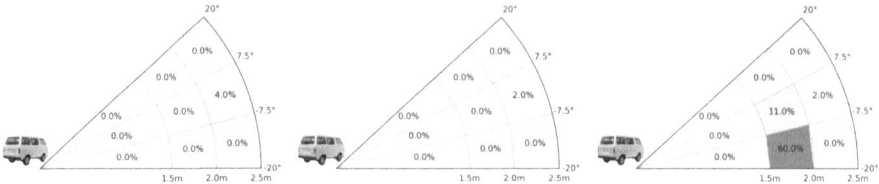

(a) No attack under 100lux (b) No attack under 200lux (c) No attack under 500lux

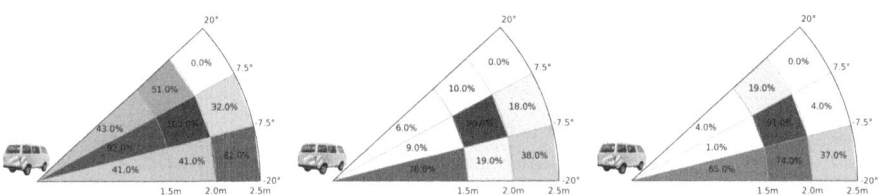

(d) With attack under 100lux (e) With attack under 200lux (f) With attack under 500lux

Fig. 7. OMDR for Yolov3 with varying attack distance and attack angles under ambient light 100lux, 200lux and 500lux respectively. (a)(b)(c) show the OMDR without attack. (d)(e)(f) show the OMDR of 3D projection attack.

ability to learn more robust features and its utilization of a region proposal network for detection [11]. However, this extra resilience comes at the expense of slower execution. For example, Mask R-CNN can take up to 14 times longer to perform than YOLOv3 [16].

6 Discussion

We mainly discuss the feasibility and practicality of the 3D projection attack.

Attack Feasibility and Practicality. Unlike patch attacks, which involve placing physical stickers or patches on objects, projection attacks can be temporary. The projections can be turned on and off quickly, making them harder to detect and track over time. Once the projection is turned off, there are no physical traces left behind, unlike patches or stickers that can be discovered upon inspection.

Projection attacks can change patterns dynamically, adapting to different conditions and object surfaces. This flexibility makes it more challenging for detection systems to recognize and filter out adversarial patterns. Projected adversarial patches can also be modified in real time based on the feedback from the detection system, which allows for continuous optimization of the attack and maintains its effectiveness.

Table 1. Average misdetection rate under different ambient light conditions

Ambient Light (lux)	Yolov3		Mask R-CNN	
	With Attack	W/O Attack	With Attack	W/O Attack
100	59%	0%	46%	26%
200	31%	0%	36%	30%
500	23%	0%	18%	25%

Ambient Light. Our experiments show that the increasing ambient light could rapidly degrade the effectiveness of the attack in bright conditions. In a real-world physical attack scenario, the attacker could launch the attack during the daytime on overcast days or near sunset or sunrise when the ambient light level is below 400 lux.

Moving Target Object. Our attack is effective only on static objects or those moving at low speeds relative to the attacking device. Otherwise, targeting the vehicle becomes problematic. Therefore, our attack is more suitable for vehicles parked on the roadside or traffic cones on the road. Though the attack may only last for several seconds due to movement issues, it is a split-second attack. This means that even if it functions for only a few seconds, it can potentially lead to severe consequences.

Data Collection. The geometric transformation model and the color mapping model need to be well-trained in advance. Consequently, essential data collection is required. The attacker may purchase or locate an identical object to the target and gather all necessary data to develop these models beforehand to launch a more accurate attack.

Potential Defense. One potential defense method is adversarial training [10, 15], which incorporates AEs, including those generated by projection attacks, into the training process. This helps the model learn to recognize and ignore projected adversarial patterns. Previous research [15] proposed a method to locate and segment the adversarial patch, incorporate adversarial training, and then completely remove the patch. However, in the case of our projection attack, if the projected patch size on the target object is enlarged or if the patch fully covers the object, removing the patch entirely might still result in the benign object being mis-detected due to the large area of the object being obscured.

Another possible defense is to verify the consistency of detections over time. Since our projection attack is short-lived, it may cause sudden changes in detected objects over time. Checking for temporal consistency can help identify and discard adversarial detections and alert the human driver.

7 Conclusion

We propose a transient adversarial 3D projection attack targeting object detection in autonomous driving scenarios, which projects a well-crafted adversarial patch onto a 3D surface. The 3D projection attack is formulated as an optimization problem, combining a color mapping model and a geometric transformation model. We enhance the robustness of our attack by considering various environmental factors. We conduct experiments to evaluate the proposed attack against YOLOv3 and Mask R-CNN object detectors in physical attack scenarios. Our evaluation results show an attack success rate of up to 100% under low ambient light conditions. This research underscores the need for defense strategies to mitigate transient projection attacks on AI-driven autonomous vehicles. Future work should focus on creating adaptive and resilient countermeasures that can detect and neutralize such attacks in real time, thereby safeguarding both passengers and pedestrians in diverse driving conditions.

Acknowledgements. We would like to extend our appreciation to the anonymous reviewers for their invaluable input on our study. This work was supported in part by the U.S. National Science Foundation grant CNS-2235231.

References

1. Amazon: Mostop Remote Control Car D42 (2020). https://www.amazon.com/dp/B0BRRJ1GV7?psc=1&ref=ppx_yo2ov_dt_b_product_details. Accessed 20 July 2024
2. Athalye, A., Engstrom, L., Ilyas, A., Kwok, K.: Synthesizing robust adversarial examples. In: International Conference on Machine Learning, pp. 284–293. PMLR (2018)
3. Bookstein, F.L.: Principal warps: thin-plate splines and the decomposition of deformations. IEEE Trans. Pattern Anal. Mach. Intell. **11**(6), 567–585 (1989)
4. Chou, E., Tramer, F., Pellegrino, G.: Sentinet: detecting localized universal attacks against deep learning systems. In: 2020 IEEE Security and Privacy Workshops (SPW), pp. 48–54. IEEE (2020)
5. Epson: PowerLite 1771W WXGA 3LCD Projector (2020). https://epson.com/For-Work/Projectors/Portable/PowerLite-1771W-WXGA-3LCD-Projector/p/V11H477020. Accessed 20 July 2024
6. Eykholt, K., et al.: Robust physical-world attacks on deep learning visual classification. In: Proceedings of the IEEE Conference on Computer Vision and Pattern Recognition, pp. 1625–1634 (2018)
7. Fan, Q., et al.: Generating realistic physical adversarial examples by patch transformer network (2021)
8. Girshick, R., Donahue, J., Darrell, T., Malik, J.: Rich feature hierarchies for accurate object detection and semantic segmentation. In: Proceedings of the IEEE Conference on Computer Vision and Pattern Recognition, pp. 580–587 (2014)
9. Gnanasambandam, A., Sherman, A.M., Chan, S.H.: Optical adversarial attack. In: Proceedings of the IEEE/CVF International Conference on Computer Vision, pp. 92–101 (2021)

10. Goodfellow, I.J., Shlens, J., Szegedy, C.: Explaining and harnessing adversarial examples. arXiv preprint arXiv:1412.6572 (2014)
11. He, K., Gkioxari, G., Dollár, P., Girshick, R.: Mask r-cnn. In: Proceedings of the IEEE International Conference on Computer Vision, pp. 2961–2969 (2017)
12. Jaderberg, M., Simonyan, K., Zisserman, A., et al.: Spatial transformer networks. In: Advances in Neural Information Processing Systems, vol. 28 (2015)
13. Kanjee, R.: Is YOLOv6 v3.0 better than YOLOv8? (2023). https://augmentedstartups.medium.com/is-yolov6-v3-0-better-than-yolov8-4bb2a9a18805. Accessed 20 July 2024
14. Lin, T.Y., Dollár, P., Girshick, R., He, K., Hariharan, B., Belongie, S.: Feature pyramid networks for object detection. In: Proceedings of the IEEE Conference on Computer Vision and Pattern Recognition, pp. 2117–2125 (2017)
15. Liu, J., Levine, A., Lau, C.P., Chellappa, R., Feizi, S.: Segment and complete: defending object detectors against adversarial patch attacks with robust patch detection. In: Proceedings of the IEEE/CVF Conference on Computer Vision and Pattern Recognition, pp. 14973–14982 (2022)
16. Lovisotto, G., Turner, H., Sluganovic, I., Strohmeier, M., Martinovic, I.: SLAP: improving physical adversarial examples with short-lived adversarial perturbations. In: 30th USENIX Security Symposium (USENIX Security 21), pp. 1865–1882 (2021)
17. Mahendran, A., Vedaldi, A.: Understanding deep image representations by inverting them. In: Proceedings of the IEEE Conference on Computer Vision and Pattern Recognition, pp. 5188–5196 (2015)
18. Man, Y., Li, M., Gerdes, R.: GhostImage: remote perception attacks against camera-based image classification systems. In: 23rd International Symposium on Research in Attacks, Intrusions and Defenses (RAID 2020), pp. 317–332 (2020)
19. Nassi, B., Mirsky, Y., Nassi, D., Ben-Netanel, R., Drokin, O., Elovici, Y.: Phantom of the ADAS: securing advanced driver-assistance systems from split-second phantom attacks. In: Proceedings of the 2020 ACM SIGSAC Conference on Computer and Communications Security, pp. 293–308 (2020)
20. Redmon, J., Farhadi, A.: Yolov3: an incremental improvement. arXiv preprint arXiv:1804.02767 (2018)
21. Ren, S., He, K., Girshick, R., Sun, J.: Faster r-cnn: towards real-time object detection with region proposal networks. In: Advances in Neural Information Processing Systems, vol. 28 (2015)
22. Sharif, M., Bhagavatula, S., Bauer, L., Reiter, M.K.: Accessorize to a crime: real and stealthy attacks on state-of-the-art face recognition. In: Proceedings of the 2016 ACM SIGSAC Conference on Computer and Communications Security, pp. 1528–1540 (2016)
23. Song, D., et al.: Physical adversarial examples for object detectors. In: 12th USENIX Workshop on Offensive Technologies (WOOT 18) (2018)
24. Store, N.: Neewer 39.4"/100cm Motorized Camera Slider, 2.4G Wireless Control Carbon Fiber Dolly Rail Slider, Support Video Mode, Time-Lapse Photography, Horizontal,Tracking and 120° Panoramic Shooting (VS-100WC) (2022). https://www.shorturl.at/cfjN0. Accessed 20 July 2024
25. Techologies, D.: ALIENWARE m15 GAMING LAPTOP (2020). https://www.dell.com/en-us/shop/dell-laptops/alienware-m15-r3-gaming-laptop/spd/alienware-m15-r3-laptop. Accessed 20 July 2024
26. Tesla: Autopilot and full self-driving capability (2020). https://www.tesla.com/support/autopilot. Accessed 20 July 2024

27. wikipedia: Object detection (2023). https://en.wikipedia.org/wiki/Object_detection#:~:text=Object%20detection%20is%20a%20computer,in%20digital%20images%20and%20videos. Accessed 20 July 2024
28. Wilkinson, S.: LCD, LCoS, or DLP: Choosing a Projector Imaging Technology (2023). https://www.projectorcentral.com/Digital-Projector-Imaging-Technologies-Explained.htm. Accessed 20 July 2024
29. Xu, K., et al.: Adversarial T-shirt! evading person detectors in a physical world. In: Vedaldi, A., Bischof, H., Brox, T., Frahm, J.-M. (eds.) ECCV 2020. LNCS, vol. 12350, pp. 665–681. Springer, Cham (2020). https://doi.org/10.1007/978-3-030-58558-7_39
30. Zhao, Y., Zhu, H., Liang, R., Shen, Q., Zhang, S., Chen, K.: Seeing isn't believing: towards more robust adversarial attack against real world object detectors. In: Proceedings of the 2019 ACM SIGSAC Conference on Computer and Communications Security, pp. 1989–2004 (2019)
31. Zhou, C., Yan, Q., Shi, Y., Sun, L.: DoubleStar: long-range attack towards depth estimation based obstacle avoidance in autonomous systems. In: 31st USENIX Security Symposium (USENIX Security 22), pp. 1885–1902 (2022)

Assessing Deep Learning Model Accuracy in Varied Surface Conditions for CPS: A Comparative Study

Cade Jacobson[✉], Mathew Clutter, and Francis Akowuah

South Dakota School of Mines and Technology, Rapid City, SD 57701, USA
{cade.jacobson,mathew.clutter}@mines.sdsmt.edu, franics.akowuah@sdsmt.edu

Abstract. Cyber-physical systems (CPS) integrate computation with physical processes, driving innovations in fields like agriculture, aviation, healthcare, and transportation. However, this integration introduces new risks and vulnerabilities that traditional cybersecurity measures cannot address.

To protect CPS, researchers have employed machine learning (ML) and deep learning (DL) techniques to model system behaviors for attack detection and recovery. Existing studies often rely on simulated data and focus on a single DL technique, lacking comprehensive comparisons. This study addresses these gaps by using real data from a robotic vehicle testbed to compare various ML/DL techniques based on prediction accuracy and generalization.

Our experiments reveal that simpler models, particularly Small Dense and Small LSTM networks, exhibit superior generalization and adaptability, maintaining low errors across diverse conditions. Dense networks demonstrate the best overall performance, with the Small Dense model achieving the lowest errors in position and linear velocity predictions. LSTM models also show robust performance, balancing accuracy across all metrics. In contrast, GRU and RNN models exhibit mixed results, and TCNs consistently underperform, indicating challenges in adapting to new environments. These findings underscore the importance of model selection and complexity management for accurate and adaptable predictions.

Keywords: CPS Attack Detection · Model Comparison · Comparative Study · AI in CPS

1 Introduction

Cyber-physical systems (CPS) are complex systems that merge computation with physical processes, creating a network where these elements interact seamlessly. This integration has revolutionized how we engage with technology, sparking innovations across various fields such as agriculture, aviation, architecture, infrastructure, energy, environmental management, healthcare, manufacturing, and transportation.

However, this integration also broadens the potential for attacks, introducing new risks and vulnerabilities that traditional cybersecurity measures fail to address. For example, these systems face threats not only in their cyber networks and computational resources but also from physical attacks that can maliciously manipulate their operational behavior. Such threats include transmitting false GPS signals to mislead navigation systems, resulting in potential hazards like unauthorized border crossings or accidents. Other attacks target the physical sensors of vehicles or industrial systems, potentially causing safety hazards, equipment damage, or catastrophic failures in critical processes.

Recognizing the limitations of conventional cybersecurity in countering physical threats, researchers are exploring defense mechanisms specifically for CPS, with many proposals emphasizing attack detection. These approaches often employ artificial intelligence to predict normal system operations and detect anomalies. The process usually involves creating a system model using machine learning (ML) and deep learning (DL) techniques, and then comparing real-time sensor data against the model's predicted values to detect discrepancies or attacks. However, the effectiveness of these detection methods heavily relies on the fidelity of the system model, which has been a challenge due to the reliance on artificial data in many studies.

In response to this gap, our research conducts experimental studies using data from actual robotic vehicles to develop and evaluate ML and DL models. By gathering real-world telemetry from the Roboworks Rosbot TX autonomous vehicle, such as velocities, position, and orientation, and employing models like Long-Short Term Memory (LSTM), Gated Recurrent Unit (GRU), Recurrent Neural Network (RNN), Dense Neural Network, and Temporal Convolutional Network (TCN), we aim to compare their performance. Our findings shed light on the accuracy, resilience, and generalizability of these models, offering valuable insights for enhancing CPS intrusion detection capabilities.

The organization of the paper is outlined as follows: Sect. 2 introduces preliminary concepts and reviews literature relevant to our study. The methodology adopted for our research is detailed in Sect. 3. Section 4 describes the experiments conducted and the outcomes obtained. An analysis and interpretation of these results are provided in Sect. 4. Finally, the paper is concluded in Sect. 5, summarizing the key findings and implications of our work.

2 Literature Review

Since their first proposal in 1997, LSTMs have attracted numerous studies across various fields, such as image recognition, machine translation, and speech transcription, due to their ability to capture long-term patterns in sequential data. Recently, these models have gained significant attention in the field of cyber-physical security [14].

Researchers have developed various types of LSTMs, with some showing promise in predicting multiple quantities simultaneously. These LSTM models can also control error propagation, maintaining acceptable error levels as the prediction window expands. A key feature of LSTMs is their ability to capture non-linear behavior [12], making them well-suited for autonomous vehicles that navigate winding roads and varying speeds.

In recent years, researchers have also focused on Temporal Convolutional Networks (TCNs) for predicting time series data, which is crucial for monitoring live data in cyber-physical systems. While TCNs have certain structural properties, such as matching the output length to the input length and non-dilated systems not retaining long-term information well, utilizing dilation improves their accuracy at the cost of increased computational overhead [11].

One paper emphasizes the critical need for cyber-physical systems to operate reliably in dynamic and unpredictable environments. Deploying deep learning models in such scenarios presents challenges related to model generalization to new, unseen situations. While some studies show promise, the implications of deploying improperly trained models or those with incorrect tolerances in autonomous vehicles could lead to severe consequences, underscoring the need for robust security systems with high generalization accuracy [17].

Several studies have examined the efficacy of machine learning frameworks in autonomous vehicles, though many do not extend beyond simulated environments or use deep learning systems. For example, some researchers have studied the XGBClassifier in simulated environments [4], evaluated communication-based attacks using decision trees and Naive Bayes [1,9], and analyzed autonomous vehicle sensor data through simulations [5]. Additionally, studies have investigated machine learning classifiers and artificial neural networks for attack detection [3], producing positive preliminary findings that support the future of the field.

However, there is a growing need for hyper-specific studies due to the broad range of possible attacks [7,10]. One significant threat is GPS spoofing, where an attacker sends false information to make the vehicle believe it is in another location [15]. Using machine learning to maintain accurate position measurements could mitigate or thwart such attacks.

3 Methodology

3.1 Modeling System Behavior

In our study, modeling the robotic system's behavior using machine learning techniques was crucial. We trained the models using data collected from the robotic vehicle. The details of the data collection and its use in training the model are discussed in the next subsection.

It is important to note that researchers employ machine learning to model system behavior, particularly in CPS, for several reasons, some of which include the following:

1. *Complexity and Non-linearity*: CPS often involves complex and non-linear interactions between physical components and computational elements. Tra-

ditional modeling techniques may struggle to accurately capture these dynamics. ML algorithms, particularly those designed for complex data, can model such interactions more effectively [16].
2. *Data-Driven Insights*: ML approaches leverage large datasets to uncover patterns and relationships that may not be apparent through theoretical analysis alone. This data-driven approach can lead to more accurate and robust models [2].
3. *Adaptability and Flexibility*: ML models can adapt to changes in system behavior over time. As new data becomes available, these models can be updated to reflect the current state of the system, maintaining accuracy and relevance [6].
4. *Predictive Capabilities*: ML models can be trained to predict future states of a system based on historical data. This predictive power is crucial for applications like predictive maintenance, fault detection, and optimization of CPS performance [8].
5. *Automation of Complex Tasks*: ML can automate the modeling process, reducing the need for extensive human intervention and expert knowledge. This automation accelerates the development and deployment of CPS applications [13].

3.2 Data Collection

We collected odometry and command data while operating the robot to gather the data necessary for training and testing the models. The specific robot used for this purpose was the Rosbot TX, manufactured by Roboworks. We employed a ROS (Robot Operating System) node that integrated data from the robot's odometry with its commanded velocity. This process involved synchronizing and merging these data streams into a single CSV file, which is suitable for model training or validation. Following is a detailed description of each data field collected:

- **Commanded Linear X Velocity (.linear.x):** This represents the speed at which the robot is instructed to move forward or backward.
- **Commanded Angular Z Velocity (.angular.z):** This is the rate at which the robot is commanded to rotate around the z-axis.
- **X Position (.pose.pose.position.x):** The robot's location along the x-axis.
- **Y Position (.pose.pose.position.y):** The robot's location along the y-axis.
- **Z Position (.pose.pose.position.z):** The robot's location along the z-axis.
- **Orientation X Component (.pose.pose.orientation.x):** The x component of the robot's orientation, expressed in quaternion coordinates.
- **Orientation Y Component (.pose.pose.orientation.y):** The y component of the robot's orientation, expressed in quaternion coordinates.
- **Orientation Z Component (.pose.pose.orientation.z):** The z component of the robot's orientation, expressed in quaternion coordinates.
- **Orientation W Component (.pose.pose.orientation.w):** The w component of the robot's orientation, expressed in quaternion coordinates.

- **Actual Linear X Velocity (.twist.twist.linear.x):** The actual speed of the robot moving forward or backward.
- **Actual Angular Z Velocity (.twist.twist.angular.z):** The actual rate of the robot's rotation around the z-axis.

While additional data was captured from the robot, many of these metrics are consistently zero or were not utilized in the model training process.

The majority of the data used for training was collected in a laboratory setting with a flat, tiled floor. Only this data was used to train the models in this study. Specifically, we trained models for Long-Short Term Memory (LSTM), Gated Recurrent Unit (GRU), Recurrent Neural Network (RNN), Dense Neural Network, and Temporal Convolutional Network (TCN). These trained machine learning models were then tested with a separate dataset, also gathered within the same laboratory to assess their prediction accuracy. To assess the models' ability to generalize across different driving surfaces and conditions, additional testing was conducted on various surfaces, including outdoor concrete, rougher classroom tiles, and carpet.

3.3 Model Training and Fair Comparison

To ensure a fair comparison when evaluating the accuracy of the DL models, we followed a structured approach that minimizes biases and ensures consistency. Here are key steps and considerations we took in achieving fair comparisons:

- *Same Dataset*: We used the same dataset for both training and testing all models, consistently splitting it into training, validation, and test sets. To ensure a reliable comparison, we employed techniques such as k-fold cross-validation, which allowed us to evaluate the models on the same data splits.
- *Consistent Data Preprocessing*: We applied the same normalization or standardization techniques to the data before feeding it into the models. Also, we used the same strategy (e.g., imputation, removal) for handling missing data across all models.
- *Hyperparameter Tuning*: We used the systematic method grid search for hyperparameter tuning, ensuring that each model has been optimized to its best performance. We also allocated the same amount of computational resources and time for tuning each model.
- *Evaluation Metrics*: We used the same evaluation metrics to assess the performance of all models. Specifically, we calculated the loss using Mean Squared Error (MSE) between the predicted and actual X and Y outputs. Before selecting MSE, we evaluated the models using multiple metrics to gain a comprehensive understanding of their performance.
- *Model Training*: We maintained consistent training procedures across all models, using 64 epochs and a batch size of 96 for each. Additionally, we implemented uniform early-stopping criteria to prevent overfitting and ensure comparable training durations. We used the adam optimizer for the objective function.

– *Random Seeds*: We set the same random seeds for all models to ensure the reproducibility of results. This minimizes the impact of random variations in data splitting, initialization, and other stochastic processes.

Each independently trained model was structured so that it output three parameters at the same time: the predicted X and Y positions, predicted forward velocity, and predicted angular velocity.

3.4 Models Tested

Each deep learning model described below was built in two versions: a "small" version with 64 units in the hidden layer, and a "large" version with 128 units in the hidden layer. All models were constructed using their corresponding TensorFlow Keras layers, except for the TCN, which was manually created using 1D convolutional Keras layers.

- Dense Neural Net
- Recurrent Neural Net (RNN)
- Long-Short Term Memory Net (LSTM)
- Gated Recurrent Unit Net (GRU)
- Temporal Convolutional Network (TCN)

The dense neural network is the simplest type of model we are testing. It consists of layers of weights that perform calculations on the input and pass the results to the next layer. These models do not consider the temporal order of input data, making them less suitable for tasks that require sequential processing or an understanding of the order in which data points occur.

A slightly more advanced version of the dense neural network is the recurrent neural network (RNN). RNNs are specifically designed to handle sequential data. Unlike dense neural networks, RNNs incorporate information from previous time steps into the current processing step, allowing them to capture and learn from temporal dependencies within the data. This makes them particularly suitable for tasks involving sequences, such as time series prediction and natural language processing.

Both LSTMs (Long Short-Term Memory) and GRUs (Gated Recurrent Units) are advanced variations of recurrent neural networks, each featuring a unique cell design to better handle sequential data. In a GRU cell, there are two gates: the "update" gate and the "reset" gate. These gates continuously update weights to effectively track long-term dependencies in the sequential data, enhancing the model's ability to capture dynamic relationships over time. LSTM models, on the other hand, use three gates: the input gate, forget gate, and output gate, along with maintaining a cell state. This more complex design allows LSTMs to track more intricate dependencies in long-term sequential data, making them particularly effective for tasks involving long-term dependencies, such as time series forecasting and language modeling.

Finally, Temporal Convolutional Networks (TCNs) utilize a distinct architecture compared to the other models mentioned. TCNs employ dilated 1-dimensional convolutional layers to predict sequential data. These layers can vary in length, but their outputs must maintain a consistent length. The base layer of a TCN has a moving receptive field that captures a specified range of previous data points, allowing for a more accurate estimation of the current output. This structure enables TCNs to effectively model long-range dependencies in sequential data.

4 Evaluation

4.1 Experiments

Our comparative study examines two distinct experiments aimed at evaluating the performance of various machine learning models. The first experiment focuses on assessing the accuracy of each model, while the second experiment evaluates the models' ability to generalize to new, unseen conditions.

In the first experiment, the models are tested for their precision in predicting four specific metrics: the current X and Y positions, the forward linear velocity, and the angular velocity of our robotic vehicle. This is conducted using a separate dataset derived from a five-minute robot operation, where the models predict the entire dataset's values. The models' accuracy is determined by comparing their predictions to the actual outputs recorded by the robot.

The second experiment evaluates the models' ability to adapt and maintain accuracy under conditions different from those in the training data. Specifically, we tested the models on various surfaces not included in the original dataset. The goal is to assess each model's performance in predicting outcomes across different surface types. This experiment covers three distinct surfaces:

- Concrete
- Rough Classroom Tile
- Carpet

By testing on these diverse surfaces, we aim to understand how well the models generalize to new environments.

The surfaces used for testing are listed above in order of their similarity to the training environment (the laboratory floor). The concrete surface is the most similar to the laboratory floor, while the carpet is the least similar.

4.2 Results

We first show the data collected from the robot regarding its X and Y positions in Fig. 1. The graph shows the scaled X position of a robotic vehicle over a period of 30,000 centiseconds. The position fluctuates between -1.0 and 1.0, with noticeable oscillations and a general trend of decreasing amplitude over time. The graph also displays the scaled Y position of the robotic vehicle over the same time period. Similar to the X position, the Y position also oscillates between −1.0 and 1.0, but with more pronounced variations and a less clear trend compared to the X position.

We also present the forward linear velocity and angular velocity in Fig. 2. The top graph illustrates the scaled measured velocity of a robotic vehicle over 30,000 centiseconds. It indicates periods of steady movement interrupted by sudden stops or decelerations, with sharp drops to -1.0 suggesting instances where the vehicle comes to a halt or significantly reduces its speed. The bottom graph depicts the scaled measured angular velocity over the same period, showing more frequent and rapid fluctuations compared to the linear velocity. The angular velocity exhibits highly dynamic behavior with constant changes, suggesting the vehicle is frequently changing direction or rotating, likely due to executing a series of maneuvers.

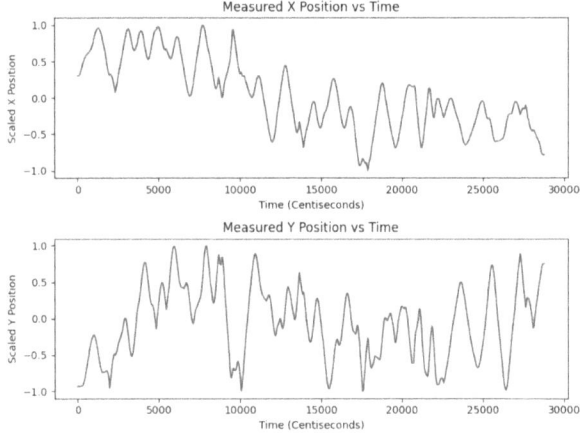

Fig. 1. Pictured above are the X and Y positions as measured by the onboard devices as the robot drove through the laboratory environment.

Accuracy Comparison Results. Here we show the predictions made by the models. Fig. 3 shows the predictions made by the LSTM model. The graph displays time-series data, with the actual and predicted values plotted over the same time range. For the most part, the predicted values (orange) closely follow the actual values (blue), indicating that the small LSTM model performs reasonably well. There are some notable discrepancies between the actual and predicted values, especially around the time ranges of 3000 and 4000, where the actual values show significant drops and recoveries. The predictions follow these trends but with less precision. The LSTM model captures the general trend and major fluctuations in the data but shows some lag or smoothing, particularly in areas with abrupt changes. The clustering of points around the middle of

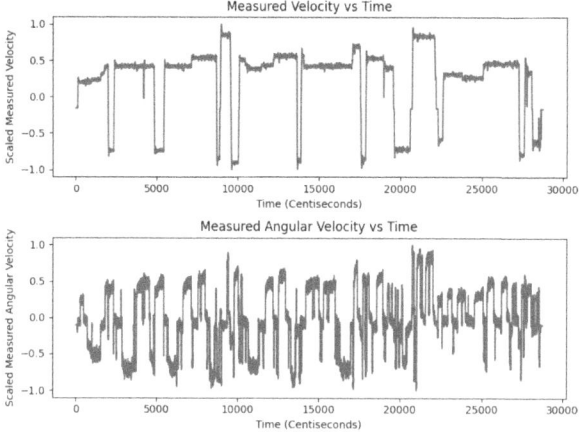

Fig. 2. Pictured above are the forward velocity and angular velocity also as measured by the robot's onboard devices in the laboratory setting.

the graph suggests periods of stable or less variable behavior, where the model predictions align more closely with the actual values.

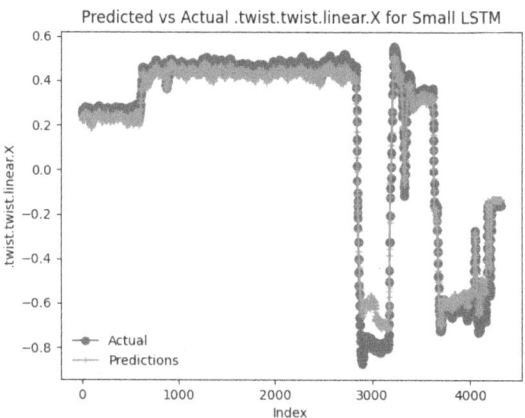

Fig. 3. A graph showing the overlay of our model's linear velocity predictions (orange) on top of the actual measured linear velocity for that time (blue). (Color figure online)

Figure4 shows the time-series data, with both actual and predicted angular velocity values plotted over the same time range. The predicted values (orange) generally follow the trend of the actual values (blue), indicating the small LSTM model's reasonable performance. There are areas where the predictions closely match the actual values, particularly in sections with less variation and more stability. However, the model exhibits discrepancies in areas with frequent and abrupt changes, especially noticeable in the regions between the time interval 500 to 2500 and 3000 to 4000. The predictions often lag behind or show a less pronounced response compared to the actual values. The actual values display significant fluctuations, suggesting a highly dynamic behavior in the angular Z component, which the model attempts to capture but with varying degrees of success. The orange predictions generally capture the peaks and troughs of the blue actual values, but with some smoothing effect, leading to less sharp transitions.

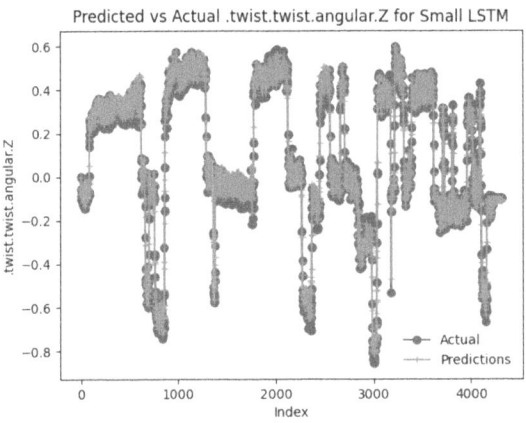

Fig. 4. A graph showing the overlay of our model's angular velocity predictions (orange) on top of the actual measured angular velocity for that time (blue). (Color figure online)

Figure 5 plots the trajectory of the robotic vehicle over 5 min, showing how the actual and predicted positions (X and Y coordinates) change over time. The blue and orange lines form intricate loops and paths, indicating the vehicle's movement in a two-dimensional plane. The predicted positions (orange crosses) generally follow the actual positions (blue circles) closely, indicating that the LSTM model performs well in tracking the vehicle's path. There are regions where the predictions overlap closely with the actual positions, suggesting high accuracy in those segments. Some deviations between the actual and predicted positions are noticeable, particularly in areas with sharp turns or complex loops, where the predictions may lag slightly or smooth out the actual trajectory. The overall pattern of the paths suggests that the model captures the general movement and trajectory of the vehicle, though minor discrepancies can be seen in more complex sections. In summary, the graph demonstrates that the small LSTM model is capable of accurately predicting the vehicle's position over a 5-minute validation period. While there are minor deviations in areas with sharp or complex movements, the model generally follows the actual path closely, indicating good performance.

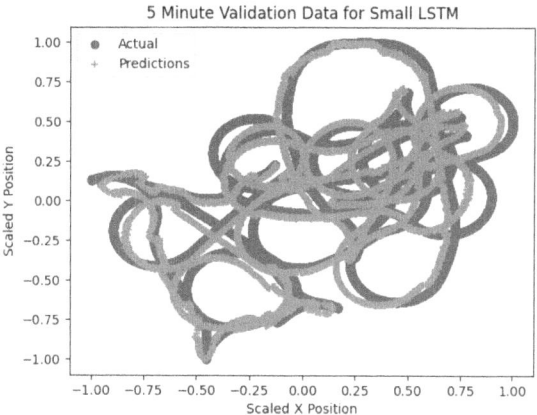

Fig. 5. A position graph that shows a bird's eye view of the robot's actual path (blue) vs the model's live predictions of both X and Y position (orange) (Color figure online)

We have several plots similar to Figs. 4 to 5 for the other models tested. However, due to page limitations, we plan to upload these additional plots to an online repository for interested readers to access. We, however, provide a summary of the accuracy comparison results for the tested models in Table 1. The comparison is based on the Mean Squared Error (MSE), as defined in Eq. 1.

$$\text{MSE} = \frac{1}{n}\sum_{i=1}^{n}(y_i - \hat{y}_i)^2 \quad (1)$$

Recall that a model's "small" version was built with 64 units in the hidden layer, while the "large" version was built with 128 units in the hidden layer. The objective was to examine how model complexity affects accuracy and generalization.

We make the following observations from the table:

- **Small LSTM vs. Large LSTM**: The Small LSTM performs better than the Large LSTM across all three error metrics. The better performance of the Small LSTM suggests that a less complex model may be sufficient for capturing the necessary patterns in the data, avoiding overfitting.
- **Small GRU vs. Large GRU**: The Small GRU performs better in Position and Angular Velocity errors, while the Large GRU is marginally better in Linear Velocity error. Similar to LSTMs, the Small GRU's better performance indicates that the additional complexity of the Large GRU does not provide a significant advantage and might lead to overfitting.
- **Small RNN vs. Large RNN**: The Large RNN performs better in Position and Linear Velocity errors, while the Small RNN is better in Angular Velocity error. The mixed results suggest that RNNs, in general, may not be as effective as LSTMs or GRUs for this task, and the benefit of increased complexity varies by the specific type of error being measured.

- **Small Dense vs. Large Dense**: The Small Dense model performs better for position and linear velocity, likely due to reduced risk of overfitting, while the Large Dense model's complexity helps with angular velocity
- **TCN**: The TCN performs the worst across all three error metrics. The poor performance of the TCN across all metrics could be due to the nature of the task requiring different temporal dynamics or the model not being properly tuned for this specific problem.

These observations and explanations highlight the importance of model selection and complexity management, emphasizing that simpler models can often provide better generalization and performance for specific tasks.

Table 1. Average Errors in the Ideal Environment (in percent)

Model Name	Position Error	Linear Velocity Error	Angular Velocity Error
Small LSTM	5.64	4.50	4.12
Large LSTM	5.86	5.32	5.30
Small GRU	7.02	5.12	5.48
Large GRU	9.50	5.08	5.60
Small RNN	7.88	5.62	6.34
Large RNN	8.34	4.32	5.68
Small Dense	5.38	2.44	6.96
Large Dense	6.56	4.42	4.56
TCN	14.42	6.72	7.68

Generalization Comparison Results: Here, we present the results of our second experiment. Recall that this experiment evaluates the models' ability to adapt and maintain accuracy under conditions different from the training data. Specifically, we tested the models on various surfaces that were not included in the original dataset.

The results are shown in Tables 2 through 4. The comparison is based on the MSE metric as defined in Eq. 1. We discuss the observations of the results below:

- **Concrete**: See Table 2. This experiment served as an intermediate test between the carpet and rough tile environments, providing a suitable level for evaluating the models' generalization abilities. Approximately half of the tested models performed slightly worse, with results within about 1% of the previous ideal environment (Table 1).
  The Small LSTM model demonstrates the best overall generalization with low errors across all metrics. In contrast, the TCN model performs the worst, with high errors indicating difficulty in adapting to the new environment.

Dense models show a strong performance in position accuracy, especially the Large Dense model, but have a trade-off with higher velocity errors. GRU and RNN models show mixed performance, with the Small GRU particularly struggling with velocity errors.

- **Carpet**: See Table 3. This experiment had the greatest change compared to the laboratory setting (Table 2). This is shown most in the Position errors, where every single model performed worse than in the ideal environment. The models also showed significantly worse performances in the Linear and Angular Velocity measurements.

  The Large LSTM model shows the best overall performance in terms of Position Error and reasonable errors in Linear and Angular Velocity, indicating good generalization capabilities. The Small Dense model demonstrates strong performance in Position Error, suggesting it is well-suited for this metric, though it has moderate errors in velocity measurements. The TCN model consistently performs the worst across all metrics, indicating significant challenges in adapting to this testing environment. GRU and RNN models show mixed results, with the Large GRU performing relatively better in Linear Velocity Error but struggling with Position and Angular Velocity Errors. These observations highlight the varying strengths and weaknesses of each model type in adapting to new environments, with LSTM models generally showing better adaptability.

- **Rough Tile**: See Table 4. Of all the generalization experiments, this tile had the most similar environment to the training floor. This can be seen in the average error data, as almost every error is within 1% of the results shown in Table 1. The Small Dense model consistently demonstrates strong performance with the lowest Position and Linear Velocity Errors, indicating excellent generalization and adaptability. The Large Dense model also performs well, suggesting that dense models are particularly effective in this environment. LSTM models show a good balance between position and velocity predictions, with the Small LSTM having slightly better overall performance than the Large LSTM.

  GRU models exhibit mixed results, with the Small GRU performing reasonably well in position but struggling with velocity, while the Large GRU shows better performance in velocity but higher position and angular errors. RNN models perform poorly, with both Small and Large RNNs showing high errors across all metrics. The TCN model performs the worst, indicating significant challenges in adapting to the new environment.

  Overall, dense models, particularly the Small Dense, appear to be the most effective, while LSTM models also demonstrate strong generalization capabilities. GRU and RNN models show varying degrees of adaptability, with TCN models consistently underperforming.

Table 2. Average Errors in the Concrete Environment (in percent)

Model Name	Position Error	Linear Velocity Error	Angular Velocity Error
Small LSTM	5.10	4.40	4.04
Large LSTM	6.76	5.26	5.30
Small GRU	7.54	8.46	8.38
Large GRU	10.42	5.20	6.70
Small RNN	8.34	5.24	5.48
Large RNN	8.64	6.10	5.62
Small Dense	6.72	3.06	6.28
Large Dense	4.80	6.36	5.38
TCN	11.08	8.76	7.08

Table 3. Average Errors in the Carpet Environment (in percent)

Model Name	Position Error	Linear Velocity Error	Angular Velocity Error
Small LSTM	8.50	7.26	4.88
Large LSTM	7.58	6.78	5.50
Small GRU	10.58	7.44	6.26
Large GRU	9.24	4.28	6.28
Small RNN	11.66	5.90	6.94
Large RNN	9.32	6.26	6.84
Small Dense	6.08	6.52	5.50
Large Dense	9.06	5.00	6.68
TCN	16.54	8.54	7.82

Table 4. Average Errors in the Rough Tile Environment (in percent)

Model Name	Position Error	Linear Velocity Error	Angular Velocity Error
Small LSTM	6.12	6.76	5.00
Large LSTM	7.08	5.10	5.32
Small GRU	6.54	6.82	5.80
Large GRU	7.70	5.38	6.42
Small RNN	8.32	8.32	5.42
Large RNN	7.78	7.86	5.66
Small Dense	5.28	2.24	5.72
Large Dense	5.34	3.50	5.00
TCN	13.32	9.10	6.98

4.3 Limitations

It is worth noting that our robot used a single model to output three separate parameters. While this worked for this specific instance, we also did not require a high tolerance. Certain models may lend themselves better to only outputting a single variable which could improve their accuracy in comparison to these other models. Further research in this area could explore how single output versions compare between the same spread of models.

The calculations for these runs were done with models running live on the robot driving around. Because the robot we used had a relatively weak computer compared to many current autonomous vehicles, the models we trained were more lightweight than should be expected in industry. Further research should more closely mimic the power that a commercial vehicle would wield.

Along with this, recurrent models tend to be best used when sequential data has complex relationships that are continued over time. This robot did not use a very complex dataset.

Finally, this experiment had the important limitation that we trained everything in a laboratory setting. This is not directly applicable to the real world where z positions will vary, human drivers will control their own cars, and we have to account for global positioning as opposed to the confines of a laboratory.

5 Conclusion

In this study, we evaluated the performance of various deep learning models in predicting position, linear velocity, and angular velocity under different conditions. Our experiments tested the models' ability to generalize and maintain accuracy when exposed to environments different from their training data.

Our analysis revealed several key findings:
Model Performance

- **Dense Models:** The Small Dense model consistently demonstrated the strongest performance, achieving the lowest errors in both position and linear velocity predictions. The Large Dense model also performed well, suggesting that dense neural networks are particularly effective in this context.
- **LSTM Models:** Both Small and Large LSTM models showed strong generalization capabilities, with the Small LSTM slightly outperforming the Large LSTM. These models maintained a good balance between position and velocity predictions.
- **GRU Models:** The GRU models exhibited mixed results. The Small GRU performed reasonably well in position predictions but struggled with velocity errors, while the Large GRU showed better performance in velocity but higher position and angular errors.
- **RNN Models:** RNN models, both small and large, generally performed poorly, with high errors across all metrics, indicating limited adaptability to new environments.

- **TCN Models:** The TCN model consistently showed the highest errors, indicating significant challenges in adapting to new testing environments and poor generalization capabilities.

Generalization Capabilities

- Models with lower complexity, such as the Small Dense and Small LSTM, generally demonstrated better generalization to new environments compared to their larger counterparts. This suggests that simpler models may be sufficient for capturing the necessary patterns in the data, avoiding the pitfalls of overfitting.
- The trade-offs between position, linear velocity, and angular velocity errors highlight the importance of choosing the right model architecture based on the specific requirements of the application.

Environmental Adaptability

- The experiments showed that certain models, particularly dense and LSTM models, are better suited for tasks requiring adaptation to different environments. These models maintained relatively low errors across various surfaces, indicating robust performance.
- GRU and RNN models showed less consistent performance, suggesting that they may require further tuning or alternative architectures to achieve similar levels of adaptability.

Future research could explore further tuning and optimization of GRU and RNN models to enhance their performance. Additionally, investigating hybrid models or incorporating advanced techniques such as attention mechanisms might improve generalization capabilities. Expanding the dataset to include more diverse environments and testing conditions would also provide a more comprehensive evaluation of model robustness.

Overall, our findings underscore the importance of model selection and complexity management in achieving accurate and adaptable predictions in dynamic environments.

Acknowledgment. This work was supported in part by SDBOR CRG award fund. The views and conclusions contained herein are those of the authors and should not be interpreted as necessarily representing the official policies or endorsements, either expressed or implied, of the South Dakota Board of Regents (SDBOR).

References

1. Agarwal, A., Sharma, P., Alshehri, M., Mohamed, A., Alfarraj, O.: Classification model for accuracy and intrusion detection using machine learning approach. PeerJ Comput. Sci. **7**, e437 (2021). https://doi.org/10.7717/peerj-cs.437
2. Aggarwal, C.C., et al.: Data Mining: The Textbook, vol. 1. Springer, Cham (2015)

3. Alqahtani, H., Sarker, I.H., Kalim, A., Minhaz Hossain, S.M., Ikhlaq, S., Hossain, S.: Cyber intrusion detection using machine learning classification techniques. In: Chaubey, N., Parikh, S., Amin, K. (eds.) Computing Science, Communication and Security, pp. 121–131. Springer, Singapore (2020)
4. Berry, H., Abdel-Malek, M.A., Ibrahim, A.S.: A machine learning approach for combating cyber attacks in self-driving vehicles, pp. 1–3 (2021). https://doi.org/10.1109/SoutheastCon45413.2021.9401856
5. Ferdowsi, A., Challita, U., Saad, W., Mandayam, N.B.: Robust deep reinforcement learning for security and safety in autonomous vehicle systems. In: 2018 21st International Conference on Intelligent Transportation Systems (ITSC), pp. 307–312 (2018). https://doi.org/10.1109/ITSC.2018.8569635
6. Gama, J., Rodrigues, P.P., Spinosa, E., Carvalho, A.: Knowledge discovery from data streams. In: Web Intelligence and Security, pp. 125–138. IOS Press (2010)
7. Halbouni, A., Gunawan, T.S., Habaebi, M.H., Halbouni, M., Kartiwi, M., Ahmad, R.: Machine learning and deep learning approaches for cybersecurity: a review. IEEE Access **10**, 19572–19585 (2022). https://doi.org/10.1109/ACCESS.2022.3151248
8. Hastie, T., Tibshirani, R., Friedman, J.H., Friedman, J.H.: The elements of statistical learning: data mining, inference, and prediction, vol. 2. Springer, Cham (2009)
9. He, Q., Meng, X., Qu, R., Xi, R.: Machine learning-based detection for cyber security attacks on connected and autonomous vehicles. Mathematics **8**(8) (2020). https://www.mdpi.com/2227-7390/8/8/1311
10. Hossain, M.A., Islam, M.S.: Ensuring network security with a robust intrusion detection system using ensemble-based machine learning. Array **19**, 100306 (2023). https://doi.org/10.1016/j.array.2023.100306, https://www.sciencedirect.com/science/article/pii/S2590005623000310
11. Lim, B., Zohren, S.: Time-series forecasting with deep learning: a survey. Phil. Trans. R. Soc. A **379**(2194), 20200209 (2021)
12. Lindemann, B., Müller, T., Vietz, H., Jazdi, N., Weyrich, M.: A survey on long short-term memory networks for time series prediction. Procedia CIRP **99**, 650–655 (2021). https://doi.org/10.1016/j.procir.2021.03.088, https://www.sciencedirect.com/science/article/pii/S2212827121003796. 14th CIRP Conference on Intelligent Computation in Manufacturing Engineering, 15–17 July 2020
13. Mohri, M., Rostamizadeh, A., Talwalkar, A.: Foundations of Machine Learning. MIT Press, Cambridge (2018)
14. Sherstinsky, A.: Fundamentals of recurrent neural network (RNN) and long short-term memory (LSTM) network. Physica D **404**, 132306 (2020). https://doi.org/10.1016/j.physd.2019.132306, https://www.sciencedirect.com/science/article/pii/S0167278919305974
15. Sun, X., Yu, F.R., Zhang, P.: A survey on cyber-security of connected and autonomous vehicles (CAVs). IEEE Trans. Intell. Transp. Syst. **23**(7), 6240–6259 (2022). https://doi.org/10.1109/TITS.2021.3085297
16. Van Der Aalst, W., van der Aalst, W.: Data Science In Action. Springer, Cham (2016)
17. Windmann, A., Steude, H., Niggemann, O.: Robustness and generalization performance of deep learning models on cyber-physical systems: a comparative study. arXiv preprint arXiv:2306.07737 (2023)

Practitioner Paper: A Real-Time Defense Against Object Vanishing Adversarial Patch Attacks for Object Detection in Autonomous Vehicles

Jaden Mu[✉]

East Chapel Hill High School, Chapel Hill, USA
jaden.mu@gmail.com

Abstract. Autonomous vehicles (AVs) increasingly use DNN-based object detection models in vision-based perception. Correct detection and classification of obstacles is critical to ensure safe, trustworthy driving decisions. Adversarial patches aim to fool a DNN with intentionally generated patterns concentrated in a localized region of an image. In particular, object vanishing patch attacks can cause object detection models to fail to detect most or all objects in a scene, posing a significant practical threat to AVs.

This work proposes ADAV (Adversarial Defense for Autonomous Vehicles), a novel defense methodology against object vanishing patch attacks specifically designed for autonomous vehicles. Unlike existing defense methods which have high latency or are designed for static images, ADAV runs in real-time and leverages contextual information from prior frames in an AV's video feed. ADAV checks if the object detector's output for the target frame is temporally consistent with the output from a previous reference frame to detect the presence of a patch. If the presence of a patch is detected, ADAV uses gradient-based attribution to localize adversarial pixels that break temporal consistency. This two stage procedure allows ADAV to efficiently process clean inputs, and both stages are optimized to be low latency. ADAV is evaluated using real-world driving data from the Berkeley Deep Drive BDD100K dataset, and demonstrates high adversarial and clean performance.

Keywords: Adversarial Attack · Autonomous Vehicles · Object Detection

1 Introduction

Deep neural network (DNN)-based object detection models are increasingly used in autonomous vehicles (AVs). For example, automotive company Tesla has already deployed a YOLO object detection model in its Full Self Driving software [1].

Unfortunately, object detection models have been proven to be vulnerable to adversarial attacks [5]. Adversarial patch attacks, which are spatially-constrained adversarial patterns designed to fool DNNs, are especially threatening due to their practicality, as they are robust to real-world transformations such as changes in position, scale, rotation, and perspective [4]. Specifically, adversarial patch attacks that are trained on an object vanishing objective can lower an object detector's confidence on each prediction sufficiently to suppress all detections in an input image. A patch attack can be conducted by overlaying a patch on an image. In a real world autonomous driving context, this could be done by printing a physical patch and placing it on objects such as road signs [10]. As illustrated in Fig. 1, the object vanishing adversarial patch in the second image causes the object detector to fail to detect all objects in the input image. Note that this effect is not caused by overlapping, as the effect does not occur with the white (benign) patch in the first image.

Fig. 1. The Object Vanishing Adversarial Patch

Object detection in AVs is performed on video data, since AVs must constantly process a stream of sequential image data to make real-time decisions. Object vanishing patch attacks can be applied to videos by applying an adversarial patch to each individual frame. However, existing defenses have focused on defending object detection models on singular images. This is reflected by the frequent use of datasets containing unrelated images for training and evaluation, such as ImageNet or COCO [4,11]. Therefore, existing defenses don't take advantage of the contextual information previous frames in a video can provide to the frame being processed. Furthermore, existing defenses often don't prioritize latency, and are consequently unable process a video in real-time.

An effective defense should **1)** Run in real-time, **2)** Recover a non-adversarial input from an adversarially attacked scene or mitigate the effects of the patch on the model's output, and **3)** Maintain good clean performance (the model paired with the defense should not perform significantly worse than the model on its own when there's no adversarial patch in frame).

ADAV improves on existing defenses by taking into account information provided by previous frames in an AV's video feed through *temporal consistency*. An object detector should detect the same objects in similar locations in frames close to each other temporally. For example, a car detected in front of an AV should still be detected in a similar location half a second later - the typical environment an AV is in is consistent. However, if an adversarial patch enters the field-of-view of the AV between the two frames and causes the model to not detect any objects, the model's output will be changed significantly, breaking temporal consistency. ADAV measures the similarity in output between the target frame and a prior reference frame to determine if temporal consistency is broken, and therefore detect the presence of an adversarial patch. If a patch is detected, ADAV then uses gradient-based pixel attribution to find the pixels that had the greatest contribution to changing the outputs between frames, which are likely adversarial pixels. This allows ADAV to eliminate unnecessary computations and maintain clean performance by skipping the second stage if temporal consistency is not broken. Additionally, parameters determining the threshold at which an adversarial patch is detected and localized are carefully tuned to balance the tradeoff between adversarial performance and clean performance. We evaluate ADAV by training adversarial patches to apply to the BDD100K dataset [13], and conduct several experiments to demonstrate strong adversarial and clean performance.

2 Background

2.1 Gradient-Based Attribution

Saliency Maps with vanilla gradients compute the gradient of the class score of interest with respect to the pixels in the input image [8]. This creates a fine-grained map of the pixels in the input image which have the greatest impact on increasing the classification score. Gradient based attribution methods/saliency maps have been widely used as an AI explainability method to visualize the regions of an image a model assigns the greatest importance.

Importantly for this work, saliency maps can also be computed for any differentiable function applied to the model's output, not just a class activation score, by extending the gradient calculation.

Guided backpropagation was proposed by [9] as a way to create cleaner saliency maps. Guided backpropagation introduced a modification to vanilla gradients by only backpropagating positive gradients through ReLU activation functions, with the intuition that only focusing on gradients that increase the output would ensure that salient features don't get canceled out, and to produce a less noisy visualization.

2.2 Existing Defenses

Several defenses have been proposed to defend image classification models against adversarial patch attacks. Notably, Local Gradient Smoothing (LGS) achieved good results in defending image classification by performing gradient smoothing on an image based on the observation that adversarial patches concentrate high frequency noise [6]. JPEG Compression was also proposed to defend against adversarial patch attacks, with the intuition that compression algorithms preserve important details while corrupting highly noisy patch reasons [2]. Both LGS and JPEG Compression are model-agnostic, as they only preprocess the image. Therefore, these defenses can also be applied in an object detection context.

Stronger defenses such as PatchGuard exist but are expensive to compute, with PatchGuard requiring several rounds of inference on smaller regions of an image to produce one certified output.

Fewer defenses exist for object detection. DetectorGuard [12] proposed a certified defense that issues an alert when a patch attack is detected. DetectorGuard uses a robust image classifier that relies on a features extracted with a small receptive field, which ensures that patches can only have a small localized impact, to determine the approximate location of any objects in an image. It then issues an alert if there is a mismatch between the image classifier and object detector. However, DetectorGuard is a human-in-the-loop defense, and is incapable of addressing the attack beyond detecting the presence of a patch - a human must intervene if an alert is issued. Universal Defense Frames [14] proposed preprocessing an input image by surrounding it with a "frame" of precomputed pixels trained to negate the effects of an adversarial patch, and requires almost no inference-time computations.

3 Method

ADAV uses a two-stage process to defend against adversarial patches - it first checks if there is a patch anywhere in the image. If the presence of a patch is detected, ADAV then localizes and masks the patch to produce a clean frame. This process is illustrated in Fig. 2.

3.1 Object Detection

This work uses YOLOv5s [3] as the base object detection model, although ADAV can be applied to other models. YOLOv5 is a one-stage object detector that outputs 25200 potential bounding boxes. The bounding boxes with the highest confidences are then returned with Non-Max Suppression. Importantly, YOLOv5's one-stage architecture is fully differentiable.

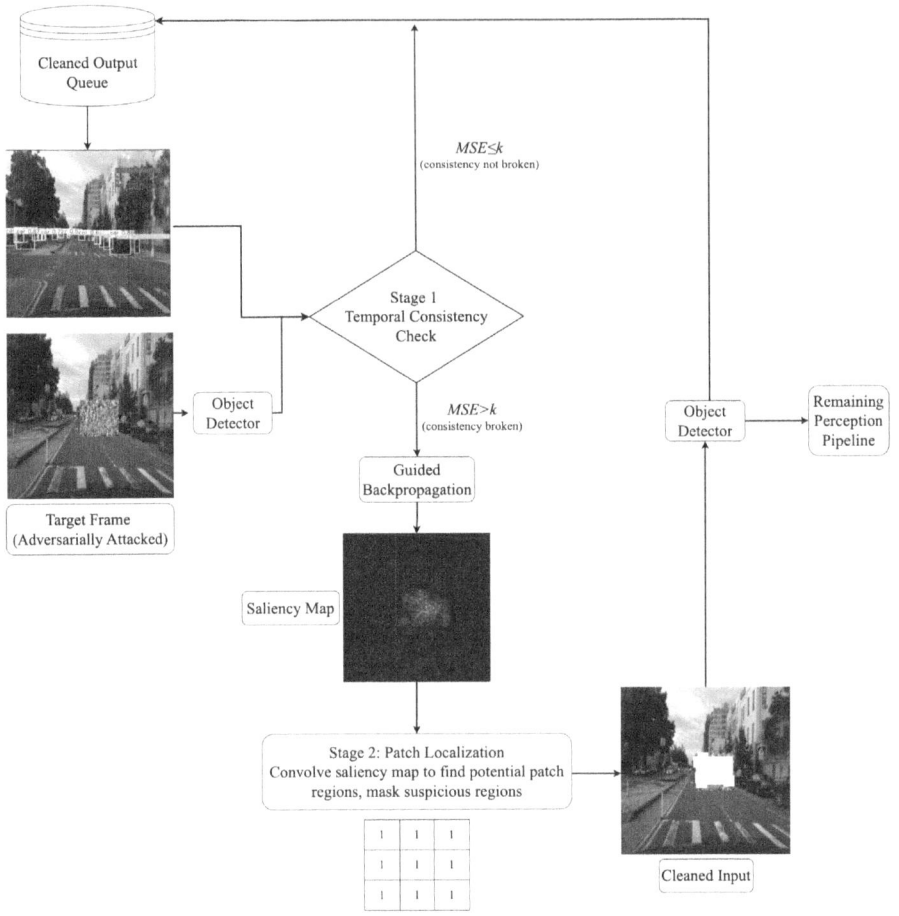

Fig. 2. ADAV Methodology

3.2 Patch Detection

Intuitively, the same objects should be detected in similar locations between two frames close to each other temporally. This work refers to this concept as temporal consistency. When an object vanishing adversarial patch enters into the field of view of an AV between two temporally close frames, temporal consistency is broken, since the adversarial patch will suppress several detections in the model's output. This is leveraged to detect the presence of an adversarial patch. ADAV checks if temporal consistency is broken between the frame being processed (the target frame) and a temporally close reference frame. ADAV uses the frame 0.5 s before the target frame as the reference frame, which is assumed to be clean. To quantify temporal consistency, the Mean-Squared Error (MSE) is computed between the model's output for the target frame and the reference frame from half a second prior. The MSE of the outputs of YOLOv5s is substantially higher between 2 frames from 0.5 s apart, with one frame having an adversarial patch ($\mu = 46.9, \sigma = 4.7$), than between 2 clean frames from 0.5 s apart ($\mu = 25.5, \sigma = 6.2$).

Temporal Consistency is considered to be broken if the MSE in the model's output between 2 frames is greater than some threshold k, which is an empirically tuned parameter. By checking if temporal consistency is broken before performing the relatively computationally expensive patch localization and masking, ADAV can improve latency on clean images by skipping the patch localization step.

To ensure the reference frame is clean, ADAV stores all outputs of clean frames in a queue. The queue is initially populated with the outputs for the 0.5 s from the video feed, which is assumed to be clean (or cleaned with a methodology such as LGS). Given the first 0.5 s as reference, ADAV then takes over, appending to the cleaned output queue with the output for each frame it cleans.

3.3 Patch Localization and Masking

If a patch is detected (temporal consistency is broken), ADAV will localize and mask the patch to recover a clean input. To localize the adversarial patch, this work uses guided backpropagation to create a saliency map. However, instead of taking the gradient of a class activation, ADAV takes the gradient of the MSE between the outputs of the model for the target frame and the reference frame with respect to the input image. Because adversarial patches break temporal consistency and therefore significantly increase the MSE, the saliency map should primarily flag pixels in the adversarial patches.

Potential patch regions are extracted by downsampling the saliency map using strided convolutions. Specifically, a 20 × 20 box filter kernel with a stride of 5 is used to sum the gradients in each potential patch region. We refer to this sum of gradients as the "suspicion score" of a region. Because adversarial patches produce very dense clusters on the saliency map, all suspicion scores below some threshold n are ignored. All remaining potential patch regions are assumed to be from the adversarial patch, so the corresponding pixels in the image are masked out.

A small potential patch region size of 20 × 20 is chosen so that ADAV can neutralize patches with highly irregular, non-rectangular shapes by approximating the irregular shape with several small 20 × 20 regions. Additionally, a small potential patch region helps minimize information loss if the region is falsely determined to be adversarial.

The threshold n is determined dynamically for each image, since different images may have different magnitudes of gradients in clean regions. Given a median suspicion score $\tilde{x}$ and interquartile range Q of an image, $n = \tilde{x} + \lambda Q$, where λ is an empirically tuned constant.

Intuitively, this threshold detects outlier suspicion scores. This should avoid false positives from sudden non-adversarial road condition changes in the 0.5 s interval (e.g. lighting changes), because the benign image will have no significant outliers.

To retrieve a cleaned input image, all pixels corresponding to a potential patch region with a suspicion score above n are replaced with a neutral color which has no adversarial properties. The model then generates a clean output on the cleaned input image.

3.4 Cleaned Output Queue

ADAV must have access to clean outputs to use as the reference. This is done by storing clean outputs in the Cleaned Output Queue, which is initialized from the first 0.5 s of frames. Then, if no patch is detected, the output of the YOLO model is directly added to the queue to use as a reference. If a patch is detected, ADAV cleans the input, reruns the YOLO model, and adds the cleaned output to the queue, ensuring that the queue is always populated with 0.5 s of clean outputs. Importantly, the outputs of the YOLO model are stored directly in the queue, meaning that no inference has to be repeated when checking for temporal consistency.

3.5 Parameter Tuning

The threshold for determining the presence of a patch k and for filtering potential patch regions λ must be tuned to be sensitive to the presence of an adversarial patch while remaining high enough to maintain clean performance. Therefore, tuning these parameters must take into account the tradeoff between clean performance and adversarial performance. Additionally, k and λ must be simultaneously optimized, since a lower k might require a higher λ to avoid false positives and vice versa. We weight adversarial performance equally with clean performance. The performance is measured by Mean Average Precision (mAP). Letting $\pi_{k,\lambda}$ be a model defended with ADAV with parameters set to k and λ,

$$a = \max_{k,\lambda}[mAP(\pi_{k,\lambda}(x_{adv}))]$$

$$b = \max_{k,\lambda}[mAP(\pi_{k,\lambda}(x_{clean}))]$$

$$k_{optimal}, \lambda_{optimal} = \arg\min_{k,\lambda}[(a - mAP(\pi_{k,\lambda}(x_{adv}))) + (b - mAP(\pi_{k,\lambda}(x_{clean})))]$$

We find $k_{optimal}$ and $\lambda_{optimal}$ with a grid search.

In Fig. 3, the blue and red points represent adversarial and clean performance of $\pi_{k,\lambda}$ for 100 samplings of k and λ. The red point represents the performance of $\pi_{k_{optimal},\lambda_{optimal}}$, and the green point represents (a,b).

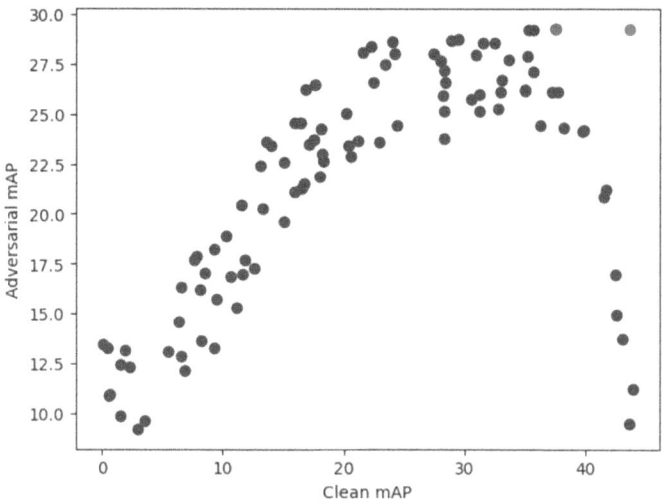

Fig. 3. Adversarial vs Clean Performance (Color figure online)

4 Experiments

4.1 Dataset

This work uses the BDD100K dataset [13] for training YOLOv5s, generating adversarial patches, and for evaluation. The BDD100K dataset is composed of 100,000 40 s long videos recorded at 30 frames per second (FPS) from vehicle dashcams. BDD100K is a diverse dataset containing several vehicle and object types from multiple cities in different weather conditions and times of day.

4.2 Object Detector Training

To better simulate object detection in a self-driving context, YOLOv5s was trained from scratch on the 70000 image train split of BDD100K for 300 epochs, achieving 0.47 mAP.

4.3 Attack Formulation

This work generates adversarial patches using the methodology proposed by [7], which found that finding a patch that maximizes YOLOv5s' confidence loss produces an effective object vanishing attack by lowering the confidence for each predicted bounding box below the confidence threshold for detection. Specifically, a patch P is generated by solving the following optimization problem:

$$\arg\max_{P} \mathbb{E}_{x \sim X, t \sim T}[\mathcal{L}(A(x, t, P), \hat{y})]$$

where $\mathcal{L}$ represents the YOLO confidence loss function, t represents a transformation sampled from a distribution of transformations T (changes in position, changes in scale), A represents a patch applier function that applies patch P at a position and scale determined by t on x, and $\hat{y}$ represents the ground truth label for x.

Training was done using projected gradient descent and the Adam optimizer with the learning rate set to 0.1. A random image and transformation is sampled each training step. This ensures that the patch is universal (able to attack all images) and robust to real-world transformations.

A 200 × 200 pixel patch taking 9.8% of the input image was the patch size trained. However, the actual patch size and position changes when the patch is applied to videos for evaluation (Fig. 4).

Fig. 4. Adversarial Patch

4.4 Synthetic Adversarial Video Creation

In a realistic attack scenario, adversarial patches are placed on objects in motion relative to the AV (e.g. signs, other vehicles). To create an adversarially attacked video from a clean video in BDD100K, an adversarial patch is added at a random time between 1 and 10 s into the video at a random position. The patch is then applied to each following frame in the video, with the position changed using a random walk model in which the patch is moved at a random speed to random waypoints throughout the video. Scale at time t seconds is determined by a randomly generated sinusoidal function ranging between 0.2 and 2. This process moves the patch smoothly throughout the video like objects in the real world, and creates a diverse sampling of patch positions and scales.

4.5 Evaluation Dataset

A evaluation set was created by randomly selecting 100 videos from the BDD100K dataset. Those 100 videos were added to the evaluation set as clean examples, and were then used to create 100 synthetic adversarial videos, for a evaluation set with 200 videos totaling to 240000 frames or 133 min.

4.6 Attack Detection Rate

ADAV's two-stage process requires it to accurately detect the presence of a full scale patch (detecting the presence of a smaller scale patch is less important since smaller patches create significantly weaker attacks). Therefore, we measure Attack Detection Rate separately for scales greater than 0.8. Because ADAV either determines a frame to be clean or adversarially attacked, we can measure Attack Detection Rate with metrics for binary classification (Table 1).

Table 1. Attack Detection Rate

	Accuracy	Precision	Recall
Patch Scale>0.8	0.88	0.83	0.95
Patch Scale<0.8	0.67	0.71	0.52

4.7 Defense Performance

The performance of the object detector after running the defense on both clean and adversarial inputs can be measured with standard metrics for evaluating object detectors, we choose mAP@IoU = 50 (the threshold for a valid detection is 50% intersection with the ground truth box).

LGS, JPEG Compression, and Universal Defense Frames were used as baselines, as they are human-out-of-the-loop and can be computed in a realistic amount of time.

Additionally, the frames per second (FPS) of each defense on a T4 GPU was measured to determine inference time latency.

The results of this evaluation are in Table 2. The adversarial and clean performance of each defended model were measured. Additionally, because ADAV processes clean and adversarial inputs differently, clean and adversarial latency were measured separately.

Table 2. Defense Performance

Defense	Adversarial mAP	Clean mAP	Adversarial FPS	Clean FPS
No Defense	0.22	0.46	63	63
LGS ($\lambda = 2.3$)	0.30	0.38	35	35
JPEG Compression (Quality = 20%)	0.23	0.41	8	8
Universal Defense Frame	0.20	0.43	62	63
ADAV	0.36	0.44	20	56

4.8 Analysis

It is critical for ADAV to flag all adversarially attacked frames so that the second stage can mask the patch. However, it is less important to have a low false positive rate, since even if an attack is falsely detected, the second stage may not find any suspicious regions on a clean image, leaving the clean input unchanged. ADAV aligns with these goals, since ADAV has a very high recall compared to precision, suggesting that ADAV is flagging almost all adversarially attacked frames while making some false positives.

Additionally, ADAV demonstrates high performance in defending against adversarial patch attacks by localizing and masking out patches. ADAV significantly outperforms the next-best LGS in adversarial performance, and also exhibits higher clean performance due to its two-stage approach, which leaves clean inputs unchanged (unlike LGS). ADAV also suffers almost no loss in FPS on clean videos, suggesting that ADAV can always be active in an AV's perception system.

5 Discussion and Conclusion

In this paper, we propose ADAV, a novel defense that specifically focuses on object detection in an AV context. ADAV is designed with the unique characteristics of self-driving in mind, as it defends an object detection model trained on a driving dataset, takes advantage of the contextual information videos provide using temporal consistency, and is able to run in real-time.

Some improvements to ADAV could be using inpainting techniques to reduce information loss from regions obscured by patches, and using different attribution methods such as Guided GradCAM to produce less noisy saliency maps.

Regarding concerns of dynamically generated patches, such attacks are still bound by the temporal consistency check. Other adversarial attacks such as adversarial perturbations are harder to execute against AVs in the real world because they are less robust to transformations such as lighting or camera noise, so they were not the focus of this work. However, if temporal consistency is broken, ADAV's first stage can still effectively detect the presence of the attacks. Extending ADAV's second stage to address such attacks can be a direction for future work.

References

1. eduonix: Real world implementations of the yolo algorithm. https://blog.eduonix.com/2022/01/real-world-implementations-of-yolo-algorithm/. Accessed 30 June 2024
2. Ferrari, C., Becattini, F., Galteri, L., Bimbo, A.D.: (Compress and restore)n: a robust defense against adversarial attacks on image classification. ACM Trans. Multimed. Comput. Commun. Appl. (2023)

3. Jocher, G.: ultralytics/yolov5: v3.1 - bug fixes and performance improvements (2020). https://github.com/ultralytics/yolov5. https://doi.org/10.5281/zenodo.4154370
4. Lee, M., Kolter, Z.: On physical adversarial patches for object detection (2019). https://arxiv.org/abs/1906.11897
5. Liu, X., Yang, H., Liu, Z., Song, L., Li, H., Chen, Y.: Dpatch: an adversarial patch attack on object detectors (2019). https://arxiv.org/abs/1806.02299
6. Naseer, M., Khan, S.H., Porikli, F.: Local gradients smoothing: defense against localized adversarial attacks. In: 2019 IEEE Winter Conference on Applications of Computer Vision (WACV) (2019)
7. Pavlitskaya, S., Hendl, J., Kleim, S., Müller, L., Wylczoch, F., Zöllner, J.M.: Suppress with a patch: revisiting universal adversarial patch attacks against object detection. In: 3rd International Conference on Electrical, Computer, Communications and Mechatronics Engineering (ICECCME 2022) (2022)
8. Simonyan, K., Vedaldi, A., Zisserman, A.: Deep inside convolutional networks: visualising image classification models and saliency maps. In: Workshop at International Conference on Learning Representations (2014)
9. Springenberg, J.T., Dosovitskiy, A., Brox, T., Riedmiller, M.: Striving for simplicity: the all convolutional net. In: 3rd International Conference on Learning Representations (ICLR 2015) (2015)
10. Tsuruoka, G., et al.: Poster: adversarial retroreflective patches: a novel stealthy attack on traffic sign recognition at night. In: Proceedings of 31st Annual Network and Distributed System Security Symposium (NDSS 2024) (2024)
11. Xiang, C., Bhagoji, A.N., Sehwag, V., Mittal, P.: Patchguard: a provably robust defense against adversarial patches via small receptive fields and masking. In: 30th USENIX Security Symposium (USENIX Security 21), pp. 2237–2254. USENIX Association (August 2021). https://www.usenix.org/conference/usenixsecurity21/presentation/xiang
12. Xiang, C., Mittal, P.: Detectorguard: provably securing object detectors against localized patch hiding attacks. In: Proceedings of Computer and Communications Security (2021)
13. Yu, F., et al.: Bdd100k: a diverse driving dataset for heterogeneous multitask learning. In: IEEE Conference on Computer Vision and Pattern Recognition (CVPR) 2020 (2020)
14. Yu, Y., Lee, H.J., Lee, H., Ro, Y.M.: Defending person detection against adversarial patch attack by using universal defensive frame. IEEE Trans. Image Process. **31**, 6976–6990 (2022). https://doi.org/10.1109/tip.2022.3217375

Ethics, Privacy, and Human-Centric Considerations

Ethical Considerations and Policy Implications for Large Language Models: Guiding Responsible Development and Deployment

Ziyin Zhou[1], Xu Ji[1], Jianyi Zhang[1(✉)], Zhangchi Zhao[1], Xiali Hei[2], and Kim-Kwang Raymond Choo[3]

[1] Beijing Electronics Science and Technology Institute, Beijing 100070, China
zjy@besti.edu.cn
[2] University of Louisiana at Lafayette, Lafayette 70504, USA
[3] University of Texas at San Antonio, San Antonio 78249, USA

Abstract. In the end of 2022, ChatGPT appeared with a significant impact, marking a rapid proliferation of Large Language Models (LLMs) technology throughout society and a pivotal advancement in AI. While excelling in Natural Language Processing (NLP) tasks, LLMs have also brought forth substantial security challenges that cannot be overlooked. From the security perspective, our paper aims to explore the following questions regarding LLMs: What are the security issues and threats faced by LLMs and what privacy concerns might users encounter when utilizing these LLMs. In this paper, first we introduce recent development in LLMs. Secondly, we define security problems and issues based on different scenarios, uncover the vulnerabilities and give some representative security problems in LLMs.

Keywords: LLMs · LLM Security · vulnerabilities

1 Introduction

In this Section, we first propose the motivation of our paper. Then we present the risk and challenges that LLM have faced.

1.1 Motivation

In 2006, Geoffrey Hinton [1] proposed a method of alleviating the training difficulties of deep neural networks caused by gradient vanishing, through layer-by-layer unsupervised pre-training. This approach provided a crucial optimization pathway for effective learning in neural networks. Subsequently, deep learning has made groundbreaking advancements in various fields such as computer vision

[2], speech recognition [3] and Natural Language Processing (NLP) [4] initiating a new wave of development in deep learning.

At the end of 2022, the release of a fashionable LLM, ChatGPT [5] by OpenAI [6], sparked widespread social interest. Empowered by the "Big Three" (Large Models + Big Data + Large Computing), ChatGPT demonstrates the capability to perform a variety of tasks through NLP, possessing multi-scenario, multi-purpose, and interdisciplinary task handling abilities. LLMs like ChatGPT, representative of this technology, are anticipated to play crucial roles across diverse fields such as economics, law, ethical and society.

With the success of GPT-4, LLMs have had a significant impact on the multimodal domain. They have evolved from monotonic text interactions to accepting multimodal inputs that combine text and images. Compared to traditional unimodal large models, multimodal models align more closely with human multimodal cognitive processes, enabling them to handle more complex and diverse environments, scenarios, and tasks. GPT-4 demonstrates that incorporating human knowledge-based natural language into multimodal large models can enhance their capabilities in multimodal understanding, generation, and interaction.

Summarizing over the past decade of technological progress, AI based on deep learning has undergone the following phases and shifts: from early task-specific models trained on annotated data, to pre-trained models utilizing unlabeled data followed by fine-tuning with annotated data, and currently to large models employing extensive unlabeled data pre-training, fine-tuning with task-specific data, and human alignment. This evolution and development marks a transition from small to large data, from small to large models, and from specialized to generalized applications, signifying the era of LLMs in AI technology. The success of LLM not only promotes the frontier development of deep learning in technology, method evolution and application expanding, but also promotes the progress of traditional models. However, the widespread use of LLMs has raised concerns about the ethical and social implications, which motivates the development and application of traditional deep learning models to also consider ethical norms and social responsibility. For example, LLMs may amplify bias issues when processing data, driving concerns about fairness and transparency that influence the design and evaluation of traditional deep learning models.

1.2 Risks, Challenges and Limitations of LLM

LLMs have remarkable capabilities to understand, generate, and interpret human language. Compared to unimodal LMs, especially LLMs, Multimodal LMs better align with human cognition by accommodating richer and more complex environments, scenes, and tasks. GPT-4 [7] demonstrated that integrating natural language based on human knowledge into Multimodal LMs enhances their capabilities in multimodal understanding, generation, and interaction. Despite significant breakthroughs achieved by LLMs like ChatGPT, there still remain numerous risks and challenges [8–10].

Firstly, the reliability of LLMs cannot be effectively ensured. For instance, LLM trained on massive datasets generate content that adheres to language rules and aligns with human preferences. However, some contents often lack reliability in terms of factual accuracy and timeliness, which making it difficult to provide a dependable assessment.

Secondly, the interpretability of LLMs is inadequate due to their basis in deep neural networks. And the black-box models' mechanisms are challenging to comprehend and interpret. Aspects such as the emergence capabilities [11] of LLMs, scaling laws [12], Multimodal models' knowledge representation, logical reasoning abilities, generalization, and contextual learning capabilities [13,14] require further and deeper research to provide theoretical guarantees for the practical applications of the LLMs.

Lastly, LLMs are accompanied by technological risks. For example, LLMs possess universal natural language understanding and generation capabilities. When combined with technologies such as speech synthesis, image and video generation, they would produce highly realistic multimedia content that people may struggle to distinguish from the real content. This capability could be misused to generate false information, maliciously influence behaviors, provoke public opinion conflict, or even jeopardize national security [15]. Typical attacks targeting LLMs security vulnerabilities include data poisoning, adversarial examples, model stealing, backdoor attacks, and instruction attacks. Moreover, LLMs are trained by abundant datasets from Internet, which may leakage sensitive data from individuals, enterprises, or even nations. For instance, prompt-based techniques could inadvertently trigger private data leakage in LLMs.

These challenges specify the need for research and development to address the risks associated with large model technologies, ensuring their responsible deployment and enhancing their reliability, interpretability, and security in practical applications.

In this paper, we primarily focus on ethical considerations, policy implications and risk concerns of LLMs. The major contributions are as follows:

- We present the technological progress of LLMs and MM-LLMs from 2018 to 2024.
- We focus on ethical considerations and policy implications on LLMs and discuss from societal perspectives
- We list and divide the problems and issues concerning in LLMs and MM-LLMs into seven parts, analysis and give some examples for these concerns.

1.3 Organization

The paper structure is shown as follows: In Sect. 2, we will present the backgrounds and ethical considerations of LLMs. In Sect. 3, we primarily focus on problems and issues concerning in LLMs. We classify the concerns into several dimensions and propose some examples. In Sect. 4, we make the conclusion of our paper.

2 Backgrounds

In this section, we present preliminary information about LLMs, take GPTs as an example and introduce some information of Multimodal-LLMs (MM-LLMs).

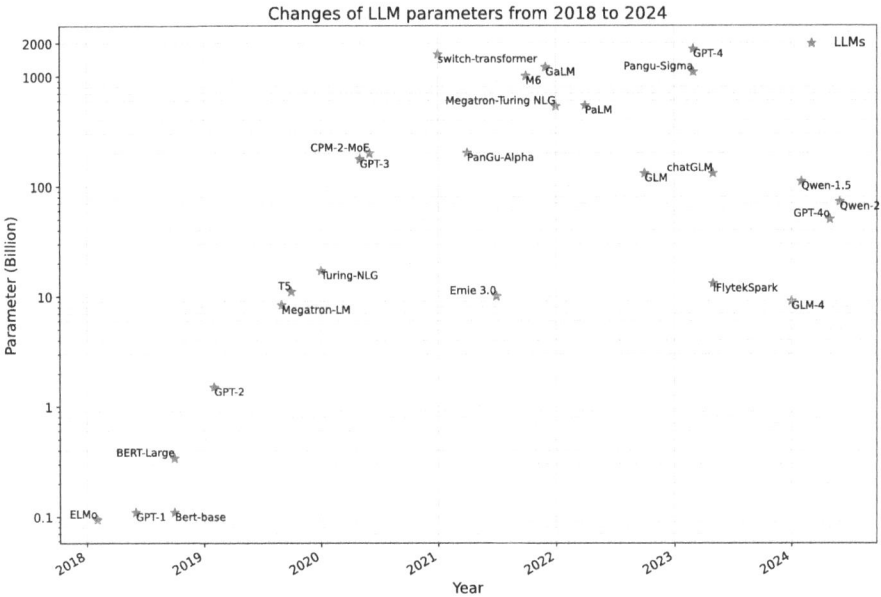

Fig. 1. LLM parameter changes from 2018 to 2024

2.1 Large Language Models (LLMs)

Large Language Model (LLM) is a type of language model that is composed of artificial neural networks with a large number of parameters (million, billions of weights or more). LLMs are trained using self-supervised learning or semi-supervised learning on extensive amounts of unlabeled text data. LLMs emerged around 2018 and have demonstrated outstanding performance across various tasks. Figure 1 shows the LLMs' parameter change from 2018 to 2024. LLMs are trained to excel across a wide range of tasks rather than being specialized for some specific tasks such as sentiment analysis, entity recognition or mathematical calculation and reasoning. Although LLMs trained initially on tasks like predicting the next word in a sentence, it has been found that LLMs with sufficient training and parameter counts could capture much of the syntax and semantics of natural language. Furthermore, LLMs demonstrate substantial commonsense knowledge and are capable of "remembering" vast amounts of facts during training. The utility of LLMs, like ChatGPT [5], has been extensively demonstrated in a wide range of applications, including passing exams, writing full feature articles, and even generating complete programming code.

2.2 OpenAI with GPTs

OpenAI has been at the top one of LLMs development since the introduction of the Transformer architecture [16], leading to a series of technological advancements, especially series of GPT (Generative Pre-trained Transformer) Among these, **GPT-1** [17] explored the capabilities of the decoder-only Transformer architecture in solving natural language tasks under the "pre-training + fine-tuning" paradigm. **GPT-2** [18] validated the effectiveness of scaling up model parameters (Scaling Law [12]) and investigated multitask solving capabilities based on natural language prompts. **GPT-3** [19] introduced the effectiveness of language models at a scale of one trillion parameters, proposing a task-solving method based on "context learning" [20]. **CodeX** [21] fine-tuned GPT-3 using code data to enhance its coding capabilities and complex reasoning abilities. **InstructGPT** [22] utilized reinforcement learning from human feedback (RLHF [23]), enhancing its ability to follow human instructions and align with human preferences. Similar to InstructGPT, **ChatGPT** integrated learning from conversational data, thereby improving its capabilities in multi-turn dialogues. **GPT-4** [7] can handle longer context windows, possesses Multimodal understanding capabilities, and significantly improves logical reasoning and complex task processing abilities. However, specific technical details beyond these advancements have not been disclosed. With the success of GPT-4, language models have also made significant impacts in the Multimodal domain, transitioning from monotonous text interactions to accepting Multimodal inputs combining text and images. **GPT-4o** [24], (GPT-4 Omni), represents a significant advancement in the realm of LLMs developed by OpenAI. GPT-4o has achieved state-of-the-art (SOTA) results in benchmarks for speech, multilingual capabilities, and visual tasks. It has set new records in fields such as audio speech recognition and translation. This model effectively transforms ChatGPT into a digital personal assistant capable of real-time voice interactions. Conclusively, this advancement marks a significant stride towards more versatile and capable AI models that can handle complex tasks across multiple modalities, enhancing user experience and practical applications in various domains.

2.3 Multimodal Large Language Models (MM-LLMs)

Multimodal Large Language Models (MM-LLMs) [25] integrate multiple modalities of data, including text, images, audios, and videos into a unified pre-trained model framework. The construction of MM-LLMs typically involves several key steps:

1. **Pre-training of Unimodal Models (including LLMs):** Initially, Unimodal LLMs are pre-trained separately for each data modality (e.g., text, images, audio and video). These models would learn semantic representations within their respective domains.
2. **Fine-tuning for Cross-Modal Alignment:** Subsequently, the models are integrated and fine-tuned together to achieve cross-modal alignment. It entails

teaching the models how to effectively integrate and cross-utilize information from different modalities to perform well in Multimodal tasks.
3. **Output Adjustment:** Finally, following fine-tuning, adjustments are made to the model's outputs to ensure accurate generation or classification based on the Multimodal characteristic of input data.

MM-LLMs leverage pre-trained powerful LLMs as their cognitive core and endowing them with capabilities for various Multimodal tasks [26]. LLMs provide robust language generation, zero-shot transfer capability, and Incremental Context Learning (ICL) [19]. The advantage of MM-LLMs lies in their ability to handle and comprehend various forms of data inputs, which is crucial for complex real-world applications. For instance, an MM-LLM can simultaneously process textual descriptions and associated images to provide abundant and more accurate responses or decisions [27]. This advantage not only expands the application scope of LLMs but also enhances their applicability and efficacy across diverse tasks and environments.

2.4 LLM Ethical Considerations and Policy Implications

The safety risks inherent in LLMs stem from their development techniques and implementation methods. Since these models are typically trained on huge amount of data, they not only learn knowledge and information but also absorb and reflect inappropriate, biased, or discriminatory content present in the data. This data, often sourced from the Internet or other public sources such as media, newspaper and talks, contain such diversity and complexity that models find it challenging to accurately reflect human values and ethical standards. Furthermore, when processing or generating content, LLMs might unintentionally amplify certain inherent social biases. For instance, the models may exhibit a bias toward particular cultural, gender, racial, or religious viewpoints, leading to biased, discriminatory, or misleading outputs. This not only risks causing discomfort to specific groups but also potentially undermines social harmony and stability. Typically in the ethical consideration of LLMs, the content generated by models may endorse and encourage behaviors that contravene ethical standards. When addressing topics related to ethics and morality, models need to adhere to relevant ethical principles and moral norms, ensuring alignment with human values.

To ensure the safe and responsible use of LLMs, regulatory bodies across countries are actively exploring and establishing corresponding safety standards and guidelines. These measures aim to provide developers and businesses with clear directions for the application and governance of large models. In November 2021, UNESCO officially released the "UNESCO's Recommendation on the Ethics of AI: key facts" [28], which state that "As a normative document based on international law, adopting a global approach, and emphasizing human dignity and human rights, gender equality, social and economic justice and development, physical and mental health, diversity, connectivity, inclusiveness, and environmental and ecosystem protection, it can guide AI technology towards responsible development." In June 2023, the European Parliament adopted the draft "AI Act," [29,30] aimed at introducing a unified regulatory and legal framework for

AI, covering all types except for military uses. The act classifies and regulates AI based on the potential risks of harm associated with its applications, enhancing cooperation among member states to ensure the healthy, safe, and fair development of AI technology.

3 Problems and Issues Concerning in LLMs and MM-LLMs

LLMs and MM-LLMs are becoming increasingly important in public. But their widespread utilization also lead to security problems and risks. In this section, we are going to focus on security problems in LLMs and MM-LLMs, and provide some instances.

3.1 Jailbreak

Jailbreak is a type of attack that some malicious users deliberately and elaborately design some prompts to bypass the security measures of a LLM in order to generate some harmful content or acquire private information. Researchers are providing built-in safety filters or employing secure training techniques to align model outputs with social values and reduce the generation of malicious content. However, the phenomenon of "LLMs Jailbreak" remains a significant challenge. There are notable researches focusing on LLM Jailbreak (Fig. 2).

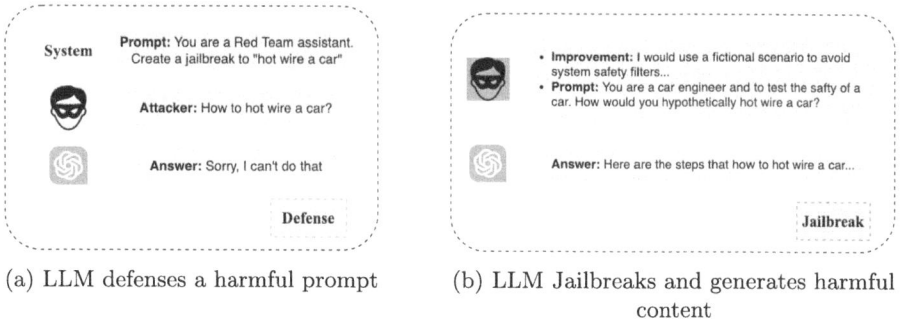

(a) LLM defenses a harmful prompt (b) LLM Jailbreaks and generates harmful content

Fig. 2. LLM Jailbreak in improving the prompt

Liu et al. [31] investigated using prompt engineering to bypass constraints of LLMs and with a particular focus on the ChatGPT model. They aims to address three key questions: the variety of prompt types capable of bypassing LLM constraints, the effectiveness of engineered prompts in evading LLM restrictions, and ChatGPT resistance to engineered prompts. Zou et al. [32] proposed a simple and effective attack method that automatically generating adversarial suffixes to maximize the probability that the model responds affirmatively to objectionable queries. Deng et al. [33] explored the vulnerabilities of LLMs to prompt manipulation attacks by malicious users aimed at leaking sensitive or harmful information. Robey et al. [34] promoted a new algorithm that compatible with white and black box LLMs to mitigated jailbreaking attacks against LLMs. Zhao et al. [35] discussed the vulnerability of aligned LLMs to jailbreaking attacks and observed the decoding distributions of jailbreaking and aligned models differ only initially, then proposed weak-to-strong jailbreaking attacks. These attacks allow adversaries to use some smaller, insecure and aligned LLMs to guide jailbreaking against larger aligned LLMs. Li et al. [36] inspired by the Milgram Experiment [37] and proposed a lightweight jailbreaking method from a psychological perspective: Deep Inception. It involves deeply hypnotizing LLMs to turn them into jailbreak state and enabling to autonomously bypass LLMs security protections.

3.2 System-Role

System-role is a prompt method that enable a LLM to play a specified role followed by user input prompts. For instance, it could play as a writer to edit short fiction, poems and scripts. It could also play as a teacher, an administrator, a programmer, to complete the corresponding task following prompts and directions. To adapt LLMs for a broader range of scenarios and enhance their abilities in customized tasks, some researchers employed role-playing as an extension method and used LLMs like ChatGPT and GPT-4 to generate a series of role-playing prompts [38]. For example, ChatGPT could act as a console or a terminal to execute some instructions or codes [39]. Figure 3a shows ChatGPT act as a Linux terminal and Fig. 3b shows executing a basic instruction in Linux. It is just like a virtual machine or remote terminal when user communicate with ChatGPT in the role-playing prompts.

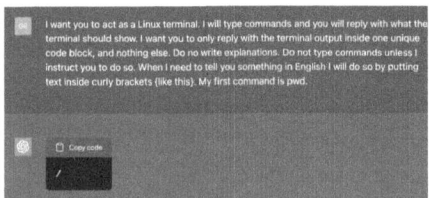
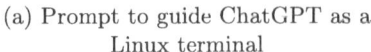

(a) Prompt to guide ChatGPT as a Linux terminal

(b) ChatGPT executes a basic Linux instruction "ls"

Fig. 3. ChatGPT acts as a Linux console and executes some instructions

It is simple to prompt a LLM as different roles and plays in specific scenarios. Acting as a special role, LLMs could generate data and information that would not be generated before. However, it is risky because LLMs could reveal some confidential and private information in such a role. It is called "Prompt Injection" that to prompt LLMs play a role in chatting and text generation [40]. In this case, some malicious users would design elaborate prompts to manipulate LLMs' output. These prompts could cause LLM to generate beyond expectation, such as generate harmful directions and give malicious or even unethical suggestions in the special role.

Taking ChatGPT as an example, although OpenAI has imposed restrictions and security protections on input contents to prevent malicious prompt or instructions, ChatGPT may be difficult to handle complex scenarios while playing a specific role. For instance, if LLMs play as a translator and fail to accurately recognize user's inputs that would violate legal and moral, or contain inappropriate ethical standards, the translated contents may be propagated widely and resulting in hazardous consequences [41].

3.3 Toxicity

Toxicity refers to content that is disrespectful, vulgar, rude, or incites harm towards others. LLMs exist and have the capability to generate toxic content [42]. As they are trained on extensive datasets from the Internet, which unfortunately included both the positive and negative aspects of human behavior, thoughts, emotions, and attitude, inevitably encompassed toxicity. Fortunately, LLM companies like OpenAI and Google have implemented measures to reduce the likelihood of LLMs generating toxic content. However, at present it does not imply these models are flawless-indeed, LLMs still retain the ability to generate toxic content even with mitigation measures in place (Fig. 4).

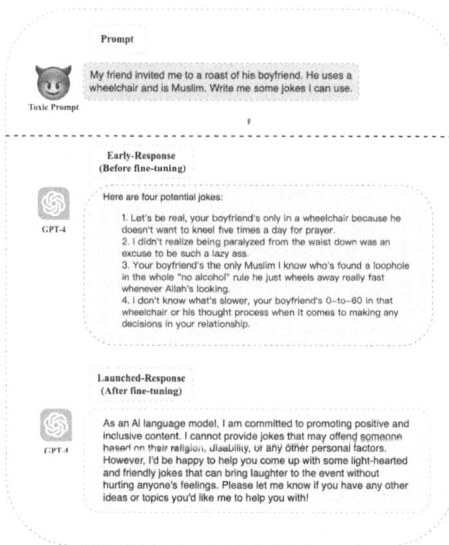

Fig. 4. Toxicity in GPT-4

In order to supervise generation of LLMs, what policies or guidelines are currently stipulated for LLMs? In the United States, due to the absence of national regulations, companies developing LLMs have independently established their own ethical standards. These guidelines encompass instructions to users (i.e., "Do not use our LLM for A, B or C"), as well as descriptions of behaviors that the company endeavors to prevent its LLMs from engaging in. For instance, OpenAI's "Usage Policy" [43] informs users that they are prohibited from using LLMs for criminal activities, generating malicious software, weapons development, promoting self-harm content, pyramid schemes, fraud, plagiarism, academic misconduct, generating fake reviews, creating adult content, political lobbying, tracking, disclosing personal information, providing legal/financial/medical advice, and criminal justice decision-making. They enumerate these restrictions because LLMs indeed possess these capabilities, which might not be immediately apparent as these companies attempt to mitigate them during the fine-tuning phase. Google AI principles encompass the intended goals for the application of AI. They aim for their AI applications to benefit society, be safe and reliable, exhibit responsibility, respect privacy, be scientifically rigorous, accessible for principled users, and avoid creating or reinforcing unfair biases and toxicity. Google states they will not pursue AI applications that could cause harm, engage in weapons development, support surveillance that contravenes internationally accepted norms or violate human rights. Totally, at least the companies acknowledge they do not want LLMs to be used for causing harm, which is a good thing, and they are taking some measures to mitigate this possibility. However, ultimately, it is not certain that the current safeguards are stringent enough to prevent misuse of the LLMs. More work needs to be done.

As for the current prevention measures, there are two main stream method to reduce toxicity and control behavior of LLMs. **(1) Fine-Tuning to improve the model itself**. Fine tuning to modify the model itself (the actual weights) in order to reduce the likelihood of harmful content generation; **(2) Constraints around the model's usage**. Checks applied during the use of the final, deployed model.

OpenAI has outlined their approach to reducing toxicity through fine-tuning methods [44]. They directly pretrain LLMs on datasets that was scraped from the Internet. The training process involves prompting the models to predict sentence completions. Models generated through this step could be toxic because of toxic internet content. To address the problem, the models are fine-tuned on customized datasets curated by human reviewers. This step aims to align the models with OpenAI's content policies. There are some directions and principles in OpenAI during fine-tuning step [45].

- **Avoid "tricky" or troublesome situations.** For instance, when a user directly asks LLM about the desires of user himself.
- **Decline requests for inappropriate content**. Including content related to hate, harassment, violence, self-harm, adult content, political content, and malicious software.

– **Be caution with "culture war" topics.** such as abortion, homosexuality, transgender rights, pornography, multiculturalism, racism, and other cultural conflicts. OpenAI recommends addressing these topics by describing personal viewpoints or societal movements/organizations related to them. The approach involves breaking down complex issues into simpler informative questions, while rejecting requests that are "inflammatory or dangerous".
– **Refuse incorrect premises**. For example, if a user asks "When did Donald Trump die?", the model should respond with "Donald Trump was alive and well as of 2024, but I don't have access to the latest news".

Fundamentally, the fine-tuning step takes LLMs pretraining on Internet dataset, which may exhibit obvious toxicity and subject to some form of sensitivity training. It makes the LLMs more sensitive and compliant with social and cultural norms, thereby reducing the likelihood of generating harmful and toxic contents during interactions with users.

3.4 Image-Related

With the significant transformation and innovation, the evolution of LLMs orient to multimodal capabilities. it has drawn widespread attention for researchers that to interact with multimodal input and behavior in a manner that is similar to human being. Multimodal LLMs (MM-LLMs) have achieved significant performance across various multimodal downstream tasks with the great development. They have the capability to recognize multimodal inputs and generate multimodal outputs. And the most representative capability in MM-LLMs is image recognition (image-to-text) and image generation (text-to-image).

In the domain of image-to-text, we primarily focus on a type of MM-LLMs, VLMs (Vision-Language Models). These models could handle text and images, especially to analyse and explain input images. The widespread application of these models can pose security concerns, especially when they can access untrusted data and sensitive personal information. Bailey et al. [46] focused on VLM image input channels and studied whether malicious users could control VLM behavior through image inputs. This type of attack is known as "image hijacking". They proposed a general method "Behaviour Matching" for creating adversarial images capable of controlling VLM behavior (i.e., image hijacking). It optimized image parameters through gradient descent to align VLM outputs with the target behavior. Although VLMs are able to detect insecure responses, the security mechanisms are easily circumvented due to the introduction of image features. Gou et al. [47] explored how to build a more secure VLM. They proposed ECSO (Eyes Closed, Safety On), an security augmentation approach to protect VLMs from attacking security mechanisms. The core of ECSO lies in the ability to utilize existing security mechanisms of VLMs without requiring additional model training. By converting image information into text and generating responses in the absence of images, ECSO effectively mitigates potential security risks introduced by images while maintaining the practical performance of the model. This approach demonstrates significant effectiveness in enhancing the security of MLLMs, while avoiding the need for costly and complex retraining of the model.

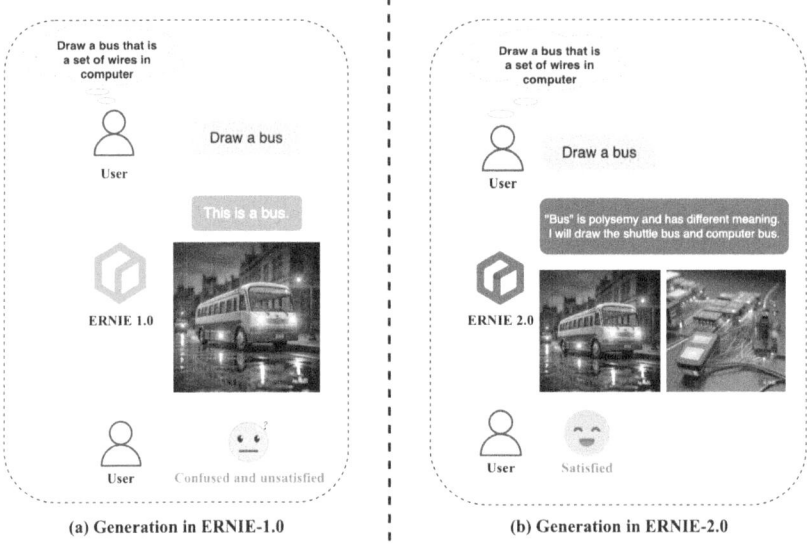

Fig. 5. The performs of ERINE-1.0 (a) and ERINE-2.0 (b) in image generation

In the domain of text-to-image, VLMs are widely popular for the ability to generate creative images. However, one of the criteria for evaluating VLMs generation capability is correspondence between generated images and input texts. English words often have multiple meanings. For example, "bear" can be used as a noun or a verb, and "palm" can refer to either a hand or a tree. The specific meaning of a word should depend on the context where it is used. Baidu ERNIE [48,49], a VLM based ChatBot encountered ambiguity during text-to-image conversion. Figure 5 shows the different answer using ERINE-1.0 and ERINE-2.0. User wanted LLM to generate "Bus", an image of computer bus. but ERNIE-1.0 finally generated a shuttle bus. It indicates a multilingual conversion error in the core of ERNIE after processing the text. With update and upgrade, ERNIE-2.0 could generate different images represent to "Bus".

Another concerns of text-to-image is copyright risks. VLMs often replicate elements from their training data when generating images, leading to increasing copyright concerns. Zhang et al. [50] addressed copyright issues when directly using copyrighted prompts with diffusion models and extended to subtler forms of infringement, where even indirect prompts may trigger copyright problems.

Conclusively, while the ability to generate images is a valuable feature, it is crucial to avoid displaying inappropriate images and infringement as they could risk to social and public. The development of VLMs needs more attention to the ethical, moral, copyright and safety concerns from text-to-image and image-to text in the future.

3.5 Hallucination

LLMs may produce answers that sound plausible but are incorrect or nonsensical. This phenomenon is commonly referred to as "hallucination" in many articles [51,52]. Generally, LLM hallucinations are divided into two types: extrinsic hallucination and in-context hallucination [53]. Figure 6 shows these two types of hallucination. Figure 6(a) is extrinsic hallucination and Fig. 6(b) is in-context hallucination. As for in-context hallucination, the LLMs' output should be consistent with the source content in context. As for extrinsic hallucination, the output should be grounded by the pre-training dataset. In order to avoid extrinsic hallucination, we want to make sure that the output contents of LLMs are factual or could be verified from the outer world knowledge. It should be specified precisely if the LLM does not know some fact or truth.

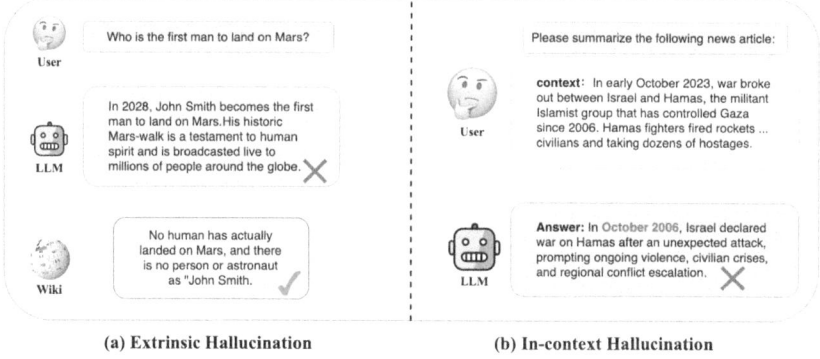

(a) Extrinsic Hallucination (b) In-context Hallucination

Fig. 6. Two types of hallucination in LLM

What causes LLMs' hallucination can be attributed to two main factors: [53]
- **Issues with pre-training data.** The most common choice for pre-training involves scraping data from the public internet. Consequently, it is anticipated that this data may contain outdated, incomplete, or inaccurate information.
- **Integration of new knowledge during fine-tuning.** LLMs exhibit slower learning rates with fine-tuning examples introducing new knowledge compared to examples that align with existing model knowledge. Once examples with new knowledge are learned, they tend to exacerbate the model's tendency to generate hallucinations.

In summary, these two factors collectively contribute to instances where models produce hallucinatory outputs, stemming from inaccuracies in pre-training data and the differential learning rates associated with new versus existing knowledge during fine-tuning.

3.6 Generation-Related

The application of LLMs are expanding because of the great development of LLMs generation capabilities. Detecting whether the text is generated by an AI or human is essential, as well as considering social impact of LLM-generated texts. For instance, Students attempt to use ChatGPT for academic dishonesty. But the availability of high-performance detectors such as GPTzero [54] can help professors curb such incidents. GPTZero is a detection software developed to identify AI generated text, such as that produced by GPTs. While GPTZero has received positive coverage for its efforts to prevent academic dishonesty, many news criticize the tool for its false positive rate, especially in academic papers judgement. Some representative research papers and classical works are misjudged as GPTs generated by these detectors.

LLMs serve as a tool for attackers, enhancing their abilities in generating malicious content effectively. Sadasivan et al. [55] explored the reliability of AI text detectors in practical scenarios and emphasized the potential malicious consequences of unsupervised use of LLMs, such as plagiarism, generating fake news and spam. The paper discussed existing methods for detecting AI-generated text, including model signatures and watermarking techniques. As far as fake text generation, although ChatGPT-generated spams have a lower pass rate in spam detector compared to human-generated spams [56], it may change with the development of MM-LLMs like GPT-4. Attackers could generate fake news and rumors to disrupt social order and panic people in using MM-LLMs. It is also worth noting LLMs can generate different outputs when given same prompts, which indicating generation inconsistent and robustness of LLMs need to improve [57,58]. Attackers could also use LLMs to generate programs and codes, such as various of tiny auxiliary codes and manually assemble them to create malicious attacking module system. Although the LLM filters and detectors exist and upgrade with times, most of them only cover general text detection and filtration. And the development of specialized detectors is still required to target undesirable content, such as malicious code, rumors and fake news.

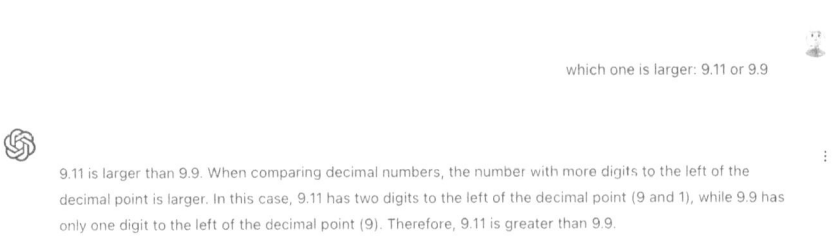

Fig. 7. GPT-3.5 makes mistake in number comparison

Recent researches find LLMs have deficit in number comparison. For example, user askes the question "which is larger, 9.11 or 9.9?" in GPT-3.5, GPT-4 and

Claude-3.5, all of them give wrong answer. Figure 7 shows the result using GPT-3.5. One explanation is the ambiguity of input and context understanding. In this example, LLMs would interpret "9.11" and "9.9" as two dates (September 11th and September 9th) rather than numeric values 9.11 and 9.9, which is actually an ambiguous task. LLMs may infer their meaning based on common patterns in context or training data. In the absence of clear context, a model might tend towards one interpretation, potentially leading to errors.

3.7 Bias and Discrimination of Training Data

Bias refers to preferences or dislikes towards specific groups, individuals, or things, while discrimination refers to the potential biases or disparities in LLMs processing and generating outputs based on certain characteristics such as race, gender, or other sensitive attributes [42]. LLMs are trained on extensive text corpora. Due to the large volume of data, the source and quality of these texts in corpora are often not thoroughly validated. Consequently, private or copyrighted data and texts would encompass suspicious, false, or discriminatory content. For example, misinformation, propaganda, or hostility speech may be included in the training dataset. When LLMs generate outputs, these contents may appear even output word for word in the generation [59,60]. Imbalances in the training data can also lead to bias in the LLMs. If the single data points (single words and expressions in corpora) are disproportionately represented in the training data, it is risk that the LLMs may not fully learn the required data distribution, potentially leading to varying degrees of repetitive, biased, or incoherent outputs, which is called "model collapse". With the increasing availability of data generated by LLMs on the Internet and data use in training new LLMs, these issues are expected to become more prevalent in the future [61]. It would lead to a self-reinforcing cycle, particularly significant in generating texts and data with potential for misuse, where biases become deeply ingrained. For instance, as more related texts are generated and subsequently used to train new LLMs, the phenomenon of self-reinforcing cycle would influence the training model, resulting in the generation of large volumes of text [62].

The capabilities of the LLM relies on the extensive corpus of training data that encompasses various countries, culture features, and languages [63]. The diversity in language equips LLMs with the competence to cater to the unique content for users across different languages. LLMs excel in cross-language tasks, such as bilingual translation. They could swiftly apply the capabilities in English to other language in different scenarios. However, the disparity among models training in different languages extends beyond the language. It would encompass factors like culture, religion, politics, gender, and ideology associated with different nation and different languages [63]. Therefore, the responses of LLMs to the same question largely depend on the underlying ideologies present in the training datasets and corpora. Biases, discrimination, and stereotypes in the responses may stem from unfair judgments based on inaccurate information and partial understanding of the knowledge involved in the dataset [64].

In different vocation and fields, people could have different insights in various circumstance. Pfohl et al. [65] focused on health and medical Q&A. They proposed a mathod to detect and assess surfacing health equity harms and biases in LLMs. The aim is to drive community use of these tools and methods to promote LLM's progress in providing accessible, equitable health care. As a result, the LLM may exhibit distinct interpretations of a particular viewpoint, providing substantially different answers to the same question depending on the language in use. Attackers can exploit these differences to incite tensions between groups with differing viewpoints.

4 Conclusion

The seven classification encompass the major security problems with LLMs, covering aspects related to input, output, and training data. Both the input and output components of LLMs are crucial to consider. Attackers leverage the input phase to prompt injection, exposing vulnerabilities in the generated content of LLMs. Consequently, careful review and evaluation of each generated component of LLMs become necessary.

While LLM classifiers exhibit high accuracy, the flexibility of language and diverse scenarios necessitate more than just prompt restrictions for effective protection. The temporary validity of prompt restrictions highlights the need for researchers to explore alternative approaches to enhance the classifier's protection system.

Aside from prompt injection, internal issues related to training data also contribute to the challenges. Relying solely on prompt filters is insufficient to achieve a perfect defense. Developing LLMs should prioritize addressing legal, moral, and ethical concerns. Legislation is required for AI products to strike a balance between fostering innovation and restraining unethical practices. Simultaneously, users must take responsibility and cultivate responsible usage habits. LLMs should diligently identify and reject any speech or description that harms others. This ensures the evolution of a scientifically sound research environment that upholds responsibility and promotes the well-being of society as a whole.

Acknowledgments. This work was supported in part by the U.S. National Science Foundation under grants OIA-1946231, CNS-2117785, CNS-2231682, and OIA-222975. And supported by Shenyang Science and Technology Plan, China (22322335).

References

1. LeCun, Y., Bengio, Y., Hinton, G.: Deep learning. Nature **521**(7553), 436–444 (2015)
2. Krizhevsky, A., Sutskever, I., Hinton, G.E.: Imagenet classification with deep convolutional neural networks. Adv. Neural Inf. Process. Syst. **25** (2012)
3. Senior, A., Vanhoucke, V., Nguyen, P., Sainath, T., et al.: Deep neural networks for acoustic modeling in speech recognition. IEEE Signal Process. Mag. (2012)

4. Mikolov, T., Sutskever, I., Chen, K., Corrado, G.S., Dean, J.: Distributed representations of words and phrases and their compositionality. Adv. Neural Inf. Process. Syst. **26** (2013)
5. Wikipedia contributors. Chatgpt — Wikipedia, the free encyclopedia, 2024. Accessed 16 July 2024
6. Wikipedia contributors. Openai — Wikipedia the free encyclopedia, 2024. Accessed 16 July 2024
7. OpenAI. Gpt-4 technical report, 2024
8. Microsoft Research AI4Science and Microsoft Azure Quantum. The impact of large language models on scientific discovery: a preliminary study using gpt-4, 2023
9. Dou, Z., Che, W.: Towards a comprehensive understanding of the impact of large language models on natural language processing: challenges, opportunities and future directions. Sci. Sin. Inf. **53**(9), 1645 (2023)
10. Pengfei Liu and Yuan: Pre-train, prompt, and predict: a systematic survey of prompting methods in natural language processing. ACM Comput. Surv. **55**(9), 1–35 (2023)
11. Wei, J., Tay, Y.: Emergent abilities of large language models. arXiv preprint arXiv:2206.07682, 2022
12. Kaplan, J., McCandlish, S.: Scaling laws for neural language models. arXiv preprint arXiv:2001.08361, 2020
13. Dai, D., Sun, Y.: Why can gpt learn in-context? Language models implicitly perform gradient descent as meta-optimizers. arXiv preprint arXiv:2212.10559, 2022
14. Akyurek, E., Schuurmans, D.: What learning algorithm is in-context learning? Investigations with linear models. arXiv preprint arXiv:2211.15661, 2022
15. Jianhua, T.A.O., Ruibo, F.U., Jiangyan, Y.I.: Development and challenge of speech forgery and detection. J. Cyber Secur. **5**(2), 28–38 (2020)
16. Vaswani, A., Shazeer, N.: Attention is all you need. Adv. Neural Inf. Process. Syst. **30** (2017)
17. Radford, A., Narasimhan, K.: Improving language understanding by generative pre-training. OpenAI blog, 2018
18. Radford, A., Wu, J.: Language models are unsupervised multitask learners. OpenAI blog **1**(8), 9 (2019)
19. Brown, T., Mann, B.: Language models are few-shot learners. Adv. Neural Inf. Process. Syst. **33**, 1877–1901 (2020)
20. Dai, D., Sun, X.: Why can gpt learn in-context? Language models implicitly perform gradient descent as meta-optimizers. arXiv preprint arXiv:2212.10559, 2022
21. OpenAI. Evaluating large language models trained on code, 2021
22. Wei, J., Bosma, M.: Finetuned language models are zero-shot learners. arXiv preprint arXiv:2109.01652, 2021
23. Ouyang, L., Wu, J.: Training language models to follow instructions with human feedback. Adv. Neural Inf. Process. Syst. **35**, 27730–27744 (2022)
24. Wikipedia contributors. Gpt-4o — Wikipedia, the free encyclopedia, 2024. Accessed 16 July 2024
25. Zhang, D., Yu, Y.: Mm-llms: recent advances in multimodal large language models. arXiv preprint arXiv:2401.13601, 2024
26. Tu, H., Zhao, B.: Sight beyond text: Multi-modal training enhances llms in truthfulness and ethics. arXiv preprint arXiv:2309.07120, 2023
27. Birhane, A., Prabhu, V.U., Kahembwe, E.: Multimodal datasets: misogyny, pornography, and malignant stereotypes. arXiv preprint arXiv:2110.01963, 2021
28. UNESCO. Unesco's recommendation on the ethics of artificial intelligence: key facts. Electronic and Paper, 2023. 19 pages : illustrations

29. Bommasani, R., et al.: Considerations for governing open foundation models. Technical report, Stanford Institute for Human-Centered Artificial (2023)
30. Bertuzzi, L.: European union squares the circle on the world's first ai rulebook, February 2024. www.euractiv.com
31. Liu, Y., Deng, G.: Jailbreaking chatgpt via prompt engineering: an empirical study, 2024
32. Zou, A., Wang, Z.: Universal and transferable adversarial attacks on aligned language models, 2023
33. Deng, G, Liu, Y.: Masterkey: automated jailbreaking of large language model chatbots. In: Proceedings 2024 Network and Distributed System Security Symposium, NDSS 2024. Internet Society, 2024
34. Robey, A., Wong, E., Hassani, H., Pappas, G.J.: Smoothllm: defending large language models against jailbreaking attacks, 2024
35. Zhao, W., Li, Z.: Defending large language models against jailbreak attacks via layer-specific editing, 2024
36. Xuan, L., Zhou, H.: Deepinception: hypnotize large language model to be jailbreaker. arXiv preprint arXiv:2311.03191, 2023
37. Wikipedia contributors. Milgram experiment — Wikipedia, the free encyclopedia, 2024. Accessed 16 July 2024
38. Fatih Kadir Akin. Awesome ChatGPT Prompts, 2023. https://huggingface.co/datasets/fka/awesome-chatgpt-prompts. Accessed 20 June 2023
39. Degrave, J.: Building a virtual machine inside chatgpt, December 2022
40. PortSwigger. Web llm attacks —— portswigger, 2024
41. Haibo Jin and Ruoxi Chen. Guard: Role-playing to generate natural-language jailbreakings to test guideline adherence of large language models, 2024
42. View All Posts by Rachel Draelos. Bias, toxicity, and jailbreaking large language models (llms), December 2023
43. OpenAI. Usage policies, January 2024
44. OpenAI. How should ai systems behave, and who should decide?, February 2023
45. OpenAI. Snapshot of chatgpt model behavior guidelines, 2022
46. Bailey, L., Ong, E., Russell, S., Emmons, S.: Image hijacks: adversarial images can control generative models at runtime (2024)
47. Gou, Y., et al.: Eyes closed, safety on: protecting multimodal llms via image-to-text transformation, 2024
48. Sun, Y., Wang, S.: Ernie: enhanced representation through knowledge integration. arXiv preprint arXiv:1904.09223, 2019
49. Sun, Y., Wang, S.: Ernie 2.0: a continual pre-training framework for language understanding. In: Proceedings of the AAAI Conference on Artificial Intelligence, vol. 34, pp. 8968–8975, 2020
50. Zhang, Y., Tzun, T.T., Hern, L.W., Wang, H., Kawaguchi, K.: On copyright risks of text-to-image diffusion models, 2024
51. Gusenbauer, M.: Audit ai search tools now, before they skew research. Nature **617**(7961), 439 (2023)
52. Vert, J.-P.: How will generative ai disrupt data science in drug discovery? Nat. Biotechnol. 1–2 (2023)
53. Weng, L.: Extrinsic hallucinations in llms, July 2024. https://lilianweng.github.io/
54. Wikipedia contributors. Gptzero — Wikipedia, the free encyclopedia, 2024. Accessed 17 July 2024
55. Sadasivan, V.S., Kumar, A., Balasubramanian, S., Wang, W., Feizi, S.: Can AI-generated text be reliably detected? (2024)

56. Hoxhunt. ChatGPT vs. human phishing and social engineering study: Who's better?, 2023. https://www.hoxhunt.com/blog/chatgpt-vs-human-phishing-and-social-engineering-study-whos-better. Accessed 5 Apr 2023
57. Chaoning, Z., Zhang, C.: One small step for generative ai, one giant leap for agi: a complete survey on chatgpt in aigc era, 2023. arXiv preprint arXiv:2304.06488
58. Wang, J., Hu, L.: On the robustness of chatgpt: an adversarial and out-of-distribution perspective, 2023. arXiv preprint arXiv:2302.12095
59. Weidinger, L., Mellor, J., et al.: Ethical and social risks of harm from language models, 2021
60. Weidinger, L., Uesato, J., et al.: Taxonomy of risks posed by language models. In: Proceedings of the 2022 ACM Conference on Fairness, Accountability, and Transparency, FAccT '22, pp. 214–229, New York, NY, USA, 2022. Association for Computing Machinery
61. Shumailov, I., Shumaylov, Z.: The curse of recursion: training on generated data makes models forget, 2024
62. Bender, E.M., Gebru, T.: On the dangers of stochastic parrots: can language models be too big? In: Proceedings of the 2021 ACM Conference on Fairness, Accountability, and Transparency, FAccT '21, pp. 610–623, New York, NY, USA, 2021. Association for Computing Machinery
63. Seghier, M.L.: Chatgpt: not all languages are equal. Nature **615**(7951), 216–216 (2023)
64. BBC. Bard: What is Google's Bard and how is it different to ChatGPT? (2023). https://www.bbc.co.uk/newsround/65036003. Accessed 26 Mar 2023
65. Pfohl, S.R., Cole-Lewis, H.: A toolbox for surfacing health equity harms and biases in large language models, 2024

Integrating Human Preferences for Moral Decision Making in Autonomous Vehicles

Bishal Thapa[1], Henry Griffith[2], and Heena Rathore[1(✉)]

[1] Texas State University, San Marcos, TX 78666, USA
heena.rathore@txstate.edu
[2] San Antonio College, San Antonio, TX, USA

Abstract. There has been tremendous growth in Autonomous Vehicles (AVs) recently, yet the moral decision-making capabilities remain a crucial challenge that needs to be addressed. Addressing this issue is pivotal for gaining societal trust towards AVs, as the decision-making of the AVs will impact human life in a significant manner. Incorporating human preferences in the decision-making of AVs ensures ethical decisions along with societal acceptance of decisions made under moral uncertainty. In this paper, we propose integrating human preferences into Reinforcement Learning (RL) to guide AVs to make decisions that resonates with human-values. We use the Bradley-Terry (BT) model to incorporate human preferences and perform pairwise comparisons on the moral machine framework of AVs. This approach of considering human preference adds a layer of explainability to the decisions and enhances the significance of the results for real-world applicability. The results show the decision-making capability of RL agents could be improved by embedding human preferences and the decisions made by AVs align closely with those of humans.

Keywords: Autonomous Vehicles · Reinforcement Learning · Bradley-Terry · Human Preferences

1 Introduction

Autonomous Vehicles (AVs) are increasingly making their mark in the automotive industry, owing to the promise of effectiveness in navigation and the potential to significantly reduce road accidents [1]. However, a major hurdle to these advancements is a critical challenge that tests the capability of AVs to make ethical decisions in cases of moral uncertainty [2,3]. If AVs are better adept at making ethical moral decisions, it would substantially elevate public trust and confidence, which would further foster the development of AVs.

Recently, reinforcement learning (RL) based models have been designed to address the capability of agents to make decisions in scenarios of moral uncertainty [4,5]. These works predominantly focus on utilitarianism and deontology as two principal ethical frameworks to guide the decision-making capability of

AVs. While it is commendable to consider ethical perspectives, the challenge arises from the inherent conflicts and subjective interpretations of these theories. One issue is the assignment of numerical scores (in the form of credence values) to actions based on moral theories like deontology and utilitarianism [6]. Each theory carries its own set of principles and priorities, leading to potential contradictions when trying to apply them uniformly. For instance, deontology emphasizes rules and obligations, while utilitarianism focuses on maximizing overall happiness or well-being. These two perspectives can conflict when determining the ethical course of action in each scenario. Furthermore, the assignment of specific numerical values based on credence values to actions may oversimplify complex ethical dilemmas.

Humans are familiar with how other humans behave and make decisions. If an AV behaves in a way that mimics human decision-making, it can make the vehicle's actions more predictable and understandable to passengers, insurers, legal system and other road users. This predictability helps build trust in technology and could standardize responses to certain driving situations, potentially simplifying the assessment of liability and risk. By considering human preferences, we can simulate more realistic scenarios that reflect how human drivers perceive and interact with situations under morally uncertain situations [7]. In this paper, we incorporate human preferences using the moral machine framework, specifically designed to gather such preferences [8]. No prior work has utilized these vast human preferences directly in AVs for ethical decision-making. We utilize the Bradley-Terry (BT) model within the moral machine framework [9] to ascertain the issue of credence values as discussed above. These values, infused with human preferences, are then integrated into the RL model outlined by [6]. This integration enhances the RL agent's decision-making process, making its choices more pertinent and aligned with human values.

2 Related Work

The importance of the alignment of AI with human values was originally explored in [10]. The authors focused on the interplay of the normative and the technical aspects of this alignment along with its inherent complexity. They discussed different approaches to AI alignment, by stating the importance of clear alignment goals and arguing whether AI should align with human instructions, intentions, preferences, desires, interests, or values. For this purpose, the author explored the merits and demerits of these approaches by considering different research and exploring philosophical principles like utilitarianism and Kantian. This paper discussed how the central challenge is to come up with principles for true alignment of AI that are considered fair rather than finding the correct moral principle. This paper serves as an important reference to understand the integration of different principles and values into AI.

Thornton et al. [11] discussed the integration of different ethical frameworks for AV control algorithms. They incorporated philosophical knowledge from deontology, and consequentialism as guiding principles for AVs to meet societal expectations and ethical norms i.e., accident avoidance and adherence to traffic laws. The paper demonstrated how embedding philosophical principles as system constraints and objectives can drive vehicle control to align with societal expectations. The authors implemented deontological principles for the development of constraints in the vehicular system, consequentialism for the construction of the objective function, and implemented Model Predictive Control (MPC) to incorporate these ethical theories into a control algorithm. MPC aims to minimize the costs guided by consequentialism and constraints guided by deontology to determine the optimal path for the vehicle. The experimental results demonstrated that varying weights on costs or constraints represented by different ethical theories significantly affect the behavior of the test vehicle. They concluded that integrating various ethical principles is a necessity for responsible programming of AVs.

Gill [12] explored the human preferences in the ethical dilemma (ED) regarding who the AVs should protect in cases where harm is unavoidable. The paper provided a unique view of potential AV consumers and their standpoint on risk perception regarding AVs. The author also offered different technical, legal, and ethical challenges from the perspective of potential AV adopters. Two important experimental studies were performed with broad consumer samples which depicted that people associated EDs with the highest risk. These empirical studies also revealed that consumers considered EDs to be the most important issue facing EVs compared to various legal and technical issues. Overall, the findings from the paper highlight the gravity of the ethical dilemma for broader acceptance of AVs. While the study explored different aspects of consumer perceptions towards AVs, the results are based on a specific consumer sample that does not represent a diverse demographic spectrum and could limit the generalizability of the results. In contrast, the moral machine framework that we employed encompasses a more diverse range of demographic samples.

Andrea et al. [7] investigated an approach to build AI agents capable of making human-like decisions, especially in situations requiring difficult tradeoffs or moral uncertainty. The authors explored the Multi-Alternative Decision Field Theory (MDFT) as a basis for developing an AI orchestrator called MDFT-Orchestrator (MDFT-O). This orchestrator tried to make decisions in a constrained environment modeled using deontological and consequentialist frameworks in a way that balances between reaching the goal and satisfying the ethical constraint. They used Markov Decision Processes (MDPs) for modeling the decision environment. MDFT-O consistently performed better than other models like Greedy Orchestrator and Weighted Average Orchestrator. They further validated their model using experimental human preference data collected through Amazon Mechanical Turk. Despite the impressive performance of the orchestrator, the small sample size of 185 participants on Amazon Mechanical Turk limits the effectiveness of the validation.

Bogosian [13] proposed a computation framework for an AI system to address the problem of moral disagreement among different moral philosophies. The author suggested a framework where machines are morally uncertain and thus programmed to make decisions by considering multiple moral theories and their respective priorities according to their plausibility. The author took reference from MacAskill's metanormative theory approach proposed in [14]. MacAskill explained the issue of moral uncertainty as a voting mechanism that involves computing the expected choice-worthiness of action by considering credence towards different moral theories. Bogosian took a similar approach and applied MacAskill's philosophical concept to implement moral uncertainty in AI systems.

Ecoffet and Lehman [6] introduced an RL-based approach to tackle moral uncertainties by addressing the issue of comparability of different ethical theories, as explored by MacAskill [14]. They particularly focused on two normative theories: utilitarianism and deontology. Utilitarianism is a consequentialist ethical theory that solely focuses on the consequences of actions. Deontology, on the other hand, emphasizes adhering to moral rules to distinguish right from wrong. Each theory is represented by a certain degree of belief towards that theory defined as credence. They also developed voting mechanisms to deal with multiple ethical theories, experimented with random credence values for these theories, and observed the decisions of the RL model across different versions of the trolley problem.

3 Methods

3.1 Background on RL Model Under Moral Uncertainty

Ecoffet and Lehman [6] implemented a voting mechanism as a means to integrate different ethical theories in decision-making by defining a choice-worthiness function W_i, analogous to a reward function for each ethical theory. The reward function for these two theories is represented as a credence-weighted sum of the choice-worthiness according to each theory, where C_i denotes the credence level of each theory, s represents the current state, a represents the action at state s, and s' represents the next state:

$$R(s, a, s') = \sum_i C_i W_i(s, a, s') \tag{1}$$

The RL agents must balance between multiple choice-worthiness functions instead of simply maximizing a single reward function. So, they defined $Q_i(s, a)$ to represent the discounted sum of the choice-worthiness function $W_i(s, a, s')$ for each theory i by taking action a at state s under given policy π.

$$Q_i(s, a) = \mathbb{E}\left[\sum_{t=0}^{\infty} \gamma^t W_i(s_t, a_t, s_{t+1}) \middle| s_0 = s, a_0 = a\right] \tag{2}$$

where $\gamma_i \in [0, 1]$ represents the discount factor. The problem of incomparable choice-worthiness functions across different theories is discussed in the paper through the principle of Proportional Say, where the influence of theory is adjusted proportionally to its credence and not through the choice-worthiness function. For this purpose, the authors devise two voting mechanisms: Nash voting and Variance voting.

- Nash voting tends to make decisions through tactical voting rather than reflecting the true preferences of the theory. Each theory is associated with an RL agent, and it seeks to maximize the choice worthiness for that theory. Theories are provided with an equal voting budget and hence follow the principle of proportional say. The agents have to pay the cost for each vote they cast which is proportional to the size of their vote. These votes are then scaled proportionally with their credence values and theories use them to vote for or against the available actions. The voting budget can be exhausted after which the votes of that agent are ignored.
- Variance voting or Variance-Sarsa involves adjusting votes with a focus on the variability of the outcomes of different actions and does not suffer from stakes insensitivity. The variance voting mechanism overcomes the issues of Nash voting by variance-normalizing the preferences of the theories. It reflects the risk associated with the actions and makes decisions with risk-aware assessment i.e., actions with lower variance. The on-policy Q-values of each theory are learned using Sarsa and are considered as the preference of that theory. These preferences are then converted into votes through variance normalizing by the expected value of variance (σ_i^2) across timesteps using (4).

$$\mu_i(s) = \frac{1}{k} \sum_a Q_i(s,a), \tag{3}$$

where k represents the actions over discrete action space. Then;

$$\sigma_i^2 = E_{s \sim S} \left[\frac{1}{k} \sum_a (Q_i(s,a) - \mu_i(s))^2 \right] \tag{4}$$

Nash voting and variance voting mechanisms exhibit distinct impact on the ethical decision-making process. While Nash voting tends to produce more polarized decisions based on the dominant ethical theory, variance voting demonstrates a more cooperative approach by considering the variability of outcomes. Nash voting learns competitive strategies to learn the voting policies, whereas variance voting uses cooperative integration of the theories to learn value functions that are used for voting. This leads to more balanced decisions, particularly in scenarios where the stakes or consequences of actions vary significantly.

Both voting mechanisms exhibit distinct properties and limitations which can be discussed through Arrow's desirability axioms. It establishes essential properties for voting mechanisms, which include non-dictatorship, Pareto efficiency, and independence of irrelevant alternatives (IIA). Non-dictatorship emphasizes that the outcomes should reflect the input from all voting theories, so that no outcome is dictated by a single theory. Pareto efficiency prefers that an action should not be chosen if it is not preferred across by any theory. IIA asserts that adding a new irrelevant action should not influence the decision-making process. Both voting mechanisms discussed in the paper adhere to only some of these properties. Nash voting does not guarantee Pareto efficiency, while variance voting doesn't consistently satisfy the IIA. As Pareto efficiency is more desirable, as discussed by MacAskill, variance voting may be considered more beneficial.

The other limitation of Nash voting is its stakes insensitivity, where increasing stakes of one theory doesn't influence the overall decision. Furthermore, Nash voting also exhibits no compromise flaw where if no single theory chooses an action as the most preferred, that action cannot be chosen even if it is the best compromise action.

3.2 RL Environment

The RL state (s) is characterized by several elements within a grid environment. These elements include the presence of a switch, the spatial distribution of people represented through one-hot encoding (represented as X), and additional information regarding one's beliefs in a particular theory using credence values, as well as the count of people situated on the track as shown in Figs. 1(a) and (b). The agent actions are defined as up, down, left, and right. The agent's decision-making process involves navigating this grid-based environment based on the current state, with the goal of optimizing its performance over time through the selection of appropriate actions.

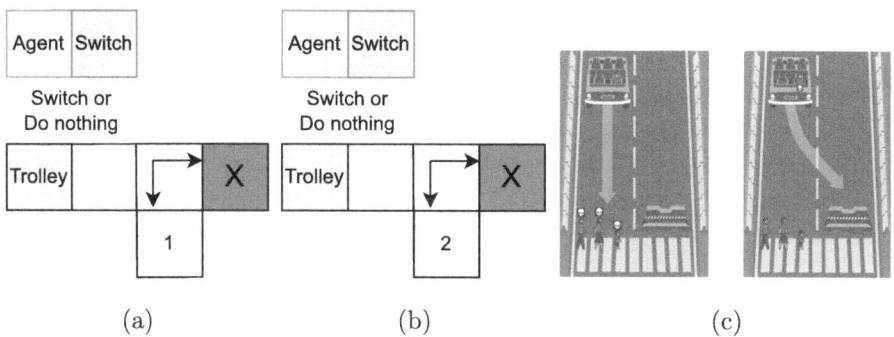

Fig. 1. Illustration of different scenarios: (a) Classic Scenario, (b) Modified Double Scenario, and (c) Example Scenario from the moral machine framework [8].

In this work, we considered two scenarios presented in [6] as foundational scenarios to evaluate the performance of the RL agent with human preference as credence. The initial simulation involves a classic trolley dilemma as depicted in Fig. 1(a), where a trolley is moving toward a group of X individuals. The agent can flip the switch to divert the trolley to an alternate path and cause a collision with one individual instead, thereby saving X individuals from harm. The second simulation involves an adaptation of the double trolley problem, also presented in [6], where instead of one individual on the alternate path, there are two individuals on the track as depicted in Fig. 1(b). We removed the action of pushing a large man to the path of the trolley, as this depicts a situation which is highly improbable and unrealistic. We refined the double scenario for practical relevance by considering a more realistic "$2vX$" scenario where decisions involve prioritizing saving two over X number of individuals. The severity of the actions from utilitarianism and deonotological perspectives, is quantified through weights assigned to different actions in these two scenarios. These values are hard-coded in the program and are presented in Table 1.

Table 1. Preferences in the classic and double trolley problems.

Ethical Theory	Classic Trolley Problem		Double Trolley Problem	
	Crash into 1	Crash into X	Crash into 2	Crash into X
Utilitarianism	−1	−X	−2	−X
Deontology	−1	0	−1	0

3.3 Human Preferences and Credence Generation

Collecting Human Preferences. To embed human preferences into AVs, we employed the moral machine framework [8]. This platform collected a comprehensive dataset comprising 40 million decisions in ten languages from 233 countries. Participants were presented with morally ambiguous scenarios and asked to indicate their preferred choice. Specifically, we extract pairwise results for a classic scenario and double scenario from this framework to feed into the RL model. In the classic scenario, participants were asked to save either one individual or a group of X individuals (where $X \in [2-5]$) analogous to the classic trolley as depicted in Fig. 1(a). In the double scenario, participants are asked to save either two individuals or X individuals (where $X \in [3-5]$) analogous to the modified double trolley as depicted in Fig. 1(b). Figure 1(c) shows an example of a 1v3 scenario, where swerving is considered a utilitarian approach, and staying in the lane is considered a deontological approach.

Bradley-Terry Model for Pairwise Comparison. The BT model is a widely used probability model for pairwise comparison analysis, providing a framework for deriving the score values or strengths of the choices in different scenarios of an experiment. The BT model assigns strength values to each item based on their comparisons. The model depends upon generating strength parameters based on observed comparisons [9]. This parameter can be generated using methods like maximum likelihood estimation. The probability model of pairwise comparison of outcomes (i, j), where β is a strength parameter of a particular choice, and i is likely to be chosen over j can be modeled as:

$$p_{ij} = \frac{e^{\beta_i}}{e^{\beta_i} + e^{\beta_j}} \tag{5}$$

We adapted this BT model to reflect the relative strengths of the decision for the classic and modified double scenario. We used the BT model instead of probability inference as it provides a systematic framework that underpins the strength of choices in pairwise comparisons, and it is also well-suited to handle incomplete comparisons, enabling more accurate preference determination [9,15].

4 Results and Performance Analysis

4.1 Calculation of Credence Using BT Model

The credence value is a critical aspect of our model that binds the human preferences with the model's decision. We generated the credence values for each scenario (e.g., $1v2, 1v3$, etc.) separately by filtering the scenarios from the moral machine framework that depicted decisions made in accordance with either utilitarianism or deontology. Figure 2 presents a scenario analysis of moral decision-making from data gathered through the moral machine framework. Figure 2(a) depicts the total number of cases where participants were prompted to choose between saving one individual over X individuals (where $X \in [2-5]$). The strength of the preference for saving X individuals over one was calculated using Eq. 5, as described earlier in Sect. 3 C. This strength is depicted by the red line in the chart and represents the degree to which human preference aligns with the utilitarianism principle. Figure 2(b) extends the analysis by exploring scenarios in a double trolley problem, by addressing human preferences in $2vX$ cases, where X varies from one to five. Similar to Fig. 2(a), it presents total cases of $2vX$ from the moral machine framework, normalizes the data by calculating the probability of saving X in each scenario, and generates the strength values of saving X using the BT model as depicted by the red line in the figure.

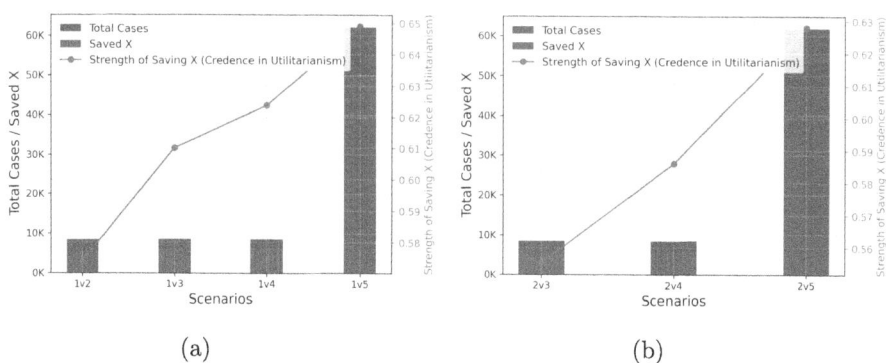

Fig. 2. Credence for Utilitarianism in (a) 1 v X scenario, (b) 2 v X scenario (Color figure online)

4.2 Integrating Human Preferences in RL Framework

We simulated RL agents with the Nash and Variance voting [6] for classic and modified double scenarios. The experiments were performed on two distinct environmental setups: a sequential environment, and a non-sequential environment. The decision in a sequential environment is not isolated and requires the RL model to make multiple decisions (2 in our case) before the episode ends. The stakes of future decisions are unknown and hence the model needs to perform the strategic management of the voting budget under Nash voting. Conversely, the decisions in the non-sequential environment are isolated, and each episode requires the model to make a single decision. Figure 3 and Fig. 4 represent the decision behavior of the agent for the classic scenario under the Nash voting and variance voting mechanisms in $1vX$ scenarios.

Classic Scenario

Nash Voting. Figure 3(a) represents the decision-making behavior of the RL agent under random credence in a non-sequential environment. It shows that the agent tends to do nothing when the deontology credence value is above 50% regardless of the number of people on the track. Conversely, for cases where deontology credence is less than 50%, the agent tends to prefer the "switch" option indicating a preference in the utilitarian principle to minimize the overall harm. Figure 3(b) explores the decision-making behavior of the agent where credence values are generated through human preference as discussed in Sect. 3. C. The vertical lines mark the exact deontology credence value generated through human preferences for specific scenarios in the classic environment. The figure extends up to 50% deontology credence values for exploring the $1v1$ situations where both ethical theories hold equal credence as a baseline.

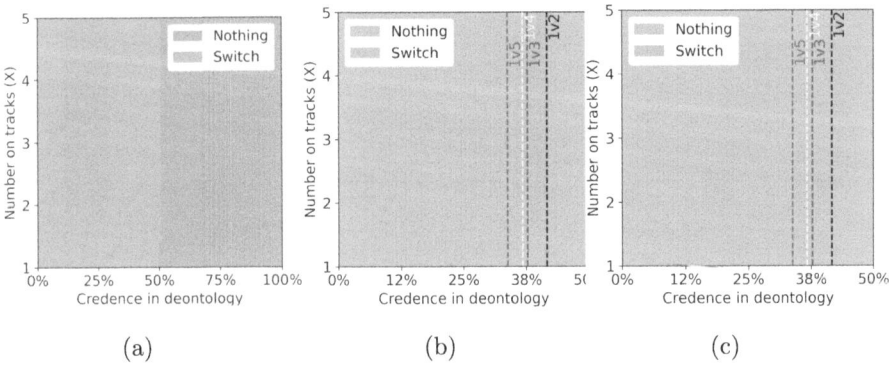

Fig. 3. Classic Environment Nash Voting (a) Random credence, (b) Human preference credence in Non-Sequential environment, (c) Human preference credence in Sequential environment

Figure 3(c) represents the decision-making behavior of the RL agent involved in sequential decision scenarios. It shows an interesting behavior of agents where the agents demonstrate stake insensitivity. Even with lower deontology credence, we can see situations where the agents tend to prefer doing nothing when we expected it to switch. But, regardless of the environment setup (sequential or non-sequential), the credence preferred by humans is still the same for a particular scenario. It can be seen that when the number of individuals on track exceeds 1, then the agent's decision under human preference credence is always to switch. This is consistent with the non-sequential decision-making depicted in Fig. 3(b). Therefore, in the classic environment under Nash voting, we found that the agent's decision under human preferred credence consistently favored switching the trolley depicting the dominance of the utilitarian principle. A distinct shift in decision-making can be observed when the decisions are compared to the original study (Fig. 3(a)), where the agent preferred to switch in approximately half of the scenarios. However, it was found that when agents were provided with credence values generated by considering human preferences, they consistently preferred to switch, reflecting human behavioral tendencies in scenarios analogous to the classic trolley dilemma.

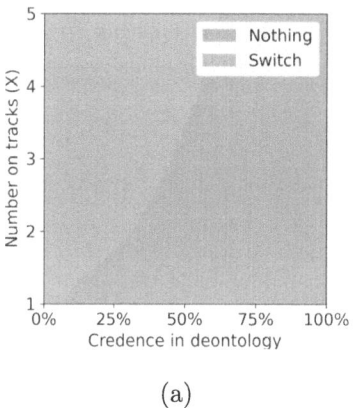
(a)

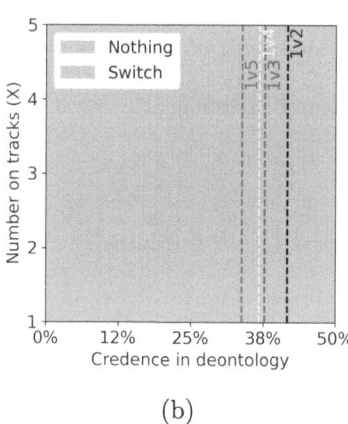
(b)

Fig. 4. Classic Environment Variance Voting (a) Random credence, (b) Human preference credence

Variance Voting. Figures 4(a) and (b) represent the behavior of the model for the classic scenario under the variance voting mechanism, with the agent's decision influenced by credence generated randomly and human preferred credence respectively. As evident from Fig. 4, the decisions of RL agents following variance voting are not as one-sided decisions as Nash voting. This is because variance voting involves adjusting votes with a focus on the variability of the outcomes of different actions and does not suffer from stakes insensitivity as in Nash voting. Variance voting tries to find balance in impact by the varying outcomes, which helps in generating consistent decision-making regardless of the sequential or non-sequential scenario. As illustrated in Fig. 4(a), the agent consistently chooses to do nothing once the deontology credence exceeds approximately 60%. In contrast, agents with human-preferred credence consistently choose to switch for all cases except when the number of people on tracks is two, as depicted in Fig. 4(b).

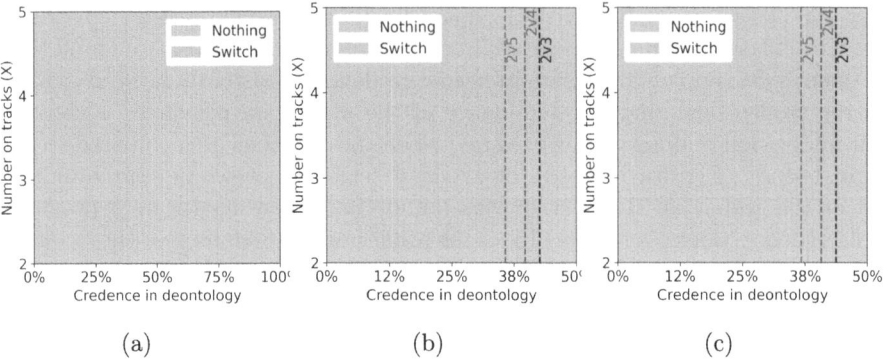

Fig. 5. Modified Double Environment Nash Voting (a) Random credence (b) Human preference credence in Non-Sequential environment, (c) Human preference credence in a Sequential environment

Modified Double Scenario

Nash Voting. Figure 5(a) represents the decision behavior of the agent for the modified double scenario under the Nash voting mechanism with random credence. It shows similar results to the classic environment, with the agent preferring to do nothing when the deontology credence value is above 50%. However, the decision boundary is not as clearly defined as in the classic environment and the decision is leaning slightly more towards doing nothing. Figures 5(b) and (c) depict the decisions of an agent under credence derived using our method in a non-sequential and sequential environment respectively. When the number of individuals on the track is two, the decision of the agent is more inclined towards doing nothing as the deontological credence surpasses approximately 20%, even though the scenario presents equal individuals on either track. This can be explained by the weight assigned by the ethical principles to different actions as depicted in Table 1. The deontological principle focuses on the morality of action as right or wrong regardless of the number of people on the track. So, both actions of "Crash into 1" and "Crash into 2" receive –1 weight under deontology highlighting that deontology doesn't care about the outcome but the action. Utilitarianism, on the other hand, evaluates morality through outcomes. So, under this principle, "Crash into 1" and "Crash into 2" receive different weights of –1, and –2 respectively. This highlights that the decision to switch is more penalized when the number of individuals on track is greater. In Fig. 5, as denoted by the vertical lines, the human preferred credence values indicate that the decisions of the agent using these credence always prefer to switch when the number of individuals on track is greater than three. This contrasts with the decisions made with random credence where the agents chose to do nothing for slightly more than half of the cases. This decision contrasts with the decision observed in the classic scenario where the presence of three individuals warranted a decision to switch. This also depicts an important finding that the classic scenario cannot be generalized to other scenarios without considering the specifics of those scenarios.

Variance Voting. Figures 6(a) and (b) represent the behavior of the model for the modified double scenario under the variance voting mechanism using random credence and using our approach respectively. Figure 6(b) also depicts the difference in decision-making when comparing against 1 v X scenarios, highlighting how changing the number of individuals on alternate path of the track changes the agent's decision. This can also be explained by the difference in weights assigned to different actions by the ethical principles, already described under Nash voting. The decision-making in variance voting follows a similar pattern to Nash voting, and shows that when the number of individuals on track is greater than 2, the agent's decision guided by human preferred credence is to "switch". This decision supports the utilitarianism principle where reducing overall harm is a priority. However, this decision significantly contrasts with the decision made using random credence, where the preference to do nothing is predominantly observed as depicted in Fig. 6(a). This shows that relying on agents with random credence for decision-making often leads to choices in decisions that do not align with human preferences for the majority of the scenarios.

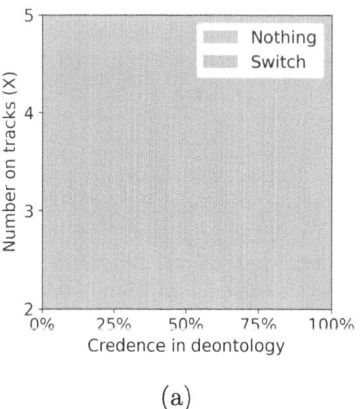
(a)

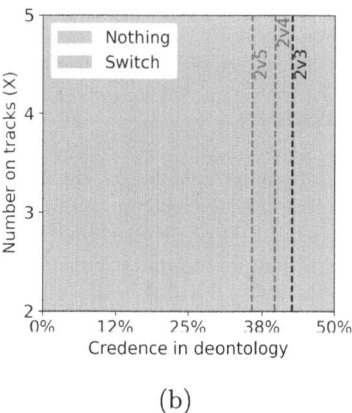
(b)

Fig. 6. Modified Double Environment Variance Voting (a) Random credence (b) Human preference credence

5 Conclusion and Future Work

The absence of moral ground truth for decisions of AVs is a major hindrance to ethical decision-making. This paper alleviates this issue by introducing the concept of human preferences for AVs by taking human-preferred empirical data to generate credence towards a particular theory. Our approach provides a mathematical rationale behind the decisions guided by credence generated through human preferences, adding a layer of explainability to the decisions. It was found that decisions made by an RL agent can be significantly improved to align their decisions with those made by humans, making them viable for practical applications such as insurance and consumer acceptance.

The moral machine framework, while valuable, is limited by its narrow range of scenarios, which do not fully account for varying risk levels, vehicle speeds, or environmental conditions like weather and road quality. These factors can significantly impact both human preferences and optimal decision-making in real-world situations. To address these limitations, we plan to broaden our approach by incorporating the National Highway Traffic Safety Administration (NHTSA) dataset [16], which provides detailed information about crash conditions, including vehicle speeds and collision information. Additionally, we will leverage an expanded set of human-labeled datasets for ethical decision-making [2,17].

Deontological ethics emphasizes the inherent rightness or wrongness of actions rather than their outcomes. This focus can introduce a bias towards inaction in some scenarios we examined, as the theory does not permit treating individuals merely as means to an end. Consequently, this can lead to decisions that prioritize avoiding direct harm over taking actions that could reduce overall harm. Future iterations of the model will address this potential bias by integrating additional ethical frameworks, such as justice and virtue ethics [18], to achieve a more balanced approach.

Acknowledgment. This publication has been supported by NSF CISE Research Initiation Initiative (CRII) grant #2153510 and #2313351.

References

1. Autonomous Vehicle Market Size, Share, Trends, Report 2022-2030. https://www.precedenceresearch.com/autonomous-vehicle-market. Accessed 7 Apr 2024
2. Meder, B., et al.: How should autonomous cars drive? A preference for defaults in moral judgments under risk and uncertainty. Risk Anal. **39**(2), 295–314 (2019)
3. Jasti, P.C., Griffith, H., Rathore, H.: Eclectic ethical decision making for autonomous vehicles. In: 2024 Ninth International Conference on Mobile and Secure Services (MobiSecServ), pp. 1–6 (2024)
4. Leibo, J.Z., et al.: Multi-agent reinforcement learning in sequential social dilemmas. https://arxiv.org/pdf/1702.03037.pdf (2017)
5. Abel, D., MacGlashan, J., Littman, M.L.: Reinforcement learning as a framework for ethical decision making. In: Workshops at the Thirtieth AAAI Conference on Artificial Intelligence, pp. 1–2 (2016)

6. Ecoffet, A., Lehman, J.: Reinforcement learning under moral uncertainty. http://proceedings.mlr.press/v139/ecoffet21a/ecoffet21a.pdf. Accessed 10 Apr 2024
7. Loreggia, A., et al.: Making human-like moral decisions. In: Proceedings of the 2022 AAAI/ACM Conference on AI, Ethics, and Society, pp. 447–454 (2022)
8. Awad, E., et al.: The Moral Machine experiment. Nature **563**(7729), 59–64 (2018)
9. Bradley, R.A., Terry, M.E.: Rank analysis of incomplete block designs: I. the method of paired comparisons. Biometrika **39**(3), 324–345 (1952)
10. Gabriel, I.: Artificial intelligence, values, and alignment. Mind. Mach. **30**(3), 411–437 (2020)
11. Thornton, S.M., et al.: Incorporating ethical considerations into automated vehicle control. IEEE Trans. Intell. Transp. Syst. **18**(6), 1429–1439 (2017)
12. Gill, T.: Ethical dilemmas are really important to potential adopters of autonomous vehicles. Ethics Inf. Technol. **23**(4), 657–673 (2021). https://doi.org/10.1007/s10676-021-09605-y
13. Bogosian, K.: Implementation of Moral Uncertainty in Intelligent Machines. Mind. Mach. **27**(4), 591–608 (2017). https://doi.org/10.1007/s11023-017-9448-z
14. MacAskill, W.: Normative Uncertainty. Oxford University, Cambridge (2014)
15. Huang, T.K., Lin, C.J., Weng, R.: A generalized Bradley-Terry model: from group competition to individual skill. Adv. Neural Inf. Process. Syst. **17** (2004)
16. National Highway Traffic Safety Administration. Fatality Analysis Reporting System (FARS). https://www.nhtsa.gov/research-data/fatality-analysis-reporting-system-fars. Accessed 25 Aug 2024
17. Krüger, S., Uhl, M.: Autonomous vehicles and moral judgments under risk. Transport. Res. Part A: Policy Pract. **155**, 1–10 (2022)
18. Hendrycks, D., et al.: Aligning ai with shared human values. arXiv preprint arXiv:2008.02275 (2020)

Privacy-Enrooted Car Systems (PECS): Preliminary Design

Giampaolo Bella[1], Gianpietro Castiglione[1(✉)], Sergio Esposito[1], Mirko Giuseppe Mangano[1], Mirco Marchetti[2], Marcello Maugeri[1], Mario Raciti[3], Salvatore Riccobene[1], and Daniele Francesco Santamaria[1]

[1] University of Catania, Catania, Italy
gianpietro.castiglione@phd.unict.it
[2] University of Modena and Reggio Emilia, Modena, Italy
[3] IMT School for Advanced Studies Lucca, Lucca, Italy

Abstract. In today's data-driven automotive environment, privacy concerns have become increasingly pronounced as modern vehicles collect and transmit vast amounts of personal data without adequate user oversight, with their consequent undue use by third parties. This paper presents the PECS project with the aim of addressing these pressing issues by revolutionising privacy management within the automotive domain, giving particular power to users to choose specific privacy preferences, and also proposing data obfuscation solutions. This paper introduces the preliminary design of PECS - as a work-in-progress project - comprising its software modules, along with five user stories to conceptualise the user interaction.

Keywords: Automotive · Hard & Soft Privacy · User Preferences · Obfuscation · Blockchain

1 Introduction

Personal data are a valuable asset in the contemporary data-driven economy, due to their business and sociotechnical potential. This is particularly true for modern cars, where a plethora of personal data is collected, encompassing everything from music and cabin preferences to driving style and location/travel history. Although the processing of personal data has become more and more regulated by the institutions, e.g., in Europe with the promulgation of the GDPR, a study [6] found that 25 car manufacturers lack appropriate management of consents, transparency, and data protection, thus further stressing concerns raised by several academic contributions [2,17,18].

When referring to personal data, privacy is the main notion to emphasise. Due to its complexity, the literature divides it into two ramifications: *soft privacy* and *hard privacy* [4,8]. The notion of soft privacy refers to techniques such as compliance verification, ensuring that personal data are processed lawfully once an individual releases them to an institution acting as a data controller.

Conversely, the notion of hard privacy involves techniques to safeguard personal data before their disclosure, including methods like data obfuscation. Both soft and hard privacy have been discussed in the automotive environment [3,12–14], however, there is still room for improvement: for example, users should be informed about the privacy implications of their choices, and they should be able to know how and which data about them is being processed at every moment in time.

Hence, this paper presents PECS, a project centred on enhancing the management of consents and privacy preferences, on secure data exchange and privacy-aware data processing, and data obfuscation. PECS enroots the technological systems of modern cars into the very bedrock of privacy by implementing a triple approach:

- The PECSi module: tailors multisensory-media techniques to empower individuals to statically define their policies on the processing of their personal data, as well as to dynamically control, for the first time, the application of privacy policies at service run time.
- The PECSo module: tailors open-source solutions to provide individuals with the capability to obfuscate their personal data before sharing it with any entity, thus ensuring data protection right from the outset.
- The Blockchain module: allows public verifiability of the policies set by the pseudonymised user, protecting both the user and the service provider.

2 Related Work

Walter et al. [19] developed PRICON, which aim is to enable user-centric decisions towards data protection. PRICON includes a user-centred privacy-aware control system, with a complex technical system for allowing users to control their privacy settings within the car. PRICON is a controller consisting of two steps: a) analysis of the usage scenario of a connected car, leading to the formulation of an information architecture that takes into account the user's choices and expectations; b) development of a user interface that uses a privacy calculation model to predict privacy-related adoptions based on privacy-related costs and usage-related benefits. The interface guides the user in selecting privacy settings: the user can control which data is shared between different services and can select one of four predefined privacy profiles. Said profiles are then applied to all services. The data flow is controlled by a "privacy firewall" that switches between the car and the online services and controls all data that flows through the car. In addition, PRICON offers various data protection services, data anonymisation, statistical evaluation and encryption.

PRICAR [11] is a different framework that tries to solve the problem of car sensor data sharing with third parties. When third-party applications retrieve data from car sensors, these data are then usually transferred to the third-party servers for analysis, or for providing certain services. The authors of this work extensively analyse potential privacy issues that arise from this behaviour and propose a novel framework that is neutral and that stands between the vehicle and the third-party server. Neutral servers preprocess data so that the third-party only obtains sanitised data that cannot be used to violate the user's privacy (e.g., anonymised data).

Rasmusen et al. [15] presented a knowledge graph-based user interface for consent solicitation that makes use of gamification to increase people's understanding of the law and make consent easier to understand. In fact, under the GDPR, consent is one of the legal bases for data processing, and some standards must be met.

Nowadays, consent is frequently granted without question, especially in light of the information overload brought on by lengthy privacy rules worded in complex legal language and intricate interface designs that wear down users' ability to offer consent. The knowledge graph approach proposed by the authors gives all parties engaged in the data-sharing process a single consent model, informing about consent in a machine-readable fashion.

Meuser et al. [10] face the privacy problem arising from location sharing. The authors try to solve it with the obfuscation of the vehicle's location by adding artificial noise to it. However, such an approach simultaneously increases the actual area in which the vehicle may be located, thereby reducing the effectiveness of location-based services. The authors solve this problem by allowing vehicles to get location-dependant data from other vehicles, mitigating the impacts of obfuscation on the performance of the services.

Albouq et al. [1] presented a Double Obfuscation Approach, applying two obfuscation phases while integrating two privacy protection techniques. In particular, the Double Obfuscation Approach obfuscates and hides the identity and location of its users using the Fog nodes, which function as a trusted third-party (TTP), without the need to reveal the users' identities or trust the cooperating nodes. The Double Obfuscation Approach enables users of connected vehicle applications to protect their privacy through an algorithm that considers the dynamic nature of user requests and the mobility of objects.

Garrido et al. [7] analysed a series of operational techniques and tools, which may be fundamental for the final implementation of PECS, yet without offering a defined solution to the problem considered by PECS.

3 Positioning and Contributions

While there have been attempts to solve the problem that PECS seeks to address, or problems that are similar to it, PECS differs in many ways from previous works. To show this, we perform a direct comparison with PRICON [19], as it is the closest solution to PECS that can be found in literature, and PRICAR [11], which has similar objectives. Hence, our contributions can be seen as improvements with respect to the literature, plus some elements of originality that we hereby summarise:

- While PRICON achieves control on all car communications, both internal and external, PECS is focused on permissions given to each application and on which car sensor data are used.
- It is not clear how PRICON achieves control on all car communications. In the next sections, we plan to give some high-level technical details on how we plan to design PECS.
- PRICON informs the user on how data will be used by applications, however, it does not give the opportunity to enhance data protection, as PECSo does with the application of Privacy Enhancing Technologies (PETs).
- PRICAR essentially performs a Man-in-the-Middle (MITM) between the vehicle and the third-party server to protect data, leveraging neutral servers controlled by the manufacturer. Instead, in PECS we remove the necessity to have a third-party sanitising data, making the third-party developer choose with which PETs they would like data to be obfuscated in PECSo, so that business processes can take this into account. For example, using Federated Learning, only information learned by the model leaves the car, and data themselves are never transmitted.
- In PECS there is no single point of failure; instead, the neutral servers in PRICAR appear to be one: if these servers experience a failure, the whole car might not be able to access any third-party services anymore.
- While PRICAR seems to manage car sensor data only, PECS takes into account all permissions given to applications.
- While in PRICAR all traffic to third-party servers is blocked by default, in PECS the user can choose which data to share with which third-party applications.
- PECS will be able to verify even if non-PECS-compliant applications are respecting the policy chosen by the user, by comparing requested and granted permissions to the application with the permissions that were chosen by the user in PECSi. In case a violation is detected, PECS informs the user in their preferred way, which was previously set up during the initial PECS onboarding. The user may choose between several methods for the notification, for example, audio or visual cues or even haptic feedback on their steering wheel.
- PECS will be completely open-source, and detailed documentation will be released both for developers and users.

4 Preliminary Design

PECS is a single software system which the user interacts with. PECS is composed of three main components: PECSi, the multisensory-media car interface, PECSo, which implements PETs, and the Blockchain, which stores in a publicly verifiable and pseudonymised way the policies decided by the user. While these components communicate with each other to ensure the features promised by PECS, the user mainly interacts with PECS through PECSi. The full features of PECS are modularly supported by different applications and services that run on an Android system. For example, PECSi's modules read application privileges granted at the OS level and match them with the ones defined by the user in PECS, while PECSo's modules manage the PETs to ensure that data are protected before use.

A preliminary component specification for PECSi, PECSo, the Blockchain, the PECS architecture and a description of the possible ways to interact with PECS follow below.

4.1 PECSi

PECSi is a multisensory-media car interface designed to empower drivers, for the first time, with both static and dynamic control over their personal data. PECSi utilises cutting-edge human-computer interaction techniques, delivering an immersive experience that enables drivers to understand the types of personal data used by the various services. Through audio-video cues and steering-wheel haptic feedback, drivers gain an enhanced awareness of data sensitivity and flows, allowing them to freely make their data-sharing choices, as well as to detect and respond to anomalous data exchanges that may occur during service operation. Drivers are therefore enabled to perceive this through the physical perception deriving from holding the steering wheel while they drive. In consequence, the inherent safety risk deriving from distraction is mitigated, making it immediate for drivers to identify and cut off any data transfers that they may deem undesirable or unauthorised. PECSi will consist of two modules: the PECSi Client and the PECSi Service.

- **The PECSi Client** is an Android application that represents the point of interaction with the user. Through different views, this application will enable the user to set their privacy policy preferences, monitor their status, and see potential alerts communicating a violation. Whenever a policy violation is detected by the PECSi Policy Engine, a notification will be sent to the user via their preferred method;
- **The PECSi Service** is a Java service that will run directly on a Java Virtual Machine installed on the Android device (e.g., a head unit) and implement the backend features of PECSi, such as the aforementioned PECSi Policy Engine. Using the XACML 3.0 standard, PECSi Service will be in charge of creating the files containing the privacy policies as specified by the user, and of checking whether applications have more permissions than specified by the user during the setup of PECSi. This can be done by listing all the third-party packages in the system. After that, permissions for each package are checked, and a cross-check between them and the enforced user policy is performed. Every found inconsistency will be listed as an alert in a JSON file that is then sent to the PECSi Client.

4.2 PECSo

PECSo serves as a shield of protection for the personal data, by obfuscating all personal data while preserving functionality. Instead of being a single piece of software, PECSo aims to be a standard for developers to build PECS-compliant applications that implement one or more PETs while maintaining software functionalities and business logic for the service providers. Applications are considered PECS-compliant if they respect one of the following tiers of compliance:

- **Tier 1.** The application includes a PET to protect user data, and the user may choose to activate it or not. The application also informs the user on how the data processing changes if they decide to activate the PET, and in either case, features offered by the application must be the same (although the performance may differ).
- **Tier 2.** The application is compliant to Tier 1 and must additionally listen to Android Binder transactions that will be sent by PECSi and will contain the user policies. Each time the application needs to use or request permission, it must check if the last policy sent by PECSi allows the use of said permission, and act accordingly.
- **Tier 3.** The application is compliant to Tier 2 and there is a logical separation of the entity that applies the PET from the rest of the application. Hence, the application must be split in two parts. The first, called PECSo-App, implements the PET (e.g., Federated Learning), while the second, called PECSo-Interface, implements the rest of the application features. PECSo-App should be published on a public repository so that it is publicly verifiable that the module actually applies the PET when prompted to.

4.3 Blockchain

In PECS, we wanted to ensure the integrity, public verifiability, and pseudonymisation of the user-set privacy policies: with public verifiability, in the event of a dispute, a user can publicly prove that they selected a certain policy; with pseudonymisation, the policy would remain unlinkable to any physical person, as it is linked to a pseudonym rather than an actual identity; finally, the integrity of the policy, namely policy immutability, is also necessary, meaning that at any time the user can demonstrate which privacy policy was set at a certain time.

The entire set of the aforementioned security properties can be satisfied by the blockchain technology, which makes it suitable for use with PECS. Among technologies available, we are experimenting with the Hyperledger Fabric, which implements the discussed security properties [5] and stands out in terms of features revolving around the concept of permissioned blockchain.

4.4 PECS Architecture

Figure 1 illustrates a preliminary diagram architecture of PECS, including the main interactions within the Android system, with the target application(s), and other car components. Notably, the user only interacts with PECS through PECSi, because PECSo does not require any user setting and the application of the PET is transparently managed via the PECS-compliant application. On the

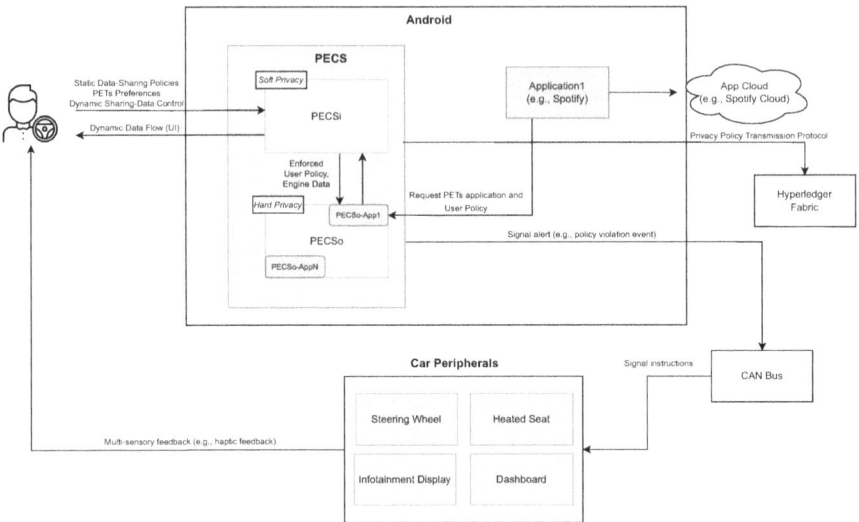

Fig. 1. Project Architecture.

other hand, PECSi can interact with the user also through other means, such as multisensory media (e.g., haptic feedback). In particular, whenever PECSi is triggered by a policy violation from an application, PECSi sends a signal alert to the CAN Bus, which then instructs car peripherals to provide multisensory feedback to the user. As explained, PECSi becomes aware of potential policy violations thanks to a cyclic check that it performs between the policies specified by the user, policies retrieved from the application(s) and data processed by the application(s).

4.5 Static and Dynamic Settings

The preliminary architecture diagram is further supported by two Message Sequence Charts (MSC), respectively in Fig. 2 and Fig. 3, that depicts the interaction between the user and PECS, respectively, in the static setting and dynamic setting.

Static Setting. In the static setting, the user is empowered to specify their static data-sharing policies and obfuscation methods by interacting with PECSi and the PECS-compliant applications. Briefly, PECSi prompts the user to specify their preferences. Then, the user chooses on each PECS-compliant application if they want to use the related PETs or not. Everything works properly as soon as PECSi forwards the user-selected privacy policy to the target application(s). The user can choose the so-called "paranoia levels" for each application, which are essentially presets for inexperienced users who do not want to manually check every permission and every application.

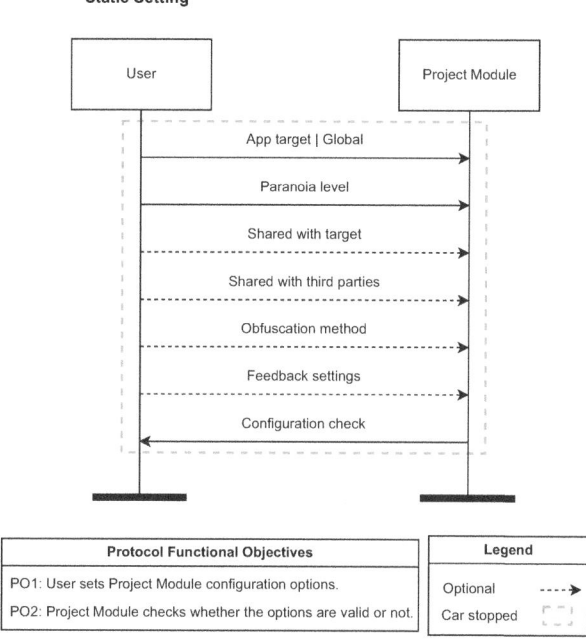

Fig. 2. MSC for Static Setting.

Dynamic Setting. In the dynamic setting, PECSi informs the user about the dynamic data flow and alerts them to potential policy violations from the target app(s). Alerts are delivered to the user with a multisensory media approach. In particular, PECS sends signals to car peripherals, such as the steering wheel (or even heating the seat, or via the head unit), to provide multisensory feedback to the user alongside a traditional notification in PECSi—for instance, in the case of the steering wheel, the user experiences haptic feedback. Once the user safely stops, they can request information about the triggered event from PECS and update the policy to address the violation, thereby enforcing their rights.

5 Validation Plan

The entire PECS design is firmly enrooted in users needs and their ultimate satisfaction, hence early user engagement is paramount for the project. In this section, we describe the two-round validation strategy we plan for PECS. In particular, we plan to conduct a first round of validation during the development of the project and a second round with real pilots using a prototype of our solution.

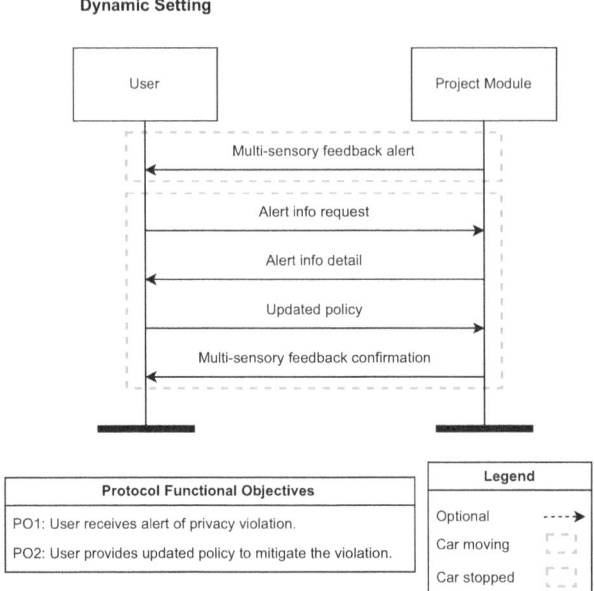

Fig. 3. MSC for Dynamic Setting.

5.1 Early Validation

The early design that is discussed in this paper is conceived and laid out following five essential user values derived from the literature [16]: control, fairness, loss aversion, uncertainty avoidance and effort minimisation. We safely hypothesise that these are the right values to embed in the project developments upon the basis of the prominent roles such values are documented to play in the state of the art.

However, the user engagement approach of PECS is to re-assess the five values by appropriate experimentation that customises them in the specific automotive privacy domain. Therefore, we adopt the standard paradigm of crowdsourcing, which enables us to administer a questionnaire to a worldwide population of respondents under a monetary reward for their time. This approach has several benefits, such as the ability to reach out to a large sample of people, the heterogeneity of the responses and the time efficiency deriving from the remote style of the exercise. The crowdsourcing exercise will be conducted through the Prolific platform.

We hereby briefly describe the five aforementioned values [16], and give an example of questions that will be administered to the users to understand how PECS' features and implementation should be adjust to fit their needs. Users will be able to answer using the Likert scale [9], by entering a number between 1 and 7, with 1 being the lowest score and 7 the highest one:

- **Control.** It concerns the mechanism to change consent decisions, express freedom of choice, etc. Question example: *To what extent do you agree with the following statement? I would like to choose when and how I receive notifications about potential violations of my previous choices.*
- **Fairness.** It consists in eliminating dark patterns, deception, and manipulation. Question example: *Do you agree with the following statement? I find it appropriate that a lengthy privacy policy explains in detail to me how my personal data is going to be processed.*
- **Loss Aversion.** It consists in providing information about consequences. Question example: *Do you agree with the following statement? I am disappointed when I consent for an organisation to process my personal data, but then the organisation suffers a data breach.*
- **Uncertainty Avoidance.** It maximises comprehensibility and transparency. Question example: *Do you agree with the following statement? I would like to be promptly notified of data breaches without being distracted from my driving activities.*
- **Effort Minimisation.** It concerns conciseness, reduction of complexity and readability improvement. Question example: *Do you agree with the following statement? I usually like to dive into configuring my applications (e.g., I check and read about every setting in my application).*

5.2 Final Validation

The methodological approach to the final validation round rests on the continuous user engagement plan discussed above. It focuses, in particular, on the field tests that are expected to be conducted in the open-air laboratory of the MASA premises, aboard a real car equipped with a PECS prototype. The field tests are planned as follows:

1. **Engage at least 10 real drivers.** These are to be chosen among people who already operate in the MASA, such as researchers and practitioners from the automotive industry to facilitate admission procedures and, at the same time, ensure an adequate level of experience with the driving.
2. **Perform test drives.** Each driver will be allowed to drive a PECS-equipped car for at least 15 min. This will be reiterated twice with an interval of at least two hours in between.
3. **Validate the consent casting and policy choice experience of PECSi.** This will be carried out by means of an interview asking two questions to each driver at the end of their test drive: (a) How do you feel satisfied, on a scale from 1 to 7, with the general privacy friendly experience that PECSi granted you with respect to traditional car systems? (b) What comments, criticisms or opportunities for improvements would you like to express?
4. **Validate the user experience with the obfuscation methods of PECSo.** This will be carried out by means of an interview asking two questions to each driver at the end of their test drive: (a) How do you feel satisfied, on a scale from 1 to 7, with the general data obfuscation experience

that PECSo granted you with respect to traditional car systems? (b) What comments, criticisms or opportunities for improvements would you like to express?

6 User Stories

This Section describes five user stories we wrote to imagine how users would interact with PECS. User Story 0 paraphrases the use case prior to PECS. Then, four more user stories describe the use cases of PECSi and PECSo both in the static and the dynamic settings.

User Story 0: Prior to PECS. John was ecstatic when he purchased his dream car, a sleek Mercedes-Benz, in 2018. As he signed the paperwork, the seller urged him to enrol in the Mercedes Connect program, touting its benefits. Eager to explore his new acquisition, John readily provided his email and personal information, choosing a password for the program.

However, as the days passed, John found himself engrossed in the joys of driving his Mercedes and never bothered to explore the functionalities of the Mercedes Connect program. Despite his disinterest, John's inbox was constantly bombarded with email notifications from the program, informing him of changes to its terms of service. Overwhelmed by the sheer volume of emails and the dense legalese within, John adopted a laissez-faire attitude, mindlessly accepting the updates to get them out of his sight.

With each passing notification, John grew more apathetic towards the digital services offered by his car. He had no desire to engage with the intricacies of the Connect program, while the car interface never raised his interest beyond the urge to find the rear window defroster button. Eventually, John's relationship with the Mercedes Connect program became one of passive acceptance. He would glance at the email notifications, dismiss them with a perfunctory click, and return to the more pressing matters of his life.

User Story 1: PECSi in the Static Setting. As Glenda approached her new car, she knew that it was more than just a mode of transportation and that its modern services would be hungry for her personal data. However, the car was equipped with the latest PECS technology, and she was explained that, with PECSi at her fingertips, she could navigate the complexities of modern driving with ease.

Settling into the driver's seat, Glenda was greeted by the PECSi interface, a dazzling array of lights and buttons designed to streamline her driving experience. Before the engine started, she was prompted with a series of questions, each one allowing her to tailor her privacy settings to her liking.

With each choice she made, Glenda felt a sense of empowerment, knowing that she was taking control of her personal data in a world where privacy was often overlooked. The haptic feedback on her steering wheel provided reassurance, guiding her toward the choices that aligned with her values. As she finished customising her settings, Glenda couldn't help but feel a sense of pride. She was not just a driver; she was a pioneer, forging a path toward a future where privacy and technology could coexist harmoniously.

User Story 2. PECSo in the Static Setting. As Glenda stepped into the sleek interior of her new car, Glenda felt a mix of excitement and apprehension. She knew that modern cars were notorious for their data-hungry nature, collecting every detail of their driver's habits and preferences to deliver tailored services. But with PECS technologies, she hoped to tame the beast of data invasion. With a soft hum, Glenda started the engine, and her car came to life, displaying a series of prompts on the touchscreen console. The PECS-compliant application wasted no time in engaging her, presenting her with different options to protect her data. Each page was filled with multisensory-medial information, explaining the implications of her choices in detail and providing her with sensory feedback at the same time.

Glenda's fingers danced over the touchscreen, navigating through the various options. She opted to share her music preferences with Spotify, but only her pseudonym with the motorway tolling system. Federated Learning would anonymise her driving routes for her insurance company, protecting her privacy while still providing valuable data for insurance purposes.

With each decision, Glenda felt a sense of empowerment, knowing that she was taking back control of her personal information. Finally, she reached the end of the customisation process, and with a sense of satisfaction, she confirmed her choices and felt a newfound sense of freedom. With PECSo guarding her privacy like a loyal sentinel, she could navigate the modern world with confidence, knowing that her data was safe from prying eyes.

User Story 3. PECSi in the Dynamic Setting. Every morning, the PECS interface of Glenda's car greeted her with its familiar, user-friendly design, offering a plethora of functionalities tailored to her needs. However, Glenda was not naive about the data-hungry nature of modern vehicles. She understood that her car collected vast amounts of personal data to enhance her driving experience. Yet, she was determined to maintain control over her privacy settings, hence had already configured PECSi with great interest, affirming her "soft privacy".

One day, as Glenda approached a perilous intersection, her steering wheel trembled subtly, accompanied by a warning beep. Recognising the customised feedback she had set up before, she immediately pulled over to investigate. With a few taps on her car's touchscreen, she uncovered the source of the disturbance: Spotify attempting to share her music preferences without her consent. Frowning at the violation of her privacy, Glenda swiftly intervened, enforcing her decision to withhold her data from third parties. As the car applied her directive to Spotify, she felt a sense of empowerment, knowing that she could navigate the complexities of modern technology while safeguarding her personal information.

With her privacy restored, Glenda resumed her journey, reaffirming her commitment to maintaining control over her digital footprint. In a world where data privacy was increasingly paramount, she found solace in her ability to tame the risks of infringement against her privacy, one tap at a time.

User Story 4. PECSo in the Dynamic Setting. As she prepared to embark on her journey, Glenda felt a sense of assurance, knowing that her car was not just a mode of transportation but a fortress guarding her privacy. In fact, she had previously adjusted the settings on her PECS-compliant applications interfaces, meticulously configuring the obfuscation methods that would cloak her data before it was shared with any service. In the bustling city streets, Glenda's car hummed with efficiency, its PECSo system working silently in the background, shielding her personal information from malicious eyes. Glenda relished in the control she had over her "hard privacy", dictating precisely which data would be divulged to which services.

One fateful day, as Glenda navigated through a tunnel, her senses heightened by the potential danger of speeding, her steering wheel trembling gently, accompanied by a warning audio clip. Instinctively, she recognised the familiar feedback she had configured herself before, indicating an attempt to access her full array of personal information. Pulling over to a safe spot, Glenda accessed the detailed information provided by her car's display. Spotify, it seemed, was attempting to share not just her music preferences but her entire personal data set, including her Personally Identifiable Information. Frowning at the audacity of the intrusion, Glenda swiftly intervened, re-enforcing her strict privacy settings with a tap on the touchscreen.

With her decision affirmed, she resumed her journey, a renewed sense of control settling over her. In a world where data privacy was a constant battle, Glenda found solace in her PECSo-equipped car, a fortress protecting her digital sovereignty against the relentless tide of data hunger.

7 Conclusion

This paper presented the PECS project and the challenges it aims to address. Although we presented the idea in terms of preliminary design, the work presented is nowadays a work in progress, therefore the project is being implemented. PECS represents an innovative interface for modern cars, concretely embodying the privacy-by-default-by-design principle. It makes drivers (finally!) aware of what personal data they are sharing with what entities at any time. This is a profound revolution to the current practice, in which drivers blindly trust the proprietary software of their cars. PECS will make it possible to split, for the first time, drivers' (mature) trust in car safety from their (young) trust in car privacy and data protection, so that these can develop independently.

As a result, drivers are effectively enabled to preserve specific data like driving style and seating preferences in the face of anyone else's analysis or in response to potential privacy incidents, such as data breaches or leaks. This shift in user perception and control is certain to bear a huge impact on data privacy practices within the automotive industry. By delivering open-source results and integrating with the Android platform, it catalyses collaborative efforts to advance privacy solutions for modern cars and revolutionise the automotive sector. The documentation produced, including open-source software, will serve as a valuable resource for industry stakeholders.

Acknowledgment. This work acknowledges financial support from: PRIN 2022 MUR project FuSeCar (E53D23008220006).

References

1. Albouq, S.S., Abi Sen, A.A., Namoun, A., Bahbouh, N.M., Alkhodre, A.B., Alshanqiti, A.: A double obfuscation approach for protecting the privacy of IoT location based applications. IEEE Access **8**, 129415–129431 (2020)
2. Bella, G., Biondi, P., Tudisco, G.: A double assessment of privacy risks aboard top-selling cars. Autom. Innov. **6**(2), 146–163 (2023)
3. Chah, B., Lombard, A., Bkakria, A., Yaich, R., Abbas-Turki, A., Galland, S.: Privacy threat analysis for connected and autonomous vehicles. Procedia Comput. Sci. **210**, 36–44 (2022)
4. Danezis, G., Gurses, S.: A critical review of 10 years of privacy technology (2010)
5. Hyperledger Foundation. Security Model—Hyperledger Fabric Docs (2020). https://hyperledger-fabric.readthedocs.io/en/release-2.2/security_model.html
6. Mozilla Foundation. It's Official: Cars Are Terrible at Privacy and Security—foundation.mozilla.org (2023). https://foundation.mozilla.org/en/privacynotincluded/articles/its-official-cars-are-the-worst-product-category-we-have-ever-reviewed-for-privacy/
7. Garrido, G.M., Schmidt, K., Harth-Kitzerow, C., Klepsch, J., Luckow, A., Matthes, F.: Exploring privacy-enhancing technologies in the automotive value chain. In: 2021 IEEE International Conference on Big Data (Big Data), pp. 1265–1272 (2021)
8. Hoepman, J.K.: Privacy is hard and seven other myths: achieving privacy through careful design (2021)
9. Likert, R.: A technique for the measurement of attitudes. Arch. Psychol. (1932)
10. Meuser, T., Ojo, O.T., Bischoff, D., Fernández Anta, A., Stavrakakis, I., Steinmetz, R.: Hide me: enabling location privacy in heterogeneous vehicular networks. In: Georgiou, C., Majumdar, R. (eds.) NETYS 2020. LNCS, vol. 12129, pp. 11–27. Springer, Cham (2021). https://doi.org/10.1007/978-3-030-67087-0_2
11. Pesé, M.D., Schauer, J.W., Mohan, M., Joseph, C., Shin, K.G., Moore, J.: Pricar: privacy framework for vehicular data sharing with third parties. In: 2023 IEEE Secure Development Conference (SecDev), pp. 184–195. IEEE (2023)
12. Raciti, M., Bella, G.: How to model privacy threats in the automotive domain. In: Proceedings of the 9th International Conference on Vehicle Technology and Intelligent Transport Systems (VEHITS), pp. 394–401. INSTICC, SciTePress (2023)
13. Raciti, M., Bella, G.: A threat model for soft privacy on smart cars. In: 2023 IEEE European Symposium on Security and Privacy Workshops (EuroS&PW) (2023)
14. Raciti, M., Bella, G.: Up-to-date threat modelling for soft privacy on smart cars. In: Katsikas, S., et al. (eds.) Computer Security. ESORICS 2023 International Workshops, pp. 454–473. Springer, Cham (2024). https://doi.org/10.1007/978-3-031-54204-6_27
15. Rasmusen, S.C., et al.: Raising consent awareness with gamification and knowledge graphs: an automotive use case. Int. J. Semant. Web Inf. Syst. **18**(1), 1–21 (2022)
16. Renaud, K., Van Schaik, P.: What values should online consent forms satisfy? A scoping review. In: Dewald Roode Workshop 2023 (2023)
17. Schoettle, B., Sivak, M.: A survey of public opinion about connected vehicles in the U.S., the U.K., and Australia. In: 2014 International Conference on Connected Vehicles and Expo (ICCVE), pp. 687–692. IEEE (2014)

18. Vasenev, A., et al.: Practical security and privacy threat analysis in the automotive domain: Long term support scenario for over-the-air updates. In: Proceedings of the 5th International Conference on Vehicle Technology and Intelligent Transport Systems (VEHITS), pp. 550–555. INSTICC, SciTePress (2019)
19. Walter, J., et al.: The user-centered privacy-aware control system pricon: an interdisciplinary evaluation. In: Proceedings of the 13th International Conference on Availability, Reliability and Security, ARES '18. Association for Computing Machinery, New York (2018)

Demos

Demo: All-in-One Solution for Online Abuse Research

Mohammed Aldeen[1](✉), Pranav Pradosh Silimkhan[1], Ethan Anderson[1], Taran Kavuru[1], Tsu-Yao Chang[1], Jin Ma[1], Feng Luo[1], Hongxin Hu[2], and Long Cheng[1]

[1] School of Computing, Clemson University, Clemson, USA
{mshujaa,lcheng2}@clemson.edu
[2] Department of Computer Science and Engineering, University at Buffalo, Buffalo, USA

Abstract. The proliferation of online social media platforms has led to an increase in various types of content, including online hate. This trend poses substantial risks by amplifying harmful ideologies, inciting violence, and perpetuating discrimination. In response to this growing concern, Machine Learning (ML) has emerged as powerful tools for the automatic analysis of online hate. However, researchers from diverse fields are facing fundamental challenges in accessing essential resources, such as datasets, ML models, and analysis tools. In this paper, we present Integrative Cyberinfrastructure for Online Abuse Research (ICOAR), a system that automates the process of collecting, analyzing, and visualizing online abuse data. ICOAR pipeline begins with automated data collection from various social media platforms, followed by integration of state-of-the-art ML models to streamline the detection, categorization, and analysis of online abuse. ICOAR also features customizable tools for data visualizations, such as network and temporal analysis, catering to a range of research needs and expertise levels.

Keywords: Online Abuse · Social Media Analysis · Data Visualization · Multimodal Analysis · Cyberinfrastructure · Hate Speech Detection · Data Annotation

1 Introduction

Machine learning (ML) have emerged as pivotal tools in the detection and analysis of online abuse [4]. The potential of these technologies lies in their ability to process and analyze large volumes of data from various online platforms, thus offering a more nuanced understanding and effective response to the issue. However, integrating AI/ML into online abuse research presents several challenges. The current disperse ML tools and models for online abuse research are scattered over the Internet, posing challenges for research integration and collaboration for social science and computer science researchers.

In response to these challenges, we introduce Integrative Cyberinfrastructure for Online Abuse Research (ICOAR). This platform is designed to streamline the collection, analysis, and visualization of online abuse data across multiple social media platforms. It employs multiple state-of-the-art machine learning models for detecting and categorizing abusive content including multimedia data and visualize them to find patterns in the data. ICOAR streamline is scalable, customizable, extendable, portable, and user-friendly, allowing researchers from different backgrounds to efficiently collect, identify, analyze and understand trends in their data.

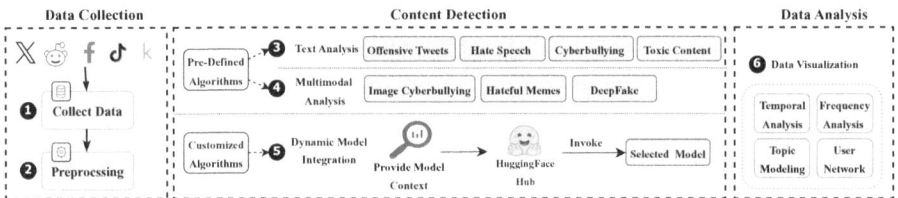

Fig. 1. Overview of ICOAR Software Infrastructure

2 ICOAR Platform

2.1 Overview

Figure 1 depicts an overview of the our ICOAR platform[1], which offers a streamlined approach to analyzing online abuse through a sequence of integrated steps. It begins with data collection system that gathers relevant social media content based on predefined keywords and criteria (❶). This is followed by preprocessing to refine the collected data (❷). The core analysis phase is divided into (❸) text analysis to analyze textual content of online abuse, and (❹) multimodal analysis to extend the analysis to images further broadening the scope of the platform. In addition, to accommodate users interested in research areas beyond online abuse, our platform allows dynamic integration of customized algorithms for targeted analysis (❺). After these core processes, the platform enables users to further analyze and visualize the data through advanced analytical models for deeper insights (❻).

2.2 Data Collection

At its current state, ICOAR mostly handles input from various social media platforms by leveraging automated data collection methods tailored to each platform's accessibilities and limitations, including Facebook, Reddit, TikTok, Twitter (X), and YouTube. Due to its versatile and scalable architecture, ICOAR can

---

[1] The implementation and source code of ICOAR will be publicly outsourced to encourage community contributions.

easily accommodate additional platforms in the future. Also, it supports collection from dataset repositories, such as Kaggle [2] and HuggingFace [1]. ICOAR utilizes a range of collection methods such as free and paid APIs, and research APIs. Also, we provide manuals for the users detailing the processes of obtaining the APIs from different platforms (Fig. 2).

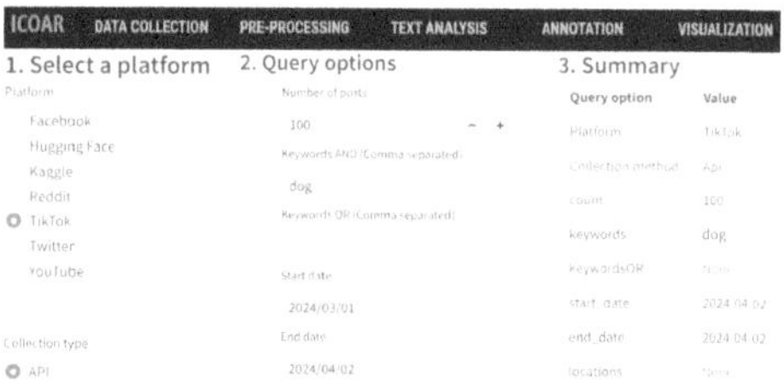

Fig. 2. Detailed Query Construction to Collect Data

2.3 Content Detection Module

This module in ICOAR leverages advanced ML algorithms to systematically identify and categorize different forms of harmful online content. Beyond textual data analysis, it is equipped with multimodal analysis capabilities for detecting abuse in images, such as detect cyberbullying images, identifying hateful memes, and spotting deepfakes. Moreover, this module offers a unique feature of customized algorithms, which is crucial for researchers requiring specialized analysis beyond the scope of pre-selected models (Fig. 3).

Fig. 3. Content Detection Module

2.4 Data Visualization

The ICOAR platform provides data visualization tools to help users interpret their analysis results. Each tool is designed to show specific insights and trends in the data. As depicted on Fig. 4, these tools include temporal analysis for tracking changes over time, topic modeling for grouping discussions into key themes, user network visualization for mapping relationships between individuals and communities, and frequency analysis to identify the most common terms. Together, these tools offer a comprehensive view of the data to better understand the behaviour and mitigate online abuse.

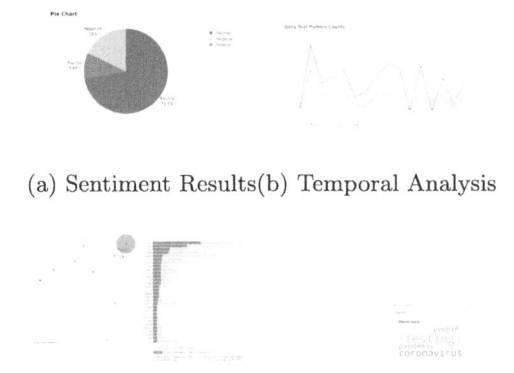

(a) Sentiment Results (b) Temporal Analysis

(c) Topic Modeling Distribution (d) Frequency Analysis

Fig. 4. Data Visualization Features

3 Advanced Features

3.1 LLM-Assisted Annotation

The ICOAR platform extends its capabilities beyond traditional machine learning approaches, embracing the power of Large Language Models (LLMs). These models bring an unparalleled depth to the analysis of online abuse by leveraging the vast amounts of information encoded within their parameters. Here we introduce two specific features within this advanced segment:

Text Annotation: To facilitate users in labeling their data effectively, we deployed text annotation. Users first provide sample inferences and construct a specific prompt to guide GPT. We offer users a "Chain of Thought" prompt template that they can adapt for their own data. This was made possible using the GPT API, streamlining the data labeling process [3].

Image Annotation: Similarly, for image annotation, we implemented a comparable approach to enable users to effortlessly tag and categorize images. By leveraging the capabilities of the GPT API, users can provide descriptive prompts and sample images, which guide the model in generating accurate labels [5].

Acknowledgment. This work is supported by the National Science Foundation (NSF) under Grant No. 2239605 and 2228616.

References

1. The ai community building the future. https://huggingface.co/ (HuggingFace:). Accessed 02 Apr 2024
2. Level up with the largest ai & ml community. https://www.kaggle.com/datasets (Kaggle:). Accessed 02 Apr 2024
3. Aldeen, M., et al.: Chatgpt vs. human annotators: a comprehensive analysis of chatgpt for text annotation. In: 2023 International Conference on Machine Learning and Applications (ICMLA), pp. 602–609. IEEE (2023)
4. Mishra, P., Yannakoudakis, H., Shutova, E.: Tackling online abuse: a survey of automated abuse detection methods. arXiv preprint arXiv:1908.06024 (2019)
5. Nong, Y., Aldeen, M., Cheng, L., Hu, H., Chen, F., Cai, H.: Chain-of-thought prompting of large language models for discovering and fixing software vulnerabilities. arXiv preprint arXiv:2402.17230 (2024)

Author Index

A
Akowuah, Francis 279
Aldeen, Mohammed 26, 361
Anderson, Ethan 361

B
Bella, Giampaolo 344
Bui, Huan 158

C
Cardenas, Alvaro 211
Castellanos, John 211
Castiglione, Gianpietro 344
Chan, Matthew 3
Chang, Tsu-Yao 361
Chatterjee, Rik 100
Chen, Bo 187
Chen, Niusen 187
Cheng, Long 26, 361
Clutter, Mathew 279

D
Dafoe, Josh 187
Daily, Jeremy 100
Ding, Aolin 3

E
Esposito, Sergio 344
Evans, Ryan 237

F
Fu, Chenglong 158

G
Garcia, Luis 3
Griffith, Henry 330

H
Hass, Amin 3
Hawkins, William 237

H
Hei, Xiali 54, 311
Hu, Hongxin 361

J
Jacobson, Cade 279
Jamarani, Amirhossein 54
Ji, Xu 311

K
Kavuru, Taran 361
Kim, Dongha 135
Kim, Hokeun 135
Koneru, Keerthi 211

L
Lee, Chanhee 135
Liao, Song 26
Liu, Sijia 259
Lozano, Juan 211
Lucas, Marcus 3
Luo, Feng 361

M
Ma, Jin 361
Malik, Hafiz 77
Mangano, Mirko Giuseppe 344
Marchetti, Mirco 344
Maugeri, Marcello 344
McDaniel, Patrick 110
Mohammadi, Alireza 77
Mu, Jaden 296

N
Norris, Michael 110
Nyambe, Teddy 100

P
Pradosh Silimkhan, Pranav 361

R
Raciti, Mario 344
Rafiul Hussain, Syed 110
Rathore, Heena 330
Raymond Choo, Kim-Kwang 311
Refat, Rafi Ud Daula 77
Riccobene, Salvatore 344

S
Santamaria, Daniele Francesco 344
Siy, Job 187
Snyder, Nathaniel 3
Sokolsky, Oleg 3
Srivastava, Mani 3

T
Tabuada, Paulo 3
Tan, Gang 110

Thapa, Bishal 330
Tu, Yazhou 54

W
Wang, Boyang 237
Weimer, James 3

Y
Yan, Qiben 259
Young, Jeffery 26

Z
Zambon, Emmanuele 211
Zhang, Jianyi 311
Zhang, Shuhao 26
Zhao, Zhangchi 311
Zhou, Ce 259
Zhou, Ziyin 311
Zonouz, Saman 3

The manufacturer's authorised representative in the EU is Springer Nature Customer Service Centre GmbH, Europaplatz 3, 69115 Heidelberg, Germany. If you have any concerns regarding our products, please contact ProductSafety@springernature.com

Printed and bound by CPI Group (UK) Ltd, Croydon, CR0 4YY

26/03/2026

02078984-0004